Detectors in Particle Physics

This textbook provides an accessible yet comprehensive introduction to detectors in particle physics. It emphasises the core physics principles, enabling a deeper understanding of the subject for further and more advanced studies. In addition to the discussion of the underlying detector physics, another aspiration of this book is to introduce the reader to practically important aspects of particle detectors, like electronics, alignment, calibration and simulation of particle detectors. Case studies of the various applications of detectors in particle physics are provided.

The primary audience is graduate students in particle or nuclear physics, in addition to advanced undergraduate students in physics.

Key Features:

- Provides an accessible yet thorough discussion of the basic physics principles needed to understand how particle detectors work.
- Presents applications of the basic physics concepts to examples of modern detectors.
- Discusses practically important aspects like electronics, alignment, calibration and simulation of particle detectors.
- Contains exercises for each chapter to further understanding.

Georg Viehhauser is a Lecturer in the Physics department at the University of Oxford, UK, and a supernumerary fellow at St. John's College, Oxford, UK. He has been working on a variety of different particle detector technologies, starting with the Forward Chamber A at the DELPHI experiment, the LKr calorimeter for NA48, the muon chambers for ATLAS, and the RICH for CLEO III. More recently, he has contributed to the construction of the ATLAS SCT and he is currently involved in the phase 2 upgrade of the ATLAS ITk, as well as the SVT for the ePIC experiment. He is one of the main organisers of the Forum on Tracking Detector Mechanics.

Tony Weidberg is a Professor of Physics at Oxford University, UK and a tutorial fellow at St. John's College. He worked on CCD readout for a scintillating fibre detector at the CERN SPS collider. He played a major role in the founding of the ATLAS experiment and the design of the ATLAS SCT. He has a wide range of experience from detector R&D, assembly and integration of complex detector systems as well as evaluating their performance. He has extensive experience in radiation hardness studies, particularly for optoelectronics and applications of reliability theory.

Both authors have a long experience in teaching undergraduate and graduate students at the University of Oxford.

Detectors in Particle Physics
A Modern Introduction

Georg Viehhauser and Tony Weidberg

CRC Press
Taylor & Francis Group
Boca Raton London New York

CRC Press is an imprint of the
Taylor & Francis Group, an **informa** business

SCOAP³

**Sponsoring Consortium for
Open Access Publishing in Particle Physics**

Designed cover image: End view of the CMS experiment (M. Hoch and M. Brice, CMS Collection, https://cds.cern.ch/record/1474902, by CC BY 4.0)

First edition published 2024
by CRC Press
2385 NW Executive Center Drive, Suite 320, Boca Raton FL 33431

and by CRC Press
4 Park Square, Milton Park, Abingdon, Oxon, OX14 4RN

CRC Press is an imprint of Taylor & Francis Group, LLC

ISBN: 978-1-032-24658-1 (hbk)
ISBN: 978-1-032-26318-2 (pbk)
ISBN: 978-1-003-28767-4 (ebk)

DOI: 10.1201/9781003287674

Typeset in Nimbus font
by KnowledgeWorks Global Ltd.

Publisher's note: This book has been prepared from camera-ready copy provided by the authors.

Contents

Preface

This book is an introduction to detectors in particle physics. It is primarily aimed at students, who want to understand the instruments that produce the data on which their research is based. This book should also be useful for physicists with little prior knowledge of detector physics who want to work on detector development, testing, commissioning or calibration/alignment.

However, we do hope that even more experienced colleagues will find one or the other topic interesting and presented with a new twist.

We will assume that the reader has a general knowledge of physics at the level of an undergraduate degree (classical mechanics and electromagnetism, quantum mechanics, special relativity, thermal physics and condensed matter physics). An introductory level knowledge of electronics will also be useful for understanding some of the discussion in the chapter on electronic signals and noise.

We consider the breadth of physics employed in the development of particle detectors one of the appealing aspects of this field.

This book is intended to bridge the gap between the terse description in summaries like the PDG [505] and much more detailed and voluminous monographs written in the field. It strives to go beyond the simple phenomenological equations that are often used for back-of-the-envelope calculations and that have become part of the folklore of the field. It wants to give the reader an introduction to the physical models that are used by the very powerful simulation programmes that have been developed to predict and model the response of particle detectors, but in which the physical models are concealed from the casual user.

This book is divided into three main parts. Chapters 2 to 5 introduce the physics and electronics concepts that are underlying the operation of all particle physics detectors. These concepts will be required to understand the workings of detectors as discussed in the latter parts of the book. To give the reader an idea of the origin of the equations presented and the underlying physics, we do provide derivations of the results in these chapters. Some of these derivations are fairly advanced, and their details are not essential for the required understanding. However, they are included to give the interested reader an opportunity and a starting point to obtain a deeper understanding of their origin, if needed. Chapter 2 will discuss the interactions of incoming particles of various types with the detector material, and the resulting response. Most modern particle detectors will produce electrical signals, and in chapter 3 we will explore their generation, and the effects of electronic noise in the signal acquisition process. Electrical signals rely on the movement of charges in electric and magnetic fields, and we will investigate this for different detector materials in chapter 4. Another signal can be scintillation light from excitation of the detector material, and this will be the topic of chapter 5.

Chapters 6 to 9 discuss how these principles lead to a signal from the detector for different types of detector materials. This part starts in chapter 6 with a look at emulsion and bubble chambers, which are largely of historical interest but have some modern applications, and then discusses detectors based on gaseous, liquid and solid detector materials, each in their own chapter (chapters 7, 8 and 9). The reader wanting to know about some but not all of these techniques could just read the relevant chapters.

The remaining chapters then discuss how these signals can be used to build experiments for particle physics, in increasing size and complexity. Chapter 10 explains how gaseous and semiconductor detectors can be combined with magnetic fields to determine the trajectories and momenta of charged particles and to reconstruct interaction vertices. In chapter 11 we will investigate the measurement of the energy deposited by an incoming electron, photon or hadron, when it is fully absorbed by a block of detector material (a calorimeter). Chapter 12 explains how different techniques can be used to identify the type of the incoming particle. Chapter 13 will give a brief introduction to how in particle physics experiments the recording of events can be instigated by a trigger, with a rate of data-taking compatible with the limitations of data processing and storage. Finally, in chapter 14 we will investigate how the technologies introduced throughout this book can be combined into powerful detectors that provide us with data for sophisticated physics studies and what some of the concerns and constraints are that drive the designs of these experiments.

Again, the reader who only wishes to learn only about specific measurement techniques can just read the relevant chapter.

In addition to a thorough introduction to the physics of particle interactions with (detector) matter and a description of the most common types of particle detectors, this book also includes introductions to practically important aspects of particle detectors, like electronics, alignment and calibration, and simulation of particle detectors. Where suitable it will demonstrate all these principles with examples of detectors that have been built, whether only at a prototype stage, or for actual particle physics experiments, although it is not aspiring to present a complete list of all types of particle detectors ever built or proposed.

At the end of each chapter there is a summary of key concepts that have been discussed in the chapter and a set of exercises that allow the reader to investigate aspects of the chapter in more detail.

We would like to thank our colleagues Wade Allison, Christoph Amelung, Giles Barr, Hugo Beauchemin, Steve Biller, Pawel Bruckman de Renstrom, Paula Collins, Louis Fayard, Alfredo Ferrari, Neville Harnew, Peter Jenni, Malcolm John, Hans Kraus, Paul Lecoq, Michel Levebre, Tim Martin, Steve McMahon, Michael Moll, Peter Phillips, Meinhard Regler, Armin Reichold, Martin Tat, Rob Veenhof, Dave Wark, Morgan Wascko, Norbert Wermes and Steve Worm, who gave us valuable inputs to this book. Any remaining mistakes are ours. We wish to thank Rebecca Hodges-Davies from Taylor and Francis for suggesting we embark on this project and Danny Kielty also from Taylor and Francis for all his help and patience throughout the process of writing this book.

1 Introduction

Detectors in particle physics detect subatomic particles. The size of particle physics experiments can range from 10s of μm to 100s of km, although typically the size of the detectors will be at the lower end of this range, with typical experiment sizes of up to a few metres. This has two important consequences: First, particle detectors are significantly larger than the Compton wavelengths (h/mc) of the particles they need to detect. It is therefore entirely justified to designate them as 'particle' detectors, and it is usually sufficient for the understanding of the detector to treat the particles as quasi-classical particles.

The second important consequence is that most of the subatomic particles cannot be directly detected, because they just don't live long enough to allow for a direct observable interaction with detector matter. Even with a significant boost, only particles with a lifetime of about 10^{-13} s and above can be detected directly, and most types of subatomic particles decay faster, in particular all particles that have strong or electromagnetic decay channels available that satisfy kinematics and conservation laws. This leaves only a limited number of particles that directly will account for signals in a particle detector.

The question how a quantum mechanical physics process results in a straight particle track in a detector has been addressed by Mott in a famous paper [381].
Direct detection here means observation of a signal generated in the detector by the particle itself.

	in standard experiments	in high granularity detectors
Leptons	$e, \mu, \nu_e, \nu_\mu, \nu_\tau$	τ
Mesons	$\pi^\pm, K^\pm, K_L$	
Baryons	p, n	$\Lambda^0, \Sigma^\pm, \Xi^{0,-}, \Omega^-$
Bosons	γ	

Table 1.1: Sub-atomic particles with a sufficiently long lifetime for interaction with a detector (anti-particles are implied, where applicable). High granularity detectors are detectors with a high density of 3D detection elements, like emulsions or bubble chambers.

Other sub-atomic particles that can interact with a detector are bound states of the two most stable baryons, the proton and the neutron (for example deuterons or αs, but also other nuclei). Finally, some mesons containing c and b quarks could potentially be directly observed in high granularity detectors, although there are no modern experiments doing this systematically.

While charged particles can leave detectable signals already in thin layers of matter (for example in a few 10s of μm of silicon), neutral particles do often require more substantial amounts of detector material. For the weakly interacting neutrinos, the interaction cross-sections are so small that the probability for detection of any one particle is very low and therefore very massive detectors are needed. In cases where the properties of a specific neutrino need to be measured (for example in collider experiments), the presence and kinematic properties of these particles can only be determined from an imbalance of the combined 4-momentum of all detected particles in the event.

All the other subatomic particles can only be observed indirectly, through their decay products, and their existence can often only be inferred from a resonance in the invariant mass reconstructed from the four-momenta of their daughter particles.

The fundamental particles that are precluded from direct observation by their basic nature are quarks and gluons, because of confinement.

DOI: 10.1201/9781003287674-1

However, the hadronisation of these particles produces jets. Because of their abundance in collider experiments (in particular at hadron colliders), detectors are needed that are capable of measuring the kinematic properties of such jets. This is often done for the jet as a whole, as the individual components are difficult to disentangle.

Jets are bundles of closely spaced parallel particles resulting from the hadronisation of a parton.

1.1 PROPERTIES TO BE MEASURED

There are cases when the only information required from the detector is evidence for the existence of particles, for example when the amount of ionising radiation needs to be measured (the classical example for the use of a Geiger-Müller particle counter). Then there are cases where the location of incoming particles needs to be observed (for example, photons in digital imaging) or their energy (for example, if the spectrum of energies from a γ source is to be measured). However, in particle physics it is usually the complete 4-momentum of individual particles that needs to be determined because of the need of full kinematic reconstruction of an event in the experiment.

The 4-momentum of each particle can be determined by measuring any combination of its 3-momentum (magnitude and direction), and the energy or its mass. The latter can be inferred if one identifies the type of particle, either by observing a process or experimental signature only possible for that particle type, or by other means, for example by combining a measurement of the speed with a measurement of the momentum.

See chapter 12.

The direction of the 3-momentum can be found from the measurement of at least two spacepoints along the particle's trajectory in 3D space. The momentum of charged particles can be inferred from the curvature of the particle's track in a magnetic field, which requires the measurement of at least three positions of the particle along its trajectory. In both cases, the space points must be separated enough to allow for a large enough lever arm to make a good measurement, and the resolution of the spatial measurement will directly affect the measurement of the kinematic variables. Any disturbance of the particle's trajectory before and between these measurements will also deteriorate the direction or momentum measurement.

See chapter 10.

The energy is typically measured by completely absorbing the particle in a sufficiently large block of matter and recording the deposited energy.

See chapter 11.

In addition to these measurements of the kinematic properties, particle physics experiments often require the rejection of backgrounds and, in cases where the raw event rate exceeds the capability of the readout to store data, fast selection of events to be recorded, which is typically based on correlations of signals in different detector elements. This selection is the task of triggers.

See chapter 13.

1.2 EXPERIMENTS IN PARTICLE PHYSICS

Particle physics experiments today are large and/or complex instruments, often comprising several different detector systems to provide improved coverage of kinematic parameters, sensitivity to different initial and final state particles, redundancy, identification of backgrounds, rate normalisation, and/or capability to cross-check and calibrate systematic effects.

This section is not intended as a complete list of particle physics experiments but gives an overview of the requirements and environments that drive the design choices of particle detectors.

At the same time, the large range of physics studies in which the detectors are employed result in very different requirements and environments.

Arguably, one of the most demanding environments for particle detectors are the experiments at the Large Hadron Collider (LHC). There, very high collision rates put high demands on speed, segmentation, read-out bandwidth, event selection and radiation hardness, and the energies of particles to be detected cover a very large range from $\sim$ GeV to 100s of GeV. The broad range of physics targeted by these experiments requires sensitivity to and discrimination of a wide range of particles.

Since the start of operation in 2008, the LHC has been undergoing continuing improvements, leading to ever higher collision rates. The experiments have pursued a concurrent set of upgrades. These upgrades divide the operation of the experiments into periods that are separated by long shutdown (LS) periods.

Table 1.2: *LHC operation periods for pp collisions. Parameters for periods after 2023 are intended. Two LS are planned during phase 2, but not shown in the table.*

	Run 1	LS1	Run 2	LS2	Run 3	LS3	Phase 2 (HL-LHC)
Period	2009–2013		2015–2018		2022–2025		2027–2041
$\sqrt{s}$ [TeV]	8		13		13.6		14
Bunch spacing [ns]	50	2013 – 2015	25	2018 – 2022	25	2025 – 2027	25
Peak luminosity [cm^{-2}s^{-1}]	8×10^{33}		2×10^{34}		2×10^{34}		10^{35}
Average number of collisions per bunch crossing	21		34		55		140–200
Integrated luminosity [fb^{-1}]	~ 30		~ 140		~ 380		3000

At the opposite end of the rate requirements are experiments searching for rare events. Here, good background rejection, shielding from cosmic backgrounds and radiopurity, and large detector masses are needed. Particle energies in these experiments are generally low ($\lesssim 1$ MeV).

Experiments in neutrino physics also require a large target mass and good background suppression. A major task is identifying the flavour of the neutrinos. Experiments studying solar neutrinos or anti-neutrinos from reactors must be sensitive to low energies ($\mathcal{O}(\text{MeV})$) while atmospheric neutrinos have energies in the range of $\sim$100 MeV to a few 10s of GeV. Very high energy neutrinos can be detected by instrumenting large volumes of ice or water. Neutrinos from accelerators have energies in the range from about 100 MeV to about 10 GeV.

In accelerators neutrinos are typically created from decays of charged pions.

Finally, particle detectors are also employed in space, where additional challenges are posed by the environment and the launch. However, the same particle detector technologies as for terrestrial experiments are being used.

In chapter 14, we will come back to these discussions and demonstrate how all these demands are met by the detectors in modern particle physics experiments.

2 Interactions of particles with matter

The biggest challenge for the detection of subatomic particles is clearly the small amount of energy these particles carry. A TeV particle has about 10^{-7} J of energy and the momentum is about 10^{-15} kgm/s, so that direct mechanical observation is not possible.

However, subatomic particles do carry the charges of the electroweak or the strong interaction, and with these charges they couple to similar charges in the detector material and transfer some or all of their energy to the detector. This energy transfer results in excitation or ionisation of atoms in the detector, which in modern particle detectors are used to create an electrical signal large enough that it can be amplified and recorded by electronic means.

Ultimately, the energy will be transferred to phonons, which in special circumstances can be detected (see section 11.2).

Ultimately, the signal creation in particle detectors always involves electromagnetic interactions. We will therefore start with a discussion of how photons interact with matter.

2.1 INTERACTION OF PHOTONS WITH MATTER

Due to the nature of the couplings in QED any interaction of an incoming photon with the charged fermions in the detector medium will result in the destruction of the photon. It is thus impossible to follow the path of a photon in a detector. The path can only be inferred if the starting point of the photon can be deduced by other means.

To maintain four-momentum conservation this is sometimes associated with the creation of another photon, but that photon does generally not retain the direction of the former.

Depending on the photon energy, there are three major interaction mechanisms.

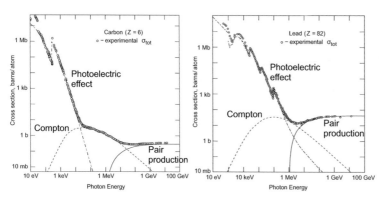

Figure 2.1: Cross-sections for the interaction of photons for a low-Z and a high-Z element (modified from [379]). In these plots we have omitted coherent scattering, photonuclear reactions and pair production in the electron fields.

Another measure for the absorption of photons in matter is the mean free path, which (in cm) is given by $\lambda = 1.66\,m_a/(\rho\sigma)$, where ρ is the mass density in g/cm^3, σ the cross-section in b, and m_a the atomic molar mass in g/mol.

In literature, the symbol A is often used for both the atomic molar mass and the dimensionless atomic mass number. We use m_a and A for the atomic molar mass (typically in g/mol), and the atomic mass number, respectively.

PHOTOELECTRIC EFFECT

At low energies (up to about 1 MeV) the dominant interaction of photons with matter is photoelectric effect. The photon gets absorbed by an atom and its energy frees an electron, which escapes with an energy $E = E_\gamma - E_{\text{bind}}$, where E_{bind} is the binding energy of the electron. At very

DOI: 10.1201/9781003287674-2

low energies this equation constitutes a threshold. Below the threshold energy for photoelectric effect the material can be transparent, if there are no molecular resonances. This energy range is called the 'optical' region.

In interactions with molecules at low energies, the energy of the photon can be used up in excitations. This type of photon absorption can occur at energies below the ionisation threshold, i.e. without the emission of an electron.

For a single atom the binding energy of the interacting electron equals the ionisation energy. For solids the relevant parameter is the 'photoelectric work function', which is defined as the minimum energy required to remove an electron from the solid to infinity. In the solid, the states of outer electrons are arranged in bands, which are occupied up to the Fermi level $-\varepsilon_F$. In addition, we also need to consider surface effects. A simple model for the surface effect in metals comprises a uniform layer of surface dipoles (often called a 'double layer'). To penetrate this double layer an additional energy W_s, depending on the surface properties, is required, so that the work function $W = -\varepsilon_F + W_s$ [73].

The minus sign here indicates that the Fermi level is lower than the energy of an electron at infinity, i.e. electrons at the Fermi level are still bound.

The work function can thus depend on surface properties like crystal orientation but also contamination.

In metals, the highest energy electrons will be in the conduction band, where they are less well bound than in a single atom, and thus the work function is typically 30–50% lower than the ionisation energy for a single atom, typically in the range of 3 to 5 eV. The lowest work functions are observed for alkalis (for example, caesium with a work function of about 1.8 eV).

$\lambda/\mu m = 1.24/(E/eV)$. Visible light is in the range from 1.75 (red) to 3.1 eV (violet).

For photon detection in semiconductors it is in principle sufficient to lift the electrons across the band gap between the valence and the conduction band, which then allows movement of the electron (and the positive hole left in the valence band) towards collecting electrodes. The small energy required for this means that semiconductor sensors are sensitive down to the infrared range of energies. For silicon the band gap is about 1.12 eV, which allows interaction of photons with wavelengths up to about 1100 nm.

However, silicon is an indirect band gap semiconductor, which means that to conserve energy and momentum additional phonons are required, and the transition will be suppressed at energies below 3.4 eV. This will be discussed in more detail in section 9.8.

In solid metals and semi-conductors collective oscillations of the free electron gas density in the valence band ('plasmons') can occur at defined energies. Photons with this energy can be absorbed without the emission of a photoelectron, and their energy used in the excitation of a plasmon. Typical plasmon energies are between 10 and 30 eV.

At energies well above the ionisation threshold the photoelectric cross-section of compounds is to good approximation given by the sum of the cross-sections of the constituent atoms.

The cross-section for photon absorption in single atoms can be calculated using time-dependent perturbation theory. In the following we will demonstrate this for the case of hydrogen. In the case of a weak field, the transition of the momentum operator, $\vec{p} \to \vec{p} - e\vec{A}(\vec{r}, t)$, with e the charge of the electron, and the vector potential describing the interacting field

For the detailed derivation see for example [162].

This is an example of "minimal substitution" and the fundamental justification comes from demanding U(1) gauge invariance [270].

$$\vec{A}(\vec{r}, t) = A_0(\omega)\, \vec{\varepsilon}\, e^{i(\vec{k} \cdot \vec{r} - \omega t)}, \tag{2.1}$$

$\vec{\varepsilon}$ describes the polarisation of the field.

can be introduced as a perturbation $\delta H(t) = (-i\hbar e/m)\vec{A}(\vec{r}, t) \cdot \vec{\nabla}$. In first order time-dependent perturbation theory the transition rate from the atomic ground state to a state n can be found from Fermi's Golden Rule

Note that Fermi's Golden Rule also implies energy conservation between the initial and final states. This is sometimes explicitly expressed using δ-functions. To improve readability we will omit these here, but energy conservation should be assumed throughout this chapter. This also removes the time-dependent term in the exponential in eq. (2.1).

$$\Gamma_{n0} = \frac{2\pi}{\hbar} |\langle \Psi_n | \delta H(t) | \Psi_0 \rangle|^2 \rho, \tag{2.2}$$

with the corresponding (hydrogen) wavefunctions Ψ_0 and Ψ_n, and ρ the density of final states. The cross-section for the transition is then found to be

$$\sigma_n = \frac{4\pi^2 \alpha \hbar^2}{m^2 \omega_n} \left| \langle \Psi_n | e^{i\vec{k}\cdot\vec{r}} \, \vec{\varepsilon} \cdot \vec{\nabla} \Psi_0 \rangle \right|^2, \qquad (2.3)$$

with $\omega_n = (E_n - E_0)/\hbar$.

$$\rho(\omega) = \frac{V m^{3/2} \sqrt{\hbar \omega}}{\pi^2 \sqrt{2} \hbar^2} = \frac{V m p}{2\pi^2 \hbar^2}.$$

α is the fine structure constant,

$$\alpha = \frac{1}{4\pi\varepsilon_0} \frac{e^2}{\hbar c}.$$

For low energy photons, the wavelength is much larger than the size of the atom, and thus we can approximate the exponential in the matrix with $1 + i\vec{k}\cdot\vec{r}$. In this approximation, the matrix element becomes

This is referred to as the 'dipole approximation'.

$$\langle \Psi_n | e^{i\vec{k}\cdot\vec{r}} \, \vec{\varepsilon} \cdot \vec{\nabla} | \Psi_0 \rangle \rightarrow -\frac{m \omega_n}{\hbar} \langle \Psi_n | \vec{\varepsilon} \cdot \vec{r} | \Psi_0 \rangle.$$

This yields, assuming random polarisation of the photon, for the integral cross-section

For random polarisation

$$\int \sigma_n \, d\omega_n \simeq \frac{4\pi^2 \alpha \omega_n}{3} \left| \langle \Psi_n | \vec{r} | \Psi_0 \rangle \right|^2.$$

$$\left| \langle \Psi_n | \vec{\varepsilon} \cdot \vec{r} | \Psi_0 \rangle \right|^2 = \frac{1}{3} \left| \langle \Psi_n | \vec{r} | \Psi_0 \rangle \right|^2.$$

It is common to introduce the 'dipole oscillator strengths'

$$f_n = \frac{2m \omega_n}{3\hbar} \left| \langle \Psi_n | \vec{r} | \Psi_0 \rangle \right|^2. \qquad (2.4)$$

The cross-section for a transition to the state n can then be written as

$$\int \sigma_n \, d\omega_n = \frac{2\pi^2 \alpha \hbar}{m} f_n.$$

In the photoelectric effect the electron does not transition to a bound state, but escapes with positive energy. We assume that the electron is non-relativistic and its energy therefore $(\hbar k_e)^2/(2m)$. For sufficiently high energies, the interaction of the electron with the nucleus can be neglected, and the emitted electron be described by a plane wave, $\Psi_e = V^{-1/2} \exp(i\vec{k}_e \cdot \vec{r})$. The matrix element then becomes

$$M_{e0} = \langle \Psi_e | e^{i\vec{k}\vec{r}} \, \vec{\varepsilon} \cdot \vec{\nabla} | \Psi_0 \rangle = \frac{1}{\sqrt{V}} \int e^{-i\vec{k}_e \cdot \vec{r}} e^{i\vec{k}\cdot\vec{r}} \, \vec{\varepsilon} \cdot \vec{\nabla} \Psi_0(\vec{r}) \, d^3 r. \qquad (2.5)$$

The cross-section for a transition to the continuum can then be obtained from

$$\sigma(\omega) = \frac{4\pi^2 \alpha \hbar^2}{m^2 \omega} \int |M_{e0}|^2 \rho(\omega) \, d\Omega.$$

The transition to the continuum is characterised by an 'oscillator strength density'

$$\frac{df}{d\omega} = \frac{2\hbar}{m\omega} \int |M_{e0}|^2 \rho(\omega) \, d\Omega. \qquad (2.6)$$

The oscillator strengths satisfy a sum rule,

$$\int \frac{df}{d\omega} \, d\omega = 1, \qquad (2.7)$$

This equation is known as the Thomas-Reiche-Kuhn (TRK) sum rule. For a detailed discussion of sum rules see for example [133].

where the oscillator strength density formally includes the discrete transitions by addition of $f_n \delta(\omega_n - \omega)$.

For many-electron atoms the position operator in the matrix elements needs to be replaced, $\vec{r} \rightarrow \sum_{i=1}^{Z} \vec{r}_i$, and the sum rule (eq. 2.7) adds up to Z instead of 1.

The matrix element in eq. (2.5), and thus the cross-section for photoelectric effect drops for increasing energy (with decreasing wavelength), as the wavelength of the incoming photon gets smaller compared to the dimension of the atom, so that in the integration of the matrix element the periods of the electric field average out. For higher energies, the shell structure of the atom becomes visible with an absorption edge for each new energetically accessible shell, with the K shell the innermost and highest energy shell. At any given energy electrons are emitted predominantly from the lowest accessible shell. Subsequently, the atoms will return to the ground state in a cascade of de-excitation steps, including the emission of fluorescence photons and Auger electrons.

Because of the complexity of shell and shielding effects, there is no closed form description of the cross-section for photoelectric effect, but typically it scales with Z^m with m between 3 and 5, and E_γ^{-n}, with n around 3. The strong decrease for the cross-section at higher energies (several 10 keV) is the reason that X-rays are highly penetrating and are useful as a diagnostic tool.

The strong dependence on Z is the reason high-Z materials (e.g. lead) are preferred shielding materials for γ radiation.

The rapid variation with Z means that X-rays are particular sensitive to heavy elements, e.g. calcium in bone.

At low energies ($\hbar\omega \ll m_e c^2$) the direction of the photoelectron emission aligns with the electric field in the incident electromagnetic wave, i.e. perpendicularly to the direction of the incoming photon. With increasing energy the emission becomes more and more forward peaked.

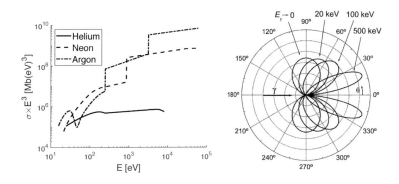

Figure 2.2: Left: Cross-section (scaled with E^3) for photoabsorption for different noble gases (data from [359]). Right: Angular distribution of photoelectrons [389].

COMPTON SCATTERING

In the intermediate energy range (about 1 MeV) Compton scattering is the dominant interaction. The cross-section for Compton scattering on a free electron can be derived from the Dirac equation and is given by the Klein-Nishina equation [310]

Compton scattering is therefore often the predominant interaction for γs from radioactive decays.

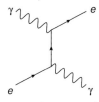

$$\sigma^{KN} = 2\pi \left(\frac{\alpha\hbar c}{m_e c^2}\right)^2 \left[\frac{1+x}{x^2}\left(\frac{2(1+x)}{1+2x} - \frac{\ln(1+2x)}{x}\right) + \frac{\ln(1+2x)}{2x} - \frac{1+3x}{(1+2x)^2}\right], \quad (2.8)$$

Figure 2.3: Tree level Feynman diagram for Compton scattering.

with $x = E_\gamma/(m_e c^2)$. For low energies the cross-section for scattering on free electrons becomes constant and equal to the cross-section for Thomson scattering,

Thomson scattering is the elastic scattering of photons by a free charged particle. It does not result in a change of frequency of the photon. It applies, as long as the wavelength of the photon is large compared to the Compton wavelength h/mc of the particle. Thomson scattering can be described by classical electromagnetism (see exercise 7).

$$\sigma^{KN} \xrightarrow[x\to 0]{} \sigma^{Th} = \frac{8\pi}{3}\left(\frac{\alpha\hbar c}{m_e c^2}\right)^2. \quad (2.9)$$

For high energies the Klein-Nishina cross-section decreases and can be approximated by

$$\sigma^{KN} \xrightarrow[x \to \infty]{} \frac{3}{8}\sigma^{Th}\frac{1}{x}\left(\ln 2x + \frac{1}{2}\right).$$

For Compton scattering on electrons bound in atoms the cross-section at high energies ($x > 1$) follows the Klein-Nishina result, and the cross-section for Compton scattering on the atom is $\sigma = Z\sigma^{KN}$. For low energies, however, binding effects are important. If the photon does not dislodge the bound electron it will scatter coherently, without energy loss. Thus the cross-section for Compton scattering decreases for low energies once the photon energy becomes comparable to, or smaller, than the energies of the bound electrons in the atom.

Photons are scattered in all directions at low energy, but are dominantly scattered in the forward direction when x becomes larger than 1.

Another elastic scattering process, Rayleigh scattering, occurs at low energies. This is elastic scattering of the photon from the atom as a whole. However, it is usually not relevant for particle detectors as its cross-section is significantly smaller than the cross-section for photoelectric effect in this energy range.

This is a general feature of photon emission at relativistic energies ('head-light' or 'searchlight' effect). Emission becomes forward peaked with a characteristic angle $\theta \sim \gamma^{-1}$ (see also exercise 4).

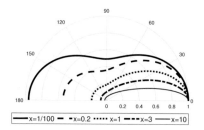

Figure 2.4: Polar angle distribution for the scattered photon in Compton scattering (the distribution is symmetric in the azimuth).

Ignoring binding effects, the scattered photon energy can be found from 4-momentum conservation, $E'_\gamma/E_\gamma = [1 + x(1 - \cos\theta)]^{-1}$, where θ is the scattering angle for the photon. The kinetic energy of the scattered electron is then

$$E_e = E_\gamma \frac{x(1 - \cos\theta)}{1 + x(1 - \cos\theta)}, \tag{2.10}$$

which has a maximum for head-on collisions of

$$E_e^{max} = \frac{2x}{1 + 2x}E_\gamma < E_\gamma. \tag{2.11}$$

The scattered electron typically loses its energy to ionisation in the detector until it comes to a stop. The scattered photon continues, until it either has another interaction in the detector, or it escapes.

PAIR PRODUCTION

Once the energy of the photon exceeds $2m_ec^2$, it can create an electron/positron pair. To maintain 4-momentum conservation it can only do this in the electric field of a charged particle, which absorbs the recoil. Pair production usually occurs in the field of a nucleus. It can also take place in the field of a shell electron, but because of the smaller charge the atomic cross-section for this type of pair production is smaller by a factor Z.

Photon absorption due to pair production starts at threshold ($2m_ec^2$) and increases with the photon energy. For high energies (i.e. full screening) the absorption coefficient tends to a constant value

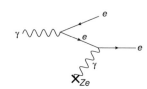

Figure 2.5: A tree level Feynman diagram for pair production.

$$\mu = n_A \sigma = \frac{7}{9}\frac{1}{X_0}, \tag{2.12}$$

The factor 7/9 in this equation is purely due to definition of X_0, which origi- nates from the description of a related process, bremsstrahlung, as discussed below.

where n_A is the volume number density of nuclei, and X_0 is called the 'ra- diation length', which we will investigate in more detail in the next sec- tion.

In pair production, the excess energy above the energy required to cre- ate the electron-positron pair is carried away as kinetic energy by the two leptons. They in term lose their energy to bremsstrahlung or ionisation, until they come to a stop. At this point the positron will annihilate with an electron in the material, with the emission of two 511 keV photons. Those can be absorbed in the detector by photoelectric effect or Compton scatter- ing, or, if the detector is of limited size, escape the detector.

Decay into two photons is required to conserve 4-momentum.

BREMSSTRAHLUNG AND THE RADIATION LENGTH

The radiation length X_0 is defined for the closely related process of bremsstrahlung, in which an electron in the field of another charged parti- cle (e.g. a nucleus) emits photons.

For large energies the energy loss of the electron per track length is proportional to the energy E of the incoming particle

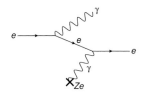

$$\frac{dE}{dx} = -\frac{E}{X_0}, \tag{2.13}$$

Figure 2.6: A tree level Feynman diagram for bremsstrahlung.

and the radiation length is the distance at which the energy of the in- coming particle has dropped by a factor $1/e$ due to the emission of bremsstrahlung. Again, the radiation length describes the radiation process at high energies, corresponding to the limit of full screening $(E \gg (\alpha Z^{1/3})^{-1} m e_c^2)$.

We can make a simple estimate of the radiation length using a semi- classical argument, based on the Weizsäcker-Williams model [498, 513]. In this approach, the electron in its frame sees the charge in the ma- terial (i.e. the nucleus) as accompanied by a cloud of virtual photons. These virtual photons scatter off the electron by Thomson scattering. By combining these two effects we can determine the photon spectrum for bremsstrahlung and hence estimate the radiation length.

For a relativistic charge with Lorentz factor γ, and an impact parame- ter b the electromagnetic fields are predominantly transverse with $E \simeq cB$. These fields can be considered as due to virtual photons. The peak electric field, when the distance between nucleus and electron is smallest, is

As the name implies, the radiation length is a distance. As we will see below, it scales with ρ^{-1}. In some situations it is useful to take out this de- pendence and use a 'reduced' radiation length $X_0' = \rho X_0$, which has units of, for example, g/cm^2. The length dimension can be restored by dividing by the mass density.
It is common to use the same designa- tion, 'radiation length', and the symbol X_0 for both properties. We do not sup- port this practice.

See exercise 3.

$$E = \frac{\gamma e}{4\pi\varepsilon_0 b^2}.$$

An observer at rest sees a field greater than half this value for a time $\Delta t = b/(\gamma c)$. The frequency spectrum of this pulse extends up to a max- imum frequency $\omega_{max} \approx (\Delta t)^{-1} = \gamma c/b$. The total energy in the pulse is $U = \varepsilon_0 E^2 V$, where V is the volume, $V \approx \pi b^2 \times b/\gamma$. Hence,

The γ factor comes from the Lorentz transformation of the electric field in the direction transverse to the boost.

The energy in the electric and magnetic fields is equal for electromagnetic waves in vacuum.

$$U \approx \frac{\gamma e^2}{16\pi\varepsilon_0 b}.$$

Assuming all this energy is radiated, its spectral density can be approximated by

$$\frac{dU(\omega)}{d\omega} \approx \frac{U}{\omega_{max}} = \frac{e^2}{16\pi\varepsilon_0 c} = \frac{\alpha\hbar}{4}.$$

The number of photons of frequency ω is given by

$$dN_\gamma(\omega) = \frac{dU(\omega)}{\hbar\omega} = \frac{\alpha}{4}\frac{d\omega}{\omega}. \tag{2.14}$$

The virtual photons are thus predominantly soft.

A more accurate calculation for the virtual photon spectrum for an electron with speed βc gives [268]

$$\frac{dN_\gamma(\omega)}{d\omega} = \frac{2\alpha}{\pi\beta^2\omega}. \tag{2.15}$$

We can now combine this flux with the Thomson cross-section for scattering of low energy photons on electrons (eq. (2.9)). The number of interactions of a particle crossing a thickness dx is $n\sigma dx$, where σ is the cross-section and $n_A = N_A\rho/m_a$ is the volume number density of nuclei, and N_A is Avogadro's constant, ρ is the mass density and m_a is the atomic molar mass. We can use this to combine eqns. (2.15) and (2.9) to calculate the energy loss

$$dE = \int_0^{\omega_{max}} \hbar\omega\, n_A\sigma dx \frac{dN_\gamma}{d\omega}d\omega =$$

$$= \int_0^{\omega_{max}} \frac{8\pi}{3}\frac{2}{\pi}\hbar\omega \frac{N_A\rho}{m_a}\frac{Z^2\alpha^3}{\omega}\left(\frac{\hbar}{m_e c\beta^2}\right)^2 dx\, d\omega.$$

As the photons are predominantly soft, their wavelengths are large compared to the size of the nucleus and the photons are coherent, so that the matrix element for scattering scales with Z, and the cross-section can be obtained by multiplication with Z^2.

We can integrate this up to the energy of the electron ($E = \hbar\omega_{max}$) and re-write it as

$$\frac{1}{\rho E}\frac{dE}{dx} = \frac{16}{3}\frac{N_A}{m_a}Z^2\alpha^3\left(\frac{\hbar}{m_e c\beta^2}\right)^2.$$

Comparing this with the definition of the radiation length (eq. 2.13) we obtain an estimate for the radiation length,

$$\frac{1}{\rho}\frac{1}{X_0} \simeq \frac{16}{3\beta^4}\frac{\hbar^2\alpha^3}{m_e^2 c^2}\frac{N_A}{m_a}Z^2. \tag{2.16}$$

The cross-sections for bremsstrahlung and pair production can be calculated from QED, but the derivation is complicated by the screening of the nuclear charge by electrons in different shells. A measure of the extent of screening is given by the ability of the momentum transfer to resolve the charge structure of the atom, given by its approximate radius $a_0 Z^{1/3}$. Considering the kinematics, full screening is achieved for $E_\gamma \gg (\alpha Z^{1/3})^{-1}m_e c^2$ [132].

The detailed calculation including shell effects yields [477, 505]

$$\frac{1}{\rho}\frac{1}{X_0} = 4\frac{\hbar^2\alpha^3}{m_e^2 c^2}\frac{N_A}{m_a}\left\{Z^2\left[L_{rad} - f(Z)\right] + ZL'_{rad}\right\}, \tag{2.17}$$

The radiation length is inversely proportional to the matrix element squared, where the matrix element $M \propto Ze^2$ for the interaction with the nuclei. For the interaction with the shell electrons, the contribution to the radiation length is $\propto Z|M|^2$, with $M \propto e^2$.

where L_{rad} and L'_{rad} can be approximated for elements with $Z > 4$ by $L_{rad} \simeq \ln(184.15 Z^{-1/3})$ and $L'_{rad} \simeq \ln(1194 Z^{-2/3})$, respectively. $f(Z)$ is a small correction $\mathcal{O}(\alpha Z^2)$. See [379] for tabulated values of L_{rad} and L'_{rad} for $Z \leq 4$ and a parameterisation of $f(Z)$.

A simple approximation for the reduced radiation length in g/cm² is $\rho X_0 \simeq 180 A / Z^2$ (good to 20% for $13 \leq Z \leq 92$).

For inhomogeneous materials the effective radiation length can be found from

$$\frac{1}{X_0} = \sum_i \frac{v_i}{X_0^i},$$

where the v_i are the volume fractions and X_0^i the radiation lengths of the components, and the sum runs over the components of the material. Similarly, for compounds

$$\frac{1}{X_0} = \sum_i \frac{m_i}{X_0^i},$$

where the m_i are the mass fractions of the elements in the compound.

INTERACTION OF PHOTONS WITH DIFFERENT MATERIALS

As shown in the previous sections, the cross-sections for the different types of photon interactions with matter have a different dependence on the atomic number of the material. Consequently, Compton scattering dominates over a wide energy range for materials with a low Z, but is confined around 1 MeV for high-Z matter.

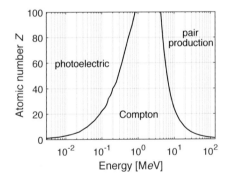

Remember that the radiation length is the scale length for bremsstrahlung and pair production in the limit of high energies.

At very high energies and material densities the cross-section for these processes are reduced, due to the Landau–Pomeranchuk–Migdal (LPM) effect [311]. Practically this is only relevant for the detection of extremely high energy cosmic rays.

Figure 2.7: Most probable interaction mechanism for photons with matter (data from [128]).

2.2 ELECTROMAGNETIC FIELDS AND MATTER

So far, we have treated electromagnetic phenomena as particles, which is in the spirit of a book on particle detectors, and arguably justified in the case of higher energy photons. However, as we will see shortly, low-energy electromagnetic interactions underlie the energy transfer of charged particles to matter. These can, to a large extent, be described by electromagnetic fields, and we therefore will now briefly revise the behaviour of such fields in the presence of matter.

Electromagnetic fields are described by Maxwell's relations, which in matter take the form of

$$\vec{\nabla} \cdot \vec{B} = 0, \qquad \vec{\nabla} \cdot \vec{D} = \rho,$$

$$\vec{\nabla} \times \vec{H} = \vec{j} + \frac{\partial \vec{D}}{\partial t}, \quad \vec{\nabla} \times \vec{E} = -\frac{\partial \vec{B}}{\partial t}, \qquad (2.18)$$

with the electric displacement field $\vec{D} = \varepsilon \vec{E}$ and the magnetic flux density $\vec{B} = \mu \vec{H}$. The proportionality constants ε and μ are the permittivity and the permeability, respectively, which can be split as $\varepsilon = \varepsilon_0 \varepsilon_r$ and $\mu = \mu_0 \mu_r$, into the value for the vacuum and a relative factor describing the effects of the medium.

We assume that these parameters are linear, isotropic, homogeneous and time independent, but as we will see later, they will depend on the frequency of the oscillation of the fields (i.e. the material is dispersive).

Introduction of the magnetic vector potential $\vec{A}$, and the electric scalar potential ϕ, defined by

$$\vec{B} = \vec{\nabla} \times \vec{A}, \text{ and } \vec{E} = -\vec{\nabla}\phi - \frac{\partial \vec{A}}{\partial t},$$

satisfies two of the Maxwell equations, but does not define the potentials unambiguously. Adoption of the Lorenz gauge

$$\vec{\nabla} \cdot \vec{A} + \varepsilon\mu\frac{\partial \phi}{\partial t} = 0$$

results in

$$-\nabla^2 \vec{A} + \varepsilon\mu\frac{\partial^2 \vec{A}}{\partial t^2} = \mu\vec{j},$$

$$-\nabla^2 \phi + \varepsilon\mu\frac{\partial^2 \phi}{\partial t^2} = \frac{\rho}{\varepsilon}, \qquad (2.19)$$

which describe wave equations with source terms $\mu\vec{j}$ and ρ/ε.

A plane wave $\vec{E}(\vec{r},t) \propto \exp[i(\vec{k}\cdot\vec{r} - \omega t)]$ is a solution to these wave equations, with a phase velocity $c = \omega/k = (\varepsilon_0\mu_0)^{-1/2}$, which gets modified by a factor $1/n = (\varepsilon_r\mu_r)^{-1/2}$ in a medium, with n the index of refraction. Absorption of a wave in the medium can be described by an imaginary part of the index of refraction, $n = n_1 + in_2$. The photoabsorption cross-section can then be written as $\sigma_\gamma = 2\omega n_2/(n_e c)$. Often the relative permeability μ_r is close to 1, and thus the absorption is attributed to a complex permittivity $\varepsilon_r = \varepsilon_1 + i\varepsilon_2$, and, for $\varepsilon_1 \gg \varepsilon_2$,

The characteristic length, with which the amplitude of the wave decays exponentially, is $\lambda_{abs} = (1/2)\mathrm{Im}k = (n_e\sigma_\gamma)^{-1}$, with n_e the number density of scattering centres, here electrons, and σ_γ the cross-section for scattering on one electron.

$$\sigma_\gamma \simeq \frac{\omega}{n_e c}\frac{\varepsilon_2}{\sqrt{\varepsilon_1}} \simeq \frac{\omega}{n_e c}\varepsilon_2, \qquad (2.20)$$

The use of the letter "n" for both the index of refraction and the number density is unfortunate, but we follow common practice here.

where the last approximation is valid for low density media (i.e. gases), and thus $\frac{\varepsilon_2}{\sqrt{\varepsilon_1}} = \frac{c}{\lambda_{abs}\omega}$.

As noted above, generally the complex permittivity will be a function of the frequency, and we will refer to $\varepsilon_r(\omega) = \varepsilon_1(\omega) + i\varepsilon_2(\omega)$ as the 'dielectric function'. As we have seen in section 2.1, the cross-section for photon interactions does change quite dramatically at low energies due to shell effects on the photoionisation cross-section, and thus we also expect significant structure for the dielectric function. To understand the origin of the variations of the index of refraction, we can use a simple classical model that is actually reasonably applicable for ionic polarisations. In the unperturbed state, the ions will be held in position by ionic bonds that can be described by a parabolic potential for not too large disturbances. An excursion of the relative position of the ions will thus result in oscillations with a frequency of ω_0. In the following we will assume a simple binary bond between two ions of charge e. In the presence of an external field $\vec{E}(t) = \vec{E}_0 \exp[i(\vec{k}\vec{r} - \omega t)]$ the dynamics of the bond separation will be described by

We assume that the rate of change of position dx/dt is small compared to the speed of light, and hence we ignore magnetic effects.

$$\mu \frac{d^2\vec{r}}{dt^2} + \mu\Gamma\frac{d\vec{r}}{dt} + \mu\omega_0^2\vec{r} = -e\vec{E}(t) = -e\vec{E}_0 e^{i(\vec{k}\cdot\vec{r}-\omega t)},$$

where μ is the reduced mass and we also have added a damping term proportional to a damping constant Γ. The solution of this equation is

In the dipole approximation the phase of the electric field at any time is effectively the same over the whole region occupied by the electron, and the spatial term in the exponential can be dropped.

The most important contribution to damping are collisions, which result in the loss of energy stored in the vibration.

$$\vec{r}(t) = -\frac{e}{\mu}\frac{1}{\omega_0^2 - \omega^2 - i\omega\Gamma}\vec{E}_0 e^{-i\omega t}.$$

This result can be used to find the polarisation

$$\vec{P} = -ne\vec{r} = \frac{ne^2}{\mu}\frac{1}{\omega_0^2 - \omega^2 - i\omega\Gamma}\vec{E}_0 e^{-i\omega t},$$

where n is the density of bonds. But $\varepsilon_r = (\varepsilon_0 E + P)/(\varepsilon_0 E)$, and thus

$$\varepsilon_r(\omega) = 1 + \frac{ne^2}{\varepsilon_0\mu}\frac{1}{\omega_0^2 - \omega^2 - i\omega\Gamma}.$$

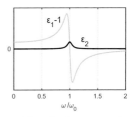

Figure 2.8: Real and imaginary index of refraction ($\Gamma = \omega/10$).

This equation describes a resonance, which in Atomic Physics is referred to as a Lorentzian. For ionic bonds typical resonance frequencies are in the infrared.

For atomic electrons similar behaviour is observed, even though they cannot be treated as classical particles with fixed equilibrium positions and linear restoring forces within the atom. Instead, their state is described by a wavefunction, which is distorted quasi-elastically by weak perturbations. The distortion can only occur to another eigenstate of the atom (including highly excited states and even the continuum), with the frequency ω_0 now describing the energy difference of the transition. The total electronic polarisation is given by the sum of the resulting resonance terms, weighted by an 'oscillator strength density' $df/d\omega$. The oscillator strength density comprises transitions to discrete atomic states (which can be formally described as additive terms $\propto f_n\delta(\omega - \omega_n)$), as well as states to the continuum. In solids discrete atomic levels rearrange into bands of states, which essentially become continuous.

This is equivalent to the oscillator strengths which we have introduced in eqs. (2.4) and (2.6).

Using the oscillator strength density, the relative permittivity is given by

$$\varepsilon_r(\omega) = 1 + \omega_p^2 \int \frac{\frac{df}{d\omega}(\omega')}{\omega'^2 - \omega^2 - i\omega\Gamma}\,d\omega'. \tag{2.21}$$

where we have used the plasma frequency ω_p, defined by

A useful dimensional approximation for the plasma frequency in eV is $28.8\sqrt{\rho z/A}$, where ρ is given in g/cm^3, and z is the effective number of free electrons per unit volume.

$$\omega_p^2 = \frac{4\pi\alpha\hbar c n_e}{m_e}. \tag{2.22}$$

Again, the sum rule $\int \frac{\mathrm{d}f}{\mathrm{d}\omega}\,\mathrm{d}\omega = 1$ applies.

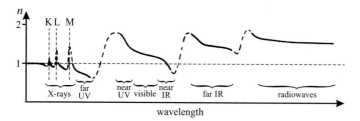

Figure 2.9: The dispersion curve of a typical transparent medium over the electromagnetic spectrum. [292].

The requirement to maintain causality between the polarisation and electric field leads to two important relations connecting the real and imaginary parts of the dielectric function (see for example [290])

These are the Kramers-Kronig relations (KKR).

$$\varepsilon_1(\omega) = 1 + \frac{2}{\pi}\,\mathrm{P}\int_0^\infty \frac{\omega'\varepsilon_2(\omega')}{\omega'^2 - \omega^2}\,\mathrm{d}\omega', \tag{2.23a}$$

P denotes the Cauchy principal value of the integral.

$$\varepsilon_2(\omega) = -\frac{2\omega}{\pi}\,\mathrm{P}\int_0^\infty \frac{\varepsilon_1(\omega') - 1}{\omega'^2 - \omega^2}\,\mathrm{d}\omega', \tag{2.23b}$$

Comparison of eq. (2.21) with the Kramers-Kronig relation eq. (2.23a) suggests a relation

$$\frac{\mathrm{d}f}{\mathrm{d}\omega}(\omega) = \frac{2}{\pi}\frac{\omega}{\omega_p^2}\varepsilon_2(\omega). \tag{2.24}$$

This is confirmed by an alternative version of the TRK sum rule eq. (2.7) that can be found for ε_2 [504]. The integral in eq. (2.23a) can be split as

$$\varepsilon_1(\omega) - 1 = \frac{2}{\pi}\,\mathrm{P}\int_0^{\omega_c} \frac{\omega'\varepsilon_2(\omega')}{\omega'^2 - \omega^2}\,\mathrm{d}\omega' + \frac{2}{\pi}\,\mathrm{P}\int_{\omega_c}^\infty \frac{\omega'\varepsilon_2(\omega')}{\omega'^2 - \omega^2}\,\mathrm{d}\omega'.$$

where ω_c is a cut-off frequency above the absorption region. For frequencies high above the absorption region, $\varepsilon_1(\omega) \to 1 - \omega_p^2/\omega^2$ and $\varepsilon_2(\omega) \to 0$. Thus the second integral goes to 0. If we now investigate this expression for $\omega \gg \omega_c$ we can ignore ω' in the denominator in the first integral and obtain

$$\frac{\omega_p^2}{\omega^2} = \frac{2}{\pi\omega^2}\int_0^{\omega_c} \omega'\varepsilon_2(\omega')\,\mathrm{d}\omega',$$

We don't need the principal value here any more because there is no singularity.

so that the sum rule becomes

$$\int_0^\infty \omega\varepsilon_2(\omega)\,\mathrm{d}\omega = \frac{\pi}{2}\,\omega_p^2. \tag{2.25}$$

But the Kramers-Kronig relations apply more generally for any complex analytical function. For example, $1/\varepsilon(\omega)$ is analytical in the upper half of the complex frequency plane, as $\varepsilon(\omega)$ is analytical and has no

See for example [504].

zeros there. Consequently,

$$\text{Re}\left(\frac{1}{\varepsilon_r(\omega)}\right) = 1 + \frac{2}{\pi}\, \text{P}\!\int_0^\infty \frac{\omega' \,\text{Im}\left(\frac{1}{\varepsilon_r(\omega')}\right)}{\omega'^2 - \omega^2}\, d\omega', \qquad (2.26a)$$

$$\text{Im}\left(\frac{1}{\varepsilon_r(\omega)}\right) = -\frac{2\omega}{\pi}\, \text{P}\!\int_0^\infty \frac{\text{Re}\left(\frac{1}{\varepsilon_r(\omega')}\right) - 1}{\omega'^2 - \omega^2}\, d\omega'. \qquad (2.26b)$$

$$\text{Re}\left(\frac{1}{\varepsilon_r(\omega)}\right) = \frac{\varepsilon_1(\omega)}{|\varepsilon_r(\omega)|^2},$$
$$\text{Im}\left(\frac{1}{\varepsilon_r(\omega)}\right) = \frac{-\varepsilon_2(\omega)}{|\varepsilon_r(\omega)|^2}.$$

$\text{Im}(1/\varepsilon_r(\omega))$ *is often referred to as 'dielectric loss function'. This is a measure of the attenuation of an electromagnetic wave in a "lossy" dielectric.*

Note that while appearing very similar, these equations are not directly derived from eq. (2.23).

A similar derivation as before yields a corresponding sum rule

$$\int_0^\infty \omega \,\text{Im}\left(\frac{-1}{\varepsilon_r(\omega)}\right) d\omega = \frac{\pi}{2}\,\omega_p^2. \qquad (2.27)$$

In principle the oscillator strengths can be calculated for a given element using time-dependent perturbation theory. However, in practice this is complicated, as it requires knowledge and processing of the electronic states of the atoms, including shell effects. A more pragmatic approach is to derive the dielectric function, and ultimately the oscillator strengths from the more conveniently available data of photon absorption using eqs. (2.20) and (2.23a).

2.3 CHARGED PARTICLE ENERGY LOSS

For charged particles the primary means of interaction with the detector material is by electromagnetic interaction. Even though the detector material is on the whole electrically neutral, it provides plenty of electrical charges at the subatomic level. Of those, the most relevant for the generation of a signal in a particle detector are electrons, which are relatively mobile and thus will readily respond to the stimulus of the passing particle, and acquire energy from the incoming particle.

We will discuss the interaction with the nuclei in the material in section 2.10.

It is instructive to look at this first using a classical approach, originally developed by Bohr [152]. We start with modelling the interaction of the incoming particle of mass M and charge ze with a single electron of mass m_e, where we make two assumptions: First, that the mass of the incoming particle is high so that its motion is not disturbed by the collision, and second, that the electron is free to move and changes in its potential energy are negligible (it behaves as a free particle). The electron will then get a radial kick

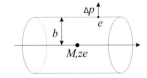

Figure 2.10: Geometry used in the derivation.

The effects of the longitudinal component cancel as the particle passes.

This simple model is based on an electrostatic interaction only. A calculation using non-relativistic quantum mechanics is discussed in exercise 6.

$$\Delta p = \int F\, dt = e\int E_\perp \frac{dt}{dx}\, dx = \frac{2z\alpha\hbar}{\beta b},$$

where we have used Gauss' law, $\int E_\perp 2\pi b\, dx = ze/\varepsilon_0$. The energy gained by an electron at radius b is then

$$\Delta E(b) = \frac{\Delta p^2}{2m_e} = \frac{2z^2\alpha^2\hbar^2}{m_e\beta^2 b^2}.$$

The energy transfer is independent of the mass of the incoming particle, and proportional to the inverse mass of the target particles. This is the reason energy transfer to more massive particles in the material, i.e. nuclei, is negligible.

If the particle passes through a slab of matter of thickness dx, with an electron density n_e, the energy loss will be

$$dE = -\Delta E(b)n_e\, 2\pi b\, db\, dx = -\frac{4\pi z^2 \alpha^2 \hbar^2 n_e}{m_e \beta^2} \frac{db}{b}\, dx.$$

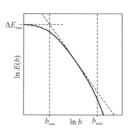

The energy loss per path length (also called the 'stopping power') is then obtained from an integration over the distance b,

$$\frac{dE}{dx} = -\frac{4\pi z^2 \alpha^2 \hbar^2 n_e}{m_e \beta^2} \ln\left(\frac{b_{max}}{b_{min}}\right), \tag{2.28}$$

Figure 2.11: Energy transfer as a function of impact parameter (after [290]).

where we have introduced physical integration boundaries b_{min} and b_{max}. These are given by the limits where the assumptions underlying this model break down.

The lower boundary b_{min} reflects the fact that the maximum energy transfer is limited and occurs in a head-on collision. If we stay within the non-relativistic approach, then $\Delta E_{max} = 2m_e v^2 = \Delta E(b_{min})$ (for $M \gg m_e$), and the stopping power becomes constant for large velocities. However, we can introduce some relativistic behaviour by using $\Delta E(b_{min}) = 2\gamma^2 m_e v^2$.

The exact expression is

$$\Delta E_{max} = \frac{2\gamma^2 m_e v^2}{1 + 2\gamma \frac{m_e}{M} + \left(\frac{m_e}{M}\right)^2}. \tag{2.29}$$

The upper boundary is given by the distance beyond which the interaction loses its ability to affect the electronic state. To estimate this, we assume that the electron is bound in an atomic orbit. If the interaction time, which can be estimated as $t(b) = b(\gamma v)^{-1}$, becomes longer than the orbital period ω^{-1} of the electron, the influence on the electronic state will be adiabatic, and no energy will be transferred.

The combined result of this classical derivation (with the relativistic cut-off energy) is thus

$$\frac{dE}{dx} = -\frac{4\pi z^2 \alpha^2 \hbar^2 n_e}{m_e \beta^2} \ln\left(\frac{\gamma^2 m_e v^3}{z \alpha c \hbar \omega}\right). \tag{2.30}$$

There are several things to note:

- The energy loss is independent of the mass of the incoming particle;
- It is proportional to the electron density of the detector material, otherwise it is only weakly depending on the detector material (through ω);
- As it originates in the interaction of the electrons in the material with the charged particle, it is proportional to $z^2 \alpha^2$;
- For low velocities ($\beta\gamma \lesssim 1$) it is dominated by the falling $1/\beta^2$ function, whereas for large velocities it will increase slowly due to the γ^2 dependence of the argument in the logarithm.

The rising section is often referred to as the 'relativistic rise', although this terminology is misleading, as the increase is really a consequence of the increased transverse reach of the electromagnetic interaction, as we will see in the next section. It only becomes relevant at relativistic energies.

A quantum mechanical treatment of the energy loss problem has been developed by Bethe [130]. This result is the basis for a widely used standard equation for the energy loss of a charged particle in matter [505]

$\frac{1}{\rho}\frac{dE}{dx}$ is called the 'mass stopping power'.

$$\frac{1}{\rho}\left\langle \frac{dE}{dx}\right\rangle = -K z^2 \frac{Z}{m_a} \frac{1}{\beta^2}\left(\frac{1}{2}\ln\frac{2m_e c^2 \beta^2 \gamma^2 \Delta E_{max}}{I^2} - \beta^2 - \frac{\delta(\beta\gamma)}{2}\right), \tag{2.31}$$

where the constant $K = 4\pi N_A \alpha^2 \hbar^2/m_e \simeq 0.3$ MeVmol^{-1}cm^2, I is the mean excitation energy (on a logarithmic scale) and $\delta(\beta\gamma)$ a correction

This equation is often referred to as the Bethe-Bloch equation. Bloch's contribution was a calculation of the mean excitation energy, which is not commonly used today, where data for this parameter exists (see below).

due to the density effect. The latter describes the reduction of the relativistic rise due to the polarisation of the detector material.

For the energy loss measured in a detector another modification of the relativistic rise is observed when the maximum energy transfer is limited to a value $E_{cut} < E_{max}$. As will be discussed in section 2.5, the energy transferred in individual interactions with the detector material has very long tails. The limit on the transferred energy cuts these tails, resulting in a weaker rise, and the energy loss becomes constant at high momentum. In practice, this occurs in thin detectors, because of the low probability of a high energy scattered electron being absorbed.

A model for the density effect has been developed by Sternheimer [466, 467], although its predictions are not very satisfactory. For a better treatment see the next section.

It is common to use the term 'energy loss' and dE/dx for the energy measured in a detector layer, although this is not strictly correct, and can lead to confusion.

This is called the 'Fermi Plateau'.

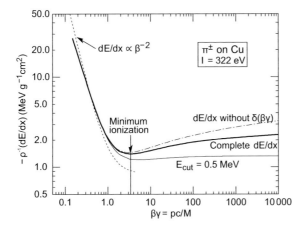

Figure 2.12: Mass stopping power as a function of $\beta\gamma$ (modified from [379]).

In the Bethe equation the only parameter that depends on the material is the mean excitation energy. While it is often treated as a parameter to fit the data, it does have a physical interpretation, and can in principle be calculated from [243]

$$\ln I = \frac{\int_0^\infty \ln(\hbar\omega)\,\sigma_\gamma(\omega)\,d\omega}{\int_0^\infty \sigma_\gamma(\omega)\,d\omega} = \frac{2}{\pi\omega_p^2}\int_0^\infty \omega\,\mathrm{Im}\left(\frac{-1}{\varepsilon(\omega)}\right)\ln(\hbar\omega)\,d\omega.$$

For most elements it can be parameterised as $I = ZI_0$, with I_0 around 10 to 12 eV. As it enters only logarithmically into the Bethe equation this variation is not significant for many applications.

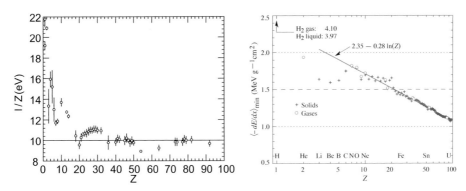

Figure 2.13: Left: Mean excitation energies divided by Z versus atomic number [379]. Right: Mass stopping power at minimum ionisation versus atomic number [505].

An important feature of the stopping power described by the Bethe equation is the minimum, which occurs at $\beta\gamma \simeq$ 3–4 with a mass stopping power of about 1–2 MeVg^{-1}cm^2. We call a particle with such a velocity a 'minimum-ionising particle' (MIP).

2.4   DIELECTRIC THEORY OF CHARGED PARTICLE ENERGY LOSS

So far, our treatment has only considered electrostatic effects, and hence has been basically non-relativistic. The complete relativistic treatment of the problem requires to solve Maxwell's equations eq. (2.18) for the source terms describing the incoming charge with velocity $\vec{\beta}c$,

$$\rho = ze\delta\left(\vec{r} - \vec{\beta}ct\right), \text{ and } \vec{j} = \vec{\beta}c\rho.$$

In the Coulomb gauge, the 4D-Fourier components of the potentials are given by [46]

We are transferring into (ω, k) space, as we will soon discuss ways to model the dielectric function in this parameter space.

$$\phi(\vec{k}, \omega) = \frac{ze}{2\pi} \frac{1/\varepsilon_r\varepsilon_0}{k^2 - \varepsilon_r\varepsilon_0\mu_0\omega^2} \delta(\omega - \vec{k}\cdot\vec{\beta}c),$$

$$\vec{A}(\vec{k}, \omega) = \frac{ze}{2\pi} \frac{\mu_0\vec{\beta}c}{k^2 - \varepsilon_r\varepsilon_0\mu_0\omega^2} \delta(\omega - \vec{k}\cdot\vec{\beta}c),$$

(2.32)

These expressions are known as the Liénard–Wiechert potentials.

Here and in the following we assume $\mu_r \simeq 1$.

with the dielectric function $\varepsilon_r = \varepsilon_r(k, \omega) = \varepsilon_1(k, \omega) + i\varepsilon_2(k, \omega)$.

The electric field generated by the incoming particle can be found from

$$\vec{E}(\vec{r}, t) = \frac{1}{(2\pi)^2} \int\int \left[i\omega\vec{A}(\vec{k}, \omega) - i\vec{k}\phi(\vec{k}, \omega)\right] e^{i(\vec{k}\cdot\vec{r} - \omega t)} \, d^3k \, d\omega.$$

The stopping power is given by the decelerating force of the electric field $\vec{E}(\vec{r}, t)$ at the location of the incoming particle

The magnetic field does not alter the energy of the incoming particle.

The multiple use of the letter E is unfortunate, but unavoidable. We denote the electric field as a vector by $\vec{E}$, and the energy of the incoming particle as E.

$$\frac{dE}{dx} = F_{\text{dec}} = ze\vec{E}(\vec{\beta}ct, t) \cdot \frac{\vec{\beta}}{\beta}.$$

After insertion of eq. (2.32), and transformation of the wave number into spherical coordinates, we integrate over the angular components, and use $\varepsilon(k, -\omega) = \varepsilon^*(k, \omega)$ to obtain

The δ-function $\delta(\omega - \vec{k}\cdot\vec{\beta}c) = (k\beta c)^{-1}\delta(\omega - k\beta c\cos\theta)$ in spherical coordinates.

The lower integration boundary for the integration over k reflects the kinematics of the interaction. Conservation of 4-momentum for an incoming particle of mass M and velocity $\vec{\beta}c$ when emitting a photon with energy $\hbar\omega$ and momentum $\hbar k$ requires

$$\frac{dE}{dx} = -\frac{z^2e^2}{2\pi^2\beta^2c^2} \int\limits_0^\infty d\omega \int\limits_{\omega/(\beta c)}^\infty dk \frac{\omega}{k} \times$$

$$\times \left[\mu_0\left(\beta^2c^2k^2 - \omega^2\right)\text{Im}\left(\frac{1}{k^2 - \mu_0\varepsilon_0\varepsilon_r(k,\omega)\omega^2}\right) - \frac{1}{\varepsilon_0}\text{Im}\left(\frac{1}{\varepsilon_r(k,\omega)}\right)\right].$$

(2.33)

$$\hbar\omega\left(1 - \frac{\hbar\omega}{2\gamma Mc^2}\right) = \hbar c\vec{\beta}\cdot\vec{k} - \frac{\hbar^2k^2}{2\gamma M},$$

which for small energy and momentum transfers ($\hbar\omega \ll \gamma Mc^2$ and $\hbar k \ll \gamma Mc\beta$) becomes

$$\omega = c\vec{\beta}\cdot\vec{k},$$

which implies $k_{min} = \omega/(\beta c)$. In practice there is also an upper integration boundary ω_{max}, which is given by eq. (2.29).

The first term in this expression is called the transverse term, as it stems from the magnetic vector potential term (in the Coulomb gauge), for which the electric field is transverse to the direction of the 3-momentum transfer, $\hbar\vec{k}$. The second term originates in the scalar potential ϕ and is often referred to as the longitudinal term, as it has the electric field parallel to the momentum transfer. It is the term relevant in the non-relativistic treatment.

The only unknown in expression (2.33) is the dielectric function, $\varepsilon_r(k, \omega)$. Unfortunately, this function is neither trivial, nor easily experimentally accessible. Its structure is determined by the quantum mechanical states of the atoms in the detector material. In addition, the phase space covered by the integral extends beyond the dispersion relation for real photons. However, simplified models, which are based on the experimentally accessible photo-absorption cross-section $\sigma_\gamma(\omega)$ have been developed to extrapolate the dielectric function into this region. One of these models, which is widely used for the accurate calculation of energy loss in particle detectors is known as the Photo-Absorption Ionisation (PAI) model [47].

At this point we will examine eq. (2.33) from a new angle and interpret this equation as an integral over collisions involving the exchange of individual photons with energy $E = \hbar\omega$ and momentum $q = \hbar k$. Because of the double integral, these photons will generally not be on mass shell, i.e. they will be virtual.

The probability density $d\sigma/dE$ for a photon exchange with energy E can be found from

$$\frac{dE}{dx} = -n_e \int_0^\infty \frac{d\sigma}{dE} E \, dE.$$

Sometimes the differential cross-section is given per atom. This just gives a different scaling factor for the cross-section. All relevant results (energy loss etc.) are unaffected by this choice as they depend on $n_e(d\sigma/dE)$.

The imaginary part of the dielectric function, $\varepsilon_2(\omega)$, for real photons (with $k = \omega/c$) can be found from the measured photo-absorption cross-section using eq. (2.20), and the real part $\varepsilon_1(\omega)$ from that using the Kramers-Kronig relation eq. (2.23a). To evaluate eq. (2.33), we have to extrapolate this over the (ω, k) plane covered by the integrals in eq. (2.33). In the PAI model the dipole approximation is assumed to be still valid for small k (in the resonance region), so that we can use $\varepsilon_r(k, \omega) \simeq \varepsilon_r(\omega)$ there. We assume that this region extends to $\hbar^2 k^2 < 2m_e\hbar\omega$. However, the loss function $\mathrm{Im}(-1/\varepsilon_r(k, \omega))$ must satisfy the sum rule [51]

$$\int_0^\infty \omega \, \mathrm{Im}\left(\frac{-1}{\varepsilon_r(k, \omega)}\right) d\omega = \frac{\pi}{2}\omega_p^2 \qquad (2.34)$$

Note that this equation is different than eq. (2.27), as ε_r depends on ω and k. It is thus valid for any slice in k.

for all k. We attribute the deficit to the absorption of the photon by stationary (quasi-free) electrons, for which non-relativistic energy and momentum conservation yields $\hbar\omega = \hbar^2 k^2/(2m_e)$, which can be expressed as a δ-function. This contribution can thus be written as $C\delta(\omega - \hbar k^2/(2m_e))$, and C can be determined using eq. (2.34). One finds

We are ignoring the Fermi motion of bound electrons.

$$C = \frac{1}{\omega} \int_0^\omega \omega' \, \mathrm{Im}\left(\frac{-1}{\varepsilon_r(\omega')}\right) d\omega',$$

and thus the loss function can be expressed as

$$\mathrm{Im}\left(\frac{-1}{\varepsilon_r(k, \omega')}\right) \simeq \Theta\left(\omega - \frac{\hbar k^2}{2m_e}\right) \mathrm{Im}\left(\frac{-1}{\varepsilon_r(\omega)}\right) +$$
$$+ \delta\left(\omega - \frac{\hbar k^2}{2m_e}\right) \frac{1}{\omega} \int_0^\omega \omega' \, \mathrm{Im}\left(\frac{-1}{\varepsilon_r(\omega')}\right) d\omega'.$$

$$\delta\left(\omega - \frac{\hbar k^2}{2m_e}\right) = \sqrt{\frac{m_e}{2\hbar\omega}} \times$$
$$\times \left[\delta\left(k - \sqrt{\frac{2m_e\omega}{\hbar}}\right) - \delta\left(k + \sqrt{\frac{2m_e\omega}{\hbar}}\right)\right].$$

After integration with this approximation the longitudinal energy loss is found to be

$$\left(\frac{dE}{dx}\right)_{long} \simeq -\frac{z^2\alpha\hbar}{\pi\beta^2 c}\int d\omega\,\omega\times$$

$$\times\left[\text{Im}\left(\frac{-1}{\varepsilon_r(\omega)}\right)\ln\frac{2m_e\beta^2 c^2}{\hbar\omega} + \frac{1}{\omega^2}\int_0^\omega \omega'\,\text{Im}\left(\frac{-1}{\varepsilon_r(\omega')}\right)d\omega'\right]. \quad (2.35)$$

In the margin: $\left(\frac{dE}{dx}\right) = \left(\frac{dE}{dx}\right)_{long} + \left(\frac{dE}{dx}\right)_{trans}$.

Essentially the only velocity dependence this expression features is the overall $1/\beta^2$ dependence which dominates the behaviour of the energy loss for low-energy projectiles. In the relativistic region it remains effectively constant. The second term in this expression is non-zero even for frequencies at which the photoabsorption cross-section, and thus $\varepsilon_2(\omega)$, has fallen to zero, hence it is the dominating term at frequencies above the absorption region.

In the margin: We introduced this term to account for the absorption of the photon by stationary (quasi-free) electrons. The integral over ω' is summing over those electrons for a given ω.

For the transverse term the integrand is largest when the denominator is small, $\omega \simeq kc/\sqrt{\varepsilon_r}$, i.e. for states close to the real photon dispersion relation, hence we simplify by replacing $\varepsilon_r(k,\omega)$ with the dielectric function for real photons $\varepsilon_r(\omega)$. The integration with this approximation over k can be done analytically and yields

In the margin: The integrals are standard, but the evaluation is elaborate, without providing much physical insight.

$$\left(\frac{dE}{dx}\right)_{trans} \simeq -\frac{z^2\alpha\hbar}{\pi\beta^2 c}\int d\omega\,\omega\times$$

$$\times\left[\text{Im}\left(\frac{-1}{\varepsilon_r(\omega)}\right)\ln\frac{1}{\sqrt{(1-\beta^2\varepsilon_1(\omega))^2 + \beta^4\varepsilon_2(\omega)^2}} + \left(\beta^2 - \frac{\varepsilon_1(\omega)}{|\varepsilon_r(\omega)|^2}\right)\Theta\right],$$

$$(2.36)$$

with $\Theta = \frac{\pi}{2} - \text{atan}\left(\frac{1-\beta^2\varepsilon_1(\omega)}{\beta^2\varepsilon_2(\omega)}\right) = \arg\left(1 - \beta^2\varepsilon_r^*(\omega)\right)$.

The transverse differential cross-section is then given by

$$n_e\left(\frac{d\sigma}{dE}\right)_{trans} = \frac{z^2\alpha}{\pi\beta^2\hbar c}\times$$

$$\times\left[\text{Im}\left(\frac{-1}{\varepsilon(E)}\right)\ln\frac{1}{\sqrt{(1-\beta^2\varepsilon_1(E))^2 + \beta^4\varepsilon_2(E)^2}} + \left(\beta^2 - \frac{\varepsilon_1(E)}{|\varepsilon(E)|^2}\right)\Theta\right].$$

The first term in this expression is the source of the relativistic rise and its saturation: In the transparent approximation ($\varepsilon_2 \simeq 0$) the argument of the logarithm becomes $(1 - v^2/u^2)^{-1/2} = \gamma'$ with the photon phase velocity $u = c/\sqrt{\varepsilon_1}$. The term depends on the density as the denominator in the logarithm becomes $|1 - \beta^2\varepsilon_r|$ in this case, and for $\beta \to 1$ tends to $|1 - \varepsilon_r| \sim n$, proportional to the density.

In the margin: It includes the screening of the electric field due to the polarisation of the material caused by the passage of the charged particle.

The second term is the only contribution to the energy loss that will be relevant at energies below the absorption region, where the loss function vanishes. It is of special interest when the material becomes transparent ($\varepsilon_2 \to 0$). In that case, Θ is close to 0, if $\beta < 1/\sqrt{\varepsilon_r}$ and jumps to close to

In the margin: The relativistic rise is a few % for solids and liquids, but can be 50%–70% in high-Z noble gases.

π above. Above this threshold the second term becomes

Below this threshold this term is an indistinguishable contribution to the transverse energy loss.

$$\left(\frac{d\sigma}{dE}\right)_{Ch} = \frac{z^2 \alpha}{n_e \hbar c} \left(1 - \frac{1}{\beta^2 \varepsilon_r}\right) \simeq \frac{z^2 \alpha}{n_e \hbar c} \sin^2 \theta_{Ch}, \qquad (2.37)$$

where $\cos\theta_{Ch} = (\beta\sqrt{\varepsilon_r})^{-1}$. This term describes the energy loss due to emission of real photons, which is called 'Cherenkov radiation', which will be discussed in more detail in section 2.8.

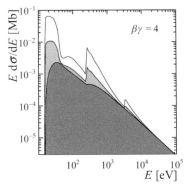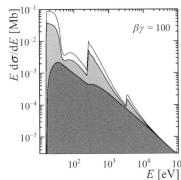

Figure 2.14: Calculated energy transfer differential cross-section for argon for $\beta\gamma = 4$ (minimum-ionising, left) and $\beta\gamma = 100$ (on Fermi plateau, right) [141]. The different shaded areas are (from top to bottom) due to distant longitudinal collisions (white - first term in eq. (2.35)), transverse collisions, (light grey - eq. (2.36)), and close longitudinal collisions (dark grey - second term in eq. (2.35). The peaks correspond to interactions with M, L and K shell electrons (from lower to higher energy).

While the PAI model has been developed for the interaction with gases, it can also be used for solids, for example silicon [276].

Another model based on dielectric theory, but specifically optimised for silicon, has been developed by Bichsel [139]. In practice, results from this model are very similar to the PAI results.

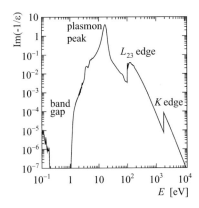

Figure 2.15: Dielectric loss function of solid silicon as a function of the photon energy [141]. Note the presence of the plasmon peak (see section 2.1) at about 17 eV, which is due to the resonant excitation of collective oscillations of electrons in the valence band. L_{23} describes X-ray transitions from 2p to 3d states.

The Bethe equation (eq. (2.31)) is a useful description of the average energy loss for charged particles, offering a simple yet reasonable parameterisation of the energy loss as a function of the projectile's velocity. However, the dielectric theory provides several benefits:

- The relativistic rise including the density effect is properly described by the theory and does not need to be added a posteriori as a correction.

- Energy loss due to Cherenkov radiation is an intrinsic aspect of the dielectric theory. It is ignored in the Bethe equation.

As we will see in section 2.8, Cherenkov energy loss is only a small fraction of the total charged particle energy loss, and can thus be ignored in many applications.

- The Bethe equation describes the mean energy loss, without providing a probability density for the energy transferred. The dielectric model interprets the energy loss as multiple exchanges of virtual

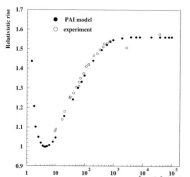

Figure 2.16: Relativistic rise of the most probable ionisation energy loss in a layer of Ar/CH$_4$ 93/7 (STP) with a thickness of 6 cm (relative to minimum-ionising energy loss). Open circles are experimental data, closed circles are from a calculation according to the PAI model [62].

photons with small energy transfers, for which a probability density can be given. This density can then be used to calculate fluctuations in the energy deposition in material slabs of finite size. This is particularly important for thin detector layers, and will be investigated in the next section.

It can be shown that, by introducing a number of approximations, the PAI model can reproduce the Bethe equation (eq. (2.31)) [46], although there is little benefit in reproducing a less accurate description.

An extension of the PAI model is the photoabsorption ionisation and relaxation (PAIR) model. In the original PAI model, it is assumed that all the energy in the transfer from the primary particle to the atoms is absorbed and converted into ionisation at the point of interaction. The PAIR model takes into account the shell structure of the atom, and that the photoelectron carries the transferred energy minus the binding energy of the given shell. After the emission of the photoelectron the atom is left in an excited state with a vacancy in the ionised shell and relaxes via the emission of fluorescent photons as well as Auger electrons. Technically, this is achieved by separating eq. (2.33) into contributions for the different shells and the subsequent simulation of atomic relaxations and δ-electrons [460].

Implementations of the PAI model (with improvements) are part of the GEANT detector simulation package [62], and of the PAIR model in the HEED software [460] (see also section 7.8). Figure 2.17 demonstrates the agreement of predictions using HEED with measured data.

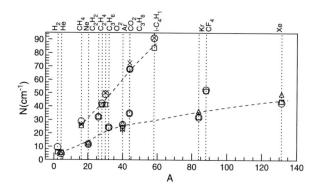

Figure 2.17: Comparison of the number of primary energy transfers for minimum-ionising particles for various gases at NTP according to calculations by HEED (circles) and various experiments [460].

2.5 ENERGY LOSS FLUCTUATIONS – STRAGGLING DISTRIBUTIONS

Once the differential cross-section $d\sigma/dE$ is known, the mean free path λ between collisions can be found from

$$\lambda^{-1} = n_e\sigma = n_e\int_{E_{min}}^{E_{max}}\frac{d\sigma}{dE}\,dE = M_0.\qquad(2.39)$$

The first moment of the differential cross-section is

$$M_1 = n_e\int_{E_{min}}^{E_{max}}E\frac{d\sigma}{dE}\,dE =$$
$$= \left\langle\frac{dE}{dx}\right\rangle.\qquad(2.38)$$

Higher moments describe the scale of the energy loss fluctuation ('straggling').

As can be seen from Figure 2.14, the differential cross-section as a function of the energy has long tails. Consequently, in each collision a varying amount of energy is transferred. As individual clusters are uncorrelated, the distance between collisions is exponentially distributed. In particular in gases the distance between collisions can be macroscopic, and can affect the reconstruction of the incoming track (see section 7.5).

Each of these discrete energy transfers is referred to as a 'cluster'.

Usually, the energy loss within a detector layer is small compared to the energy of the incoming particle, and because individual collisions are uncorrelated, the probability distribution for the number of collisions k within a distance x is given by a Poisson distribution

$$p(k,x) = \frac{\langle k\rangle^k}{k!}e^{-\langle k\rangle},$$

where the expectation value $\langle k\rangle = x/\lambda = xM_0$.

As a consequence of the statistical fluctuations of the energy transmission the energy deposited in a finite slab of detector material varies from event to event. In a thick layer the distribution of the deposited energy will acquire a Gaussian shape, in accordance with the central limit theorem. For thin layers the distribution of energy loss for a large number of incoming particles with identical energy is skewed, with a long tail due to rare high-energy clusters. Such a distribution is called a 'straggling distribution'.

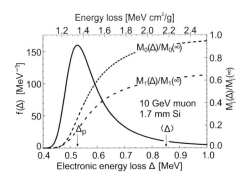

Figure 2.18: Straggling distribution. Δ_p: most probable energy loss, $\langle\Delta\rangle$: energy loss from Bethe equation, M_0 and M_1: cumulative 0th moment (mean number of collisions) and 1st moment (mean energy loss) [505].

For thin absorbers the average $\langle dE/dx\rangle$ is not a very useful parameter for characterising energy loss spectra. The most probable value Δ_p of the energy loss distribution is usually significantly smaller than the mean energy loss.

The probability density for a particle to lose an amount of energy E in a single collision is given by

$$f^{(1)}(E) = \frac{1}{M_0}n_e\frac{d\sigma}{dE}(E).\qquad(2.40)$$

In principle, the probability for an energy loss of Δ in k collisions can be found from a recurrent convolution

$$F(\Delta,k) = \int_0^\Delta f^{(k-1)}(\Delta-E)\, f^{(1)}(E)\, dE. \qquad (2.41)$$

The probability for an energy loss of Δ in k collisions is the convolution of the distribution to lose $\Delta - E$ in $k - 1$ collisions with the distribution to lose the remainder in one collision. However, we are usually interested in the energy loss distribution for a given detector thickness, not for a fixed number of collisions. Different approaches to find this are being used [140].

An early analytic approach was developed by Landau [324], and later modified by Vavilov [485]. The computations involved are quite complicated and, due to the approximations used, generally the results of the Landau calculations do not predict real straggling distributions very well. However, the general shape of the solutions are widely used for parameterising measured distributions, which led to straggling distributions generally being referred to (not very accurately) as 'Landau distributions'.

A better approach is again by convolution. We start with dividing the thickness of the detector into thin slabs of thickness δx, for each of which the probability $\langle k \rangle = \delta x / \lambda$ for a collision is small (typically a few percent in practical computations). The probability for an energy loss of Δ in this slab is then

$$F(\Delta,\delta x) = (1 - \langle k \rangle) f^{(0)}(\Delta) + \langle k \rangle f^{(1)}(\Delta) + \mathcal{O}\left(\langle k \rangle^2\right) \simeq$$

$$\simeq \left(1 - \frac{\delta x}{\lambda}\right) \delta(\Delta) + n_e \delta x \frac{d\sigma}{dE}(E). \quad (2.42)$$

where $f^{(0)}(\Delta)$ is the probability for an energy loss of Δ without a collision, which is obviously a delta function at $\Delta = 0$. From this, the material thickness can be built up by convolution, using

$$F(\Delta,x_1+x_2) = \int_0^\Delta F(\Delta-E,x_1) F(E,x_2)\, dE. \qquad (2.43)$$

The discrete nature of the ionisation loss and the associated cluster size distribution is particularly observable in gaseous detectors, where the lower interaction density along the track leads to a wide separation of the collision events.

2.6 EFFECTS OF CHARGED PARTICLE ENERGY LOSS

Usually, the virtual photon exchange between the incoming charged particle and the detector matter will result in the change of state of an electron in the material. Depending on the transferred energy this electron can be lifted to a higher bound state or, in a solid, into the conduction band, or it can escape altogether as a free particle. We have seen that if the transferred energy is sufficient, it will be predominantly inner electrons which will be affected. If the excited electron escapes with sufficient energy it can result in further excitations and ionisation of the detector material.

At the same time the electron will leave a vacancy, which will be filled in relaxation processes that involve the emission of X-rays and/or Auger electrons. The latter will typically lose their energy quickly in further collisions with the detector matter, whereas the former, due to the longer absorption lengths for photons in the X-ray regime, have a higher chance to escape the detector volume without further interaction.

The result of all the response effects is typically a number of ionisation and excitation events for each collision. Both of these effects can be made use of in particle detectors. Free charges can be moved in electric fields towards readout electrodes. The resulting induced currents on the electrodes can be used to record an electronic signal. Excited states will decay with the emission of photons ('scintillation'), and if the detector material is made transparent for these scintillation photons, they can be guided to dedicated photon detectors.

See chapter 4.

See chapter 5.

Usually, the secondaries are created in close proximity to the original collision. In gases the distance between the collisions is larger, and the clusters will display macroscopic separation.

In the extreme case, the interaction can result in the transfer of a sizeable amount of energy in a single collision, which gets transmitted to a single electron that then travels for macroscopic distances in the detector material. Such an electron is called a 'δ-electron'. δ-electron emission with enough energy to be distinguishable from the original track is rare (in argon only 5×10^{-4} of the ionisation electrons have an energy above 10 keV [148]). Nevertheless, they can be an important contribution to the degradation of the location of the primary track.

The average number of ionisation pairs created within a length L of detector material is given by

$$\langle N_{\mathrm{ip}} \rangle = \frac{\left\langle \frac{\mathrm{d}E}{\mathrm{d}x} \right\rangle L}{\langle W \rangle},$$

where the use of mean values reflects the fact that the number of ionisation pairs, like the energy loss, is subject to straggling, while W, the energy required to create the pair, will be subject to variations in the sharing of energy loss due to ionisation and excitation.

In gases $\langle W \rangle$ does depend on the energy of the incoming particle, but reaches a constant value for energies above 1 keV. The value in the high-energy limit depends on the material, and is usually about twice as large as the ionisation potential I. It is similar for incoming electrons and photons, but larger by about 15% for α particles hitting molecular gases.

For a given energy deposition, the number of ionisation pairs will be given by the distribution of the energy required to create the pair, W. If that would be the same for all collisions, no additional fluctuations to this number would be introduced, due to energy conservation. Without this constraint, the number of ionisation pairs will be characterised by a Poisson counting statistics and the variance is $\langle N_{\mathrm{ip}} \rangle$. In practice, the situation is somewhere in between, which is customarily described by [242],

$$\sigma^2 = F \langle N_{\mathrm{ip}} \rangle, \qquad (2.44)$$

Even though the ionisation charge in a cluster is produced closely together, the separation between clusters is usually lost when the charges are drifting towards a readout electrode, due to diffusion in the gas (see section 4.1).

Because of the large transferred energy this energy loss will be predominantly due to the second term in eq. (2.35).

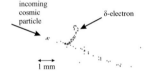

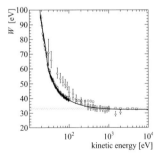

Figure 2.19: Charge deposition from a cosmic particle track in a He/iso-C_4H_{10} 80/20 gas mixture. Note the ionisation clusters and the presence of a δ-electron [179].

Figure 2.20: Energy per ionisation pair as a function of the energy of the incoming particle for CO_2 [141]. The full line shows calculation using GARFIELD. The shaded band indicates the high-energy limit.

where F is called the 'Fano factor'. Its value is between 0 (no fluctuations) and 1 (Poisson distribution). At high incident energies the value of the Fano factor is constant, but increases toward unity as the initial energy of an incident particle becomes close to the ionisation potential of the material because at low electron energies the non-ionising collisions become more likely.

The Fano factor is relevant whenever we measure an energy by measuring the ionisation charge in a detector, for example in the measurement of dE/dx (see section 12.3), or in photon spectroscopy (see section 11.1).

The fluctuations of the energy deposition between ionisation and excitations become smaller for decreasing values of $\langle W \rangle$. The Fano factor empirically correlates with $\langle W \rangle / I$, with I the ionisation energy, as [167]

$$F = 0.188 \frac{\langle W \rangle}{I} - 0.15.$$

In mixtures of gases where one of the components has excited states with energies exceeding the ionisation threshold of another component, excitation energy in the former can be transferred to ionisation of the latter, which will boost ionisation loss and hence reduce W and F ('Jesse effect' [293]). In mixtures without such an energy transfer, they are, to good approximation given by the values in the pure gases, weighted by their respective concentrations. Accurate predictions for these two parameters in gases can be made using the energy-dependent collision cross-sections for the electrons generated in the ionisation processes with molecules in the gas. Such an approach is taken in the GARFIELD simulation software, using the underlying MagBoltz code, with good success [445].

In semiconductors $\langle W \rangle$ is much less dependent on the energy of the incoming particle (only at the level of a few %), because the energy loss process is dominated by collective lattice effects. W is correlated with the band gap energy E_g, and its value at high energies can be empirically parameterised as [222]

$$\langle W \rangle = 2E_g + 1.43\,eV.$$

In semiconductors it is the Fano factor that shows a much larger energy dependence for reasons not fully understood. It is also temperature-dependent with higher values at elevated temperature. Generally, Fano factors for semiconductors are lower than for gases.

Table 2.1: High-energy values for W and F for selected gases and semiconductors [222].

Gas	W [eV]	W/I	F
He	42.3	1.72	0.21
Ne	36.6	1.70	0.13
Ar	26.4	1.68	0.14
Kr	24.1	1.72	0.17
Xe	21.9	1.80	0.13-0.17
CH_4	29.0	2.23	0.27
C_2H_6	24.9	2.13	0.25

Solid	W [eV]	E_g [eV]	F
Si	3.62	1.12	0.06
Ge	2.96	0.67	0.06
GaAs	4.2	1.43	0.14

In practical silicon detectors energy resolution is often not dominated by ionisation statistics, but by other effects like lattice imperfections etc. Measurement of the Fano factor in silicon is therefore difficult.

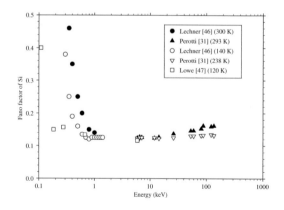

Figure 2.21: Fano factor for Silicon as a function of energy ([222] and references therein).

2.7 RANGE OF CHARGED PARTICLES IN MATTER

When a beam of particles with kinetic energy T hits a slab of matter, the particles will lose energy at a rate given by the stopping power $\langle dE/dx \rangle$, as discussed previously in this chapter. If the slab is thick and dense enough, the particles will lose all their kinetic energy and come to a stop. The distance at which this happens is called the 'range' of the particles. Because of the fluctuations in the stopping power discussed in previous sections, not all the particles will stop at the same distance and the particle density in the beam will go to zero over some finite length ('range straggling'). In many cases the relevant criterion is the essential absence of particles at a certain distance, and thus it is common to describe the range either as the distance at which the particle density has decreased to a small value (e.g. 5%), or, alternatively, using the 'practical range' R_p, which is given by the distance where the tangent with the maximum slope to the density function becomes zero.

The measurement of the range can be used to estimate the initial energy of the particle. This is used for example in large neutrino experiments to estimate the energy of muons which stop in the detector volume.

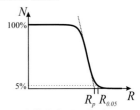

Figure 2.22: Definitions for the range of charged particles in matter.

For high energy incoming particles the range straggling will be small compared to the total distance travelled. Also, because of the increased energy loss at low velocities, the energy deposition is largest shortly before the end point of the particle track. This results in an energy loss profile that peaks towards the end of the range, with a fairly sharp fall-off. This profile for energy deposition is known as the 'Bragg peak' [161] and is, for example, exploited in particle beam therapy [387]. It allows for a high dose to be deposited locally in the region where we wish to kill cancerous cells, while having low doses on the surrounding healthy tissue.

This profile is different than for a photon beam, where the intensity decays exponentially with depth.

The range of particles as a function of T scales again with $\beta\gamma$. At high energies two regimes can be distinguished. Below relativistic energies ($\beta\gamma \lesssim 1$) the energy lost will be from the β^{-2}-dependent part of the charged particle energy loss. Above that, it will be dominated by the relativistic rise, resulting in a slower growth of the range with $\beta\gamma$. At lower energies ($\beta\gamma \lesssim 0.1$) the stopping power is smaller again, a fact that is not correctly reflected in the parameterisation eq. (2.31) of the Bethe equation, but which can be reproduced with the correct treatment of ionisation in MagBoltz.

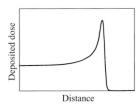

Figure 2.23: Typical dose deposition as a function of depth for a proton beam.

An empirical parameterisation that works for a large range of energies is [313]

$$\rho R_{0.05}(T) = AT\left(1 - \frac{B}{1+CT}\right), \tag{2.45}$$

with

$$A = \left(1.06Z^{-0.38} + 0.18\right) \times 10^{-3} \, \text{g/(cm}^2\,\text{keV)},$$
$$B = 0.22Z^{-0.055} + 0.79,$$
$$C = \left(1.3Z^{0.3} + 0.21\right) \times 10^{-3} \, (\text{keV})^{-1},$$

with Z for mixed materials and compounds replaced by its average value, weighted over the mass fractions.

This allows for an estimate of the range of δ-electrons. Eq. (2.45) predicts a range of 80 μm for a 1 keV electron and 1.6 mm for 10 keV in argon gas. In liquids or solids the range will be about three orders of magnitude smaller because of the density.

As discussed above, in Argon $\sim 5 \times 10^{-4}$ of the electrons from charged particle energy loss have an energy above 10 keV.

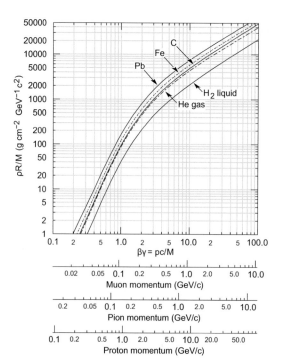

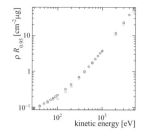

Figure 2.24: Range of heavy charged particles in different materials (modified from [379]). M is the mass of the incoming particle, and ρ the density of the absorber.

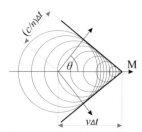

Figure 2.25: Measurements (squares) and MagBoltz calculations (circles) of the practical range of electrons in methane (at atmospheric pressure) [141].

2.8 CHERENKOV RADIATION

In section 2.4 we have seen that part of the transverse energy loss can be attributed to a term that turns on at a threshold given by $\beta_{min} = 1/n$. While the interaction of the incoming particle with the charges in the detector medium can be usually described by the exchange of virtual photons, real photons in the form of Cherenkov radiation can be emitted, if the material is transparent ($\varepsilon_2 \to 0$).

A simple picture of this effect is a luminar boom. The particle is seen as the source of spherical electromagnetic waves, which expand with the speed of light in the medium, c/n. If the particle moves faster than this, the spherical waves will build up to a conical wavefront with an opening angle given by $\cos\theta_{Ch} = (c/n)/v = (n\beta)^{-1}$.

The requirement for optical transparency is usually only satisfied in the optical region (at energies below the threshold for photoionisation), at photon energies of a very few eV. In this region the index of refraction $n \simeq 1 - \varepsilon_1$ increases with energy.

From eq. (2.37) we can see that the energy lost to Cherenkov radiation is given by

Figure 2.26: Cherenkov cone.

Eq. (2.46) is known as the Frank-Tamm equation [469].

$$\frac{dE}{dx} = \frac{z^2 \alpha \hbar}{c} \int\limits_{\beta > n(\omega)^{-1}} \left(1 - \frac{1}{\beta^2 n(\omega)^2}\right) \omega \, d\omega. \tag{2.46}$$

The energy density of the radiation is proportional to the frequency. The integral stays finite, as the index of refraction approaches 1 for $\omega \to \infty$.

As the index of refraction increases with energy the number of photons with a given wavelength is shifted towards the short wavelength end of the spectrum. This, together with the spectral sensitivity of the human eye, is the reason Cherenkov radiation from nuclear reactors immersed in water appears blue.

The number of photons $N = E/\hbar\omega$, and is then

$$\frac{dN}{dx} = \frac{z^2\alpha}{\hbar c} \int_{E_{\min}}^{E_{\max}} \left(1 - \frac{1}{\beta^2 n(E_\gamma)^2}\right) dE_\gamma \simeq$$

$$\simeq 370z^2 \int_{E_{\min}}^{E_{\max}} \sin^2\theta_{\mathrm{Ch}}(E_\gamma)\, dE_\gamma\,(eV)^{-1}\mathrm{cm}^{-1}, \quad (2.47)$$

where E_γ is the photon energy.

The amount of energy lost to Cherenkov radiation is only a small fraction of the charged particle energy loss, but it is the long range of the Cherenkov photons in transparent detectors, and the correlation of their direction with the direction and speed of the incoming particle that leads to this effect being used in a whole range of particle detectors.

The integral extends over the energies where photons are observed, i.e. where Cherenkov radiation is emitted, but also where the photon detector is sensitive.

This expression does not depend explicitly on the density. However, typically the index of refraction, and thus the Cherenkov angle and the number of photons emitted per unit track length, is larger for denser materials.

If we assume detection of Cherenkov radiation in the spectral range of visible light ($E_\gamma \simeq 1$ eV), the energy loss is about 1 keV $\times z^2 \sin^2\theta_C$ per cm, about 10^{-3} of the total charged particle energy loss in a solid.

2.9 TRANSITION RADIATION

As discussed in the previous section, in a material with index of refraction $n > 1$ Cherenkov radiation is emitted with a defined angle $\theta_{\mathrm{Ch}} = \mathrm{acos}(1/n\beta)$. However, if the material is contained within a limited thickness $L \simeq \lambda$, the superposition of radiation from emission sites along the limited track length of the incoming particle in the thin slab of material will lead to diffraction of the Cherenkov radiation, and thus of a broadening of the emission angle [292, 48].

The difference of the phases of the radiation emitted at the start of the radiator (point A) and the end (point B), when it hits the line BC, is given by $\Phi(\theta) = \Phi_A - \Phi_B = (2\pi/\lambda)[L\cos\theta - L/(\beta n)]$. The diffraction spreads the emission in θ around θ_{Ch}, which we account for by using a double differential $d^2N/(d\omega d\Omega)$. The transition from a large slab of material to a thin one can then be described by

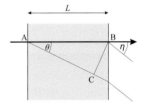

Figure 2.27: Geometry for Cherenkov radiation emission in a thin slab of material.

$$\frac{d^2N}{d\omega d\Omega} = \frac{z^2\alpha L}{2\pi c}\sin^2\theta\,\delta\left(\cos\theta - \frac{1}{\beta n}\right) \to \frac{z^2\alpha}{2\pi c}\frac{L^2}{\lambda}\sin^2\theta\left(\frac{\sin(\Phi(\theta)/2)}{\Phi(\theta)/2}\right)^2.$$

The broadening by the diffraction has the interesting consequence that real photons can get produced even if the Cherenkov condition $\beta n \geq 1$ is not met.

Outside of the thin layer is vacuum, and thus the radiation gets refracted when it exits the layer ($\sin\eta = n\sin\theta$). Hence we obtain

The interference pattern is similar to that from a finite slit ($\propto \mathrm{sinc}^2$), because in both cases the pattern can be seen as the result of the superposition of a continuous, uniformly distributed oscillator manifold with linearly changing phase.

We ignore the effect of reflections.

$$\frac{d^2N}{d\omega d\Omega'} \simeq \frac{z^2\alpha}{\pi^2\omega n^2}\sin^2\eta\,\frac{\sin^2\left[\frac{\omega L}{2c}\left(\sqrt{n^2 - \sin^2\eta} - \frac{1}{\beta}\right)\right]}{\left(\sqrt{n^2 - \sin^2\eta} - \frac{1}{\beta}\right)^2}. \quad (2.48)$$

$d\Omega' = 2\pi\sin\eta\,d\eta$.

The $\sin^2(\Phi(\theta)/2)$ factor in the numerator can be interpreted as the interference of radiation emitted from the front and back faces of the dielectric. It goes to zero for $L \to 0$, as necessary.

However, eq. (2.48) cannot be complete, as it would predict the emission of photons even in the case that the slab would be filled with vacuum

($n = 1$). We can correct this by applying the superposition principle. Introducing the slab of material has replaced a corresponding layer of vacuum and we have to subtract the equivalent amplitude for this layer, with $n = 1$ [48],

$$\frac{d^2N}{d\omega d\Omega'} = \frac{z^2\alpha}{\pi^2\omega n^2}\sin^2\eta\,\sin^2\left[\frac{\omega L}{2c}\left(\sqrt{n^2-\sin^2\eta}-\frac{1}{\beta}\right)\right] \times$$

$$\times\left(\frac{1}{\sqrt{n^2-\sin^2\eta}-\frac{1}{\beta}}-\frac{1}{\cos\eta-\frac{1}{\beta}}\right)^2. \quad (2.49)$$

The interference term remains the same as we are replacing a slab of vacuum with the equivalent slab of dielectric.

The emission of photons described by this equation is called 'transition radiation' (TR), as it requires the transition of the charged particle between layers of dielectric material and vacuum.

The first term in the bracket in eq. (2.49) is singular for the Cherenkov condition in the dielectric. It is relevant for radiation in the optical frequency region [150]. There $\varepsilon_1 > 1$ and $\varepsilon_2 \simeq 0$. In the resonance region above that the strong absorption prevents the propagation of real photons, but at very large energies (X-rays) the absorption length becomes macroscopic again. This is the region where the second term becomes important, which is singular for very small angles and β close to 1. In that region

$$n(\omega) = 1 - \frac{\omega_p^2}{\omega^2}.$$

The plasma frequency ω_p is defined in eq. (2.22).

In these conditions we can also approximate

$$\sin\eta \simeq \eta, \cos\eta \simeq 1 - \frac{\eta^2}{2}, \frac{1}{\beta} \simeq 1 + \frac{1}{2\gamma^2}.$$

Eq. (2.49) then becomes

$$\frac{d^2N}{d\omega d\Omega'} \simeq \frac{4z^2\alpha\eta^2}{\pi^2\omega}\sin^2\left[\frac{\omega L}{4c}\left(\frac{\omega_p^2}{\omega^2}+\eta^2+\frac{1}{\gamma^2}\right)\right] \times$$

$$\times\left(\frac{1}{\frac{\omega_p^2}{\omega^2}+\eta^2+\frac{1}{\gamma^2}}-\frac{1}{\eta^2+\frac{1}{\gamma^2}}\right)^2.$$

Here we have also approximated one factor of n^2 with 1.

If we remove the interference term we obtain the flux from a single face transition

$$\frac{d^2N}{d\omega d\Omega'} \rightarrow \frac{z^2\alpha\eta^2}{\pi^2\omega}\left(\frac{1}{\frac{\omega_p^2}{\omega^2}+\eta^2+\frac{1}{\gamma^2}}-\frac{1}{\eta^2+\frac{1}{\gamma^2}}\right)^2.$$

We have assumed vacuum outside of the dielectric slab. However, dielectric properties of the material outside can be accommodated by adding another factor ω_p^2/ω with the plasma frequency for the outer material in the denominator of the second term of this equation. We use the small angle approximation $d\Omega' \simeq 2\pi\theta\,d\theta$.

Integration over the energy gives the angular distribution of the single face radiated energy density ($I = N\hbar\omega$)

$$\frac{dI}{d\eta} = \frac{z^2\alpha\hbar\omega_p}{2}\frac{\eta^3}{\left(\frac{1}{\gamma^2}+\eta^2\right)^{5/2}}.$$

As expected, the emission of X-ray transition radiation photons is close to the direction of the incoming particle, with the maximum at $\eta_{\max} = \sqrt{3/2}\,\gamma^{-1}$.

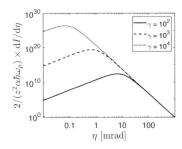
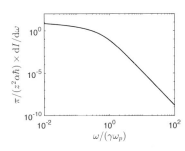

Figure 2.28: Left: Single face TR intensity as a function of angle. Right: Single face TR intensity as a function of energy.

Integration over the angle yields

$$\frac{\mathrm{d}I}{\mathrm{d}\omega} = \frac{z^2 \alpha \hbar}{\pi} \left[\left(1 + 2\frac{\omega^2}{\gamma^2 \omega_p^2}\right) \ln\left(1 + \frac{\gamma^2 \omega_p^2}{\omega^2}\right) - 2 \right].$$

This expression diverges for $\omega \to 0$. This is an artefact of the approximations used for n.

The intensity as a function of the energy scales with $\gamma \omega_p$ and drops sharply beyond $\gamma \omega_p \simeq 1$. To achieve significant intensities in the X-ray regime, where long distance propagation of the photons is possible, does require a significant Lorentz boost ($\gamma \gtrsim 100$) for typical values of the plasma frequency.

The total radiated energy at a single face is given by $I = z^2 \alpha \gamma \hbar \omega_p / 3$, increasing linearly with γ. The total number of photons diverges for $\omega \to 0$, which is a consequence of our approximations in this region. However, in practical applications there will be a low energy cut-off, which we can use for the start of the integration. In this way we obtain $N(\omega > 0.15 \gamma \omega_p) \simeq 0.5 z^2 \alpha$ for a single face transition [71]. The small photon yield per face transition is the reason practical TR detectors comprise a large number of dielectric layers.

There is no threshold for the turn-on of transition radiation.

Again, the energy lost to transition radiation is a small fraction of the charged particle energy loss. It is useful for particle detectors because of the macroscopic penetration capabilities of X-ray photons, and the dependence on the γ factor of the incoming particle.

At this point we need to take into account interference and absorption. The former is described by the interference term. One implication of this term is that the dielectric layer needs to be sufficiently deep so that radiation can develop. The length scale for which this happens is called the 'formation zone', and is given by

$$L_f = \frac{4c}{\omega} \left(\frac{\omega_p^2}{\omega^2} + \eta^2 + \frac{1}{\gamma^2} \right)^{-1} \simeq \frac{\lambda \gamma^2}{3\pi},$$

The interference term is then $4\sin(L/L_f)$.

where we used the characteristic values $\omega \simeq \gamma \omega_p$ and $\eta \simeq \gamma^{-1}$ for the approximation. A consequence of the interference is that the transition radiation intensity for a foil does not continue to linearly increase with γ, but saturates around $\gamma \simeq \sqrt{3\pi L/\lambda}$. While it appears therefore beneficial to increase the thickness of the dielectric, and also to increase the number of interfaces to increase the number of photons produced, in practice this increases absorption, and thus for practical detection of transition radiation a suitable compromise has to be sought.

The formation zone is typically a few 10 μm.

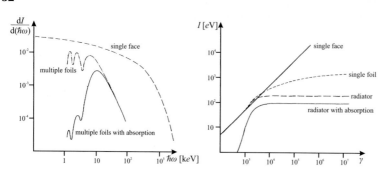

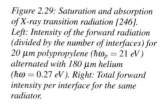

Figure 2.29: Saturation and absorption of X-ray transition radiation [246]. Left: Intensity of the forward radiation (divided by the number of interfaces) for 20 μm polypropylene ($\hbar\omega_p = 21$ eV) alternated with 180 μm helium ($\hbar\omega = 0.27$ eV). Right: Total forward intensity per interface for the same radiator.

2.10   MULTIPLE SCATTERING

So far we have discussed the energy loss of charged particles due to in-elastic collisions with the electrons in the detector material, which ultimately generates the signal we require for observation. However, charged particles also interact with the nuclei in the detector. While such collisions generally do not lead to a detectable energy transfer, they can have significant effects on the particle trajectory, which has the inconvenient consequence of degrading the knowledge of the original direction of the particle.

Depending on the number of scattering events, we distinguish three regimes: Single scattering, plural scattering (from 2 to about 20 scatters) and multiple scattering (more than 20 scatters).

Single scattering

In principle this type of scattering is described by the Rutherford scattering differential cross-section

$$\frac{\mathrm{d}\sigma}{\mathrm{d}\Omega} = \left(\frac{z^2 Z^2 \alpha \hbar c}{2\beta cp}\right)^2 \frac{1}{\sin^4\left(\frac{\chi}{2}\right)} \simeq \left(\frac{2z^2 Z^2 \alpha \hbar c}{\beta cp}\right)^2 \frac{1}{\chi^4},$$

which diverges for very small scattering angles. However, in matter the reach of the interaction is reduced by screening of the nuclear charge by the shell electrons. This reduces long-distance, small-angle scatters and effectively introduces a lower cut-off angle χ_a, with the differential cross-section becoming

$$\frac{\mathrm{d}\sigma}{\mathrm{d}\Omega} \simeq \left(\frac{2z^2 Z^2 \alpha \hbar c}{\beta cp}\right)^2 \frac{1}{(\chi^2 + \chi_a^2)^2}. \qquad (2.50)$$

Various calculations of the screening angle exist, but an important model by Molière [375] uses the Thomas-Fermi model to obtain

$$\chi_a^2 = \chi_0^2 \left(1.13 + 3.76 \left(\frac{zZ\alpha}{\beta}\right)^2\right),$$

with

$$\chi_0 = \frac{\lambdabar}{a} = \frac{1}{0.885} \frac{\hbar}{p} \frac{Z^{1/3}}{a_0},$$

where a is the Thomas-Fermi atomic radius, $a_0 = \hbar/(m_e c \alpha)$ the Bohr radius, and p and β the momentum and speed of the incoming particle.

We assume that the scattering is symmetric around the direction of the incoming particle, i.e. on spherically symmetric atoms, or, for multiple scattering, that asymmetric scatters are randomly oriented.

It is common practice in the discussion of multiple scattering to denote the polar angle for single scattering by χ, and for multiple scattering by θ.

This is sometimes called the 'screening angle'.

$\mathrm{d}\Omega = \sin\chi \, \mathrm{d}\chi \mathrm{d}\phi \simeq \chi \mathrm{d}\chi \mathrm{d}\phi.$

Plural scattering

There is no quantitative model for this, but it is usually treated as a transition between the two other cases and results are interpolated.

Multiple scattering

Multiple scattering distributions are the result of a large number (>20) of collisions. In principle, one would assume that in accordance with the central limit theorem, the distribution of the overall scattering angle after multiple scattering is Gaussian. However, the single distribution scattering distribution is not falling quickly enough for large scattering angles for this to fully apply, and the observed distribution is dominated by a central Gaussian, with non-Gaussian tails at larger scattering angles, due to the rare large-angle scattering events.

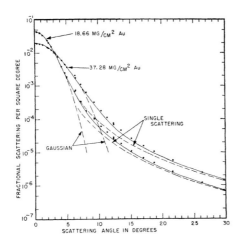

Figure 2.30: Angular scattering distribution of 15.7 MeV electrons from thick and thin gold foils [278].

The probability density function $f(\theta,t)$ that a particle has a direction given by the polar angle θ after passing a thickness t of the material must satisfy a transport equation

$$\frac{\partial f(\theta,t)}{\partial t} = n_A \int \left[f(\theta - \chi, t) - f(\theta, t) \right] \frac{d\sigma}{d\chi} \, d\chi, \qquad (2.51)$$

where $d\sigma/d\chi$ is the single scattering cross-section given in eq. (2.50) after integration of ϕ, and n_A is the number of nuclei per unit volume. In the Molière theory of multiple scattering [376] a characteristic single scattering angle

$$\chi_c^2 = 4\pi n_A t \left(\frac{zZ\alpha\hbar c}{\beta c p} \right)^2$$

The physical meaning of χ_c is that on average a particle observes one scatter with a scattering angle larger than χ_c.

is used to define the parameter

$$b = \ln\left(\frac{\chi_c^2}{1.167 \chi_a^2} \right),$$

In the discussion here we have ignored energy loss of the incoming particle, but Molière also describes a modification of χ_a and χ_c to account for this [376]. The remainder of the treatment is the same.

where the argument in the brackets is the effective number of collisions in the target. The solution to eq. (2.51) can now be found as a power series in B^{-1}, where B is defined as the solution to the equation $b = B - \ln B$. The

Typically, B is around 5 for thin absorbers and increases logarithmically with target thickness to about 20.

distribution of projected scattering angles can then be expressed by the first three elements of the series

$$f(\theta')\,\mathrm{d}\theta' \simeq \left(f^{(0)}(\theta') + \frac{f^{(1)}(\theta')}{B} + \frac{f^{(2)}(\theta')}{B^2} \right)\mathrm{d}\theta', \qquad (2.52)$$

with $\theta' = \theta_{\mathrm{proj}}/(\chi_c\sqrt{B})$ a reduced scattering angle in units of $\chi_c\sqrt{B}$, and the distribution functions $f^{(n)}(\theta')$ defined by

$$f^{(n)}(\theta') = \frac{2}{\pi n!}\int_0^\infty y e^{-y^2/4}\left[\frac{y^2}{4}\ln\left(\frac{y^2}{4}\right)\right]^n \cos(y\theta')\,\mathrm{d}y.$$

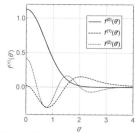

Figure 2.31: The first three distribution functions $f^{(n)}(\theta')$.

These distribution functions are generally not trivial, but the first (and dominating) one is simply a Gaussian,

$$f^{(0)}(\theta') = \frac{2}{\sqrt{\pi}} e^{-\theta'^2} = \frac{2}{\sqrt{\pi}} e^{-\frac{\theta_{\mathrm{proj}}^2}{\chi_c^2 B}}.$$

Numerical values for the distribution functions have been given in Molière's original paper [376] and in a follow-up paper by Bethe [131]. Simple empirical representations are given in [248].

In θ_{proj} the standard deviation of this Gaussian is $\chi_c\sqrt{2B}$. The other terms in eq. (2.52) describe the non-Gaussian tails of the distribution.

Usually, the direction of the scattered particle is given in 3D space, and is symmetric around the direction of the incoming particle, so that multiple scattering distributions are depending on the polar angle θ_{space}. However, it is often the projection onto a plane containing the direction of the incoming particle that is relevant. In that projection the direction of the scattered particle is given by an angle θ_{proj}.

That plane, for example, can be the bending plane of a magnetic spectrometer.

The non-Gaussian tails of the multiple scattering distribution are usually very small (at the level of 10^{-3}), and thus can be neglected for the estimation of errors in tracking in particle physics experiments. It is therefore common to use the width of the Gaussian core as the sole measure to parameterise the effects of multiple scattering on tracking. Early on [433] it was noticed that the scaling by $(N_A/A)Z^2$ invites the use of the radiation length X_0 as a scale variable to describe multiple scattering. Highland [286] developed a parameterisation for this, for which Lynch and Dahl [353] have found improved parameter values which are today generally accepted as a standard description of multiple scattering [505]

In particle therapy the tails cannot be neglected.

Multiple scattering is not a radiative process. Nevertheless, the radiation length is still a useful parameter here, as the underlying electromagnetic interactions are similar.

It is amusing that Lynch and Dahl in their paper propose a parameterisation they consider better than the Highland form (and is not relying on the radiation length). However this parameterisation has not won general acceptance.

$$\theta_0 = \theta_{\mathrm{proj}}^{\mathrm{stdev}} = \frac{1}{\sqrt{2}}\theta_{\mathrm{space}}^{\mathrm{stdev}} =$$

$$= \frac{13.6\,\mathrm{MeV}}{\beta cp} z \sqrt{\frac{x}{X_0}}\left[1 + 0.088\log_{10}\left(\frac{z^2}{\beta^2}\frac{x}{X_0}\right)\right], \qquad (2.53)$$

where x is the distance travelled through the material.

$\theta_{\mathrm{space}}^{\mathrm{stdev}}$ is the standard deviation of the 3D spatial scattering angle distribution, which for small angles can be approximated by

$$P(\theta)\,\mathrm{d}\Omega =$$

$$= \frac{2\theta}{\theta_{\mathrm{space}}^{\mathrm{stdev}\,2}}\exp\left[-\left(\frac{-\theta}{\theta_{\mathrm{space}}^{\mathrm{stdev}}}\right)^2\right]\mathrm{d}\theta.$$

Molière theory (and other descriptions of multiple scattering) makes predictions of the scattering angle but not for the displacement of the particle after the passage through a given slab of material. For the simulation of the impact of multiple scattering on tracking performance the position of the particle along its trajectory must be determined. Two main approaches are used: In 'detailed' simulations a complete sequence of individual scatters is generated. The results from such a calculation is accurate, but due to computational demands this method is typically limited to

$\theta_{\mathrm{proj}}^{\mathrm{stdev}}$ is the standard deviation of the angle in a projection of this distribution onto a plane containing the direction of the incoming particle.

a few hundred scatters. The second approach are 'condensed' simulations, which make use of a multiple scattering model like Molière theory. The track is divided into segments (steps) that are still long compared to the mean free path, and the displacement due to multiple scattering for each segment based on the underlying multiple scattering theory is computed. The algorithms used for this are not exact and the main source of errors in these Monte Carlo codes. In particular, results can depend on the step size, and only become stable if these are sufficiently small, thus becoming computationally demanding. More recently, 'mixed' approaches are being explored, which combine condensed simulations for small scattering angles with a limited number of detailed calculations of large angle scatters [248].

For an overview of the methods employed by GEANT see ref. [478].

2.11 INTERACTIONS OF ELECTRONS WITH MATTER

Electrons as charged particles will lose energy in interactions with shell electrons similarly to other charged particles. In principle, this interaction follows the same principles as outlined previously for other charged particles, with adjustment for the kinematics and interaction cross-sections.

In addition, as electrons have a low mass they do have sizeable cross-section for bremsstrahlung.

CRITICAL ENERGY

If we compare the energy loss by bremsstrahlung, eq. (2.13), to the energy loss to ionisation and excitation as, for example, given in eq. (2.31), we note that the radiative energy loss scales dominantly with Z^2 compared to Z, as it is due to an interaction with the nuclei rather than the shell electrons. The radiative energy loss increases with the energy E of the incoming particle. Hence it will dominate at high energies. The energy at which the radiative energy loss exceeds the loss to ionisation and excitation is called the critical energy E_c.

The exact definition of E_c is ambiguous. Two definitions exist, one where the ionisation energy loss equals the radiative energy loss at the critical energy,

$$\left(\frac{dE}{dx}\right)^{ion}_{E=E_c} = \left(\frac{dE}{dx}\right)^{rad}_{E=E_c},$$

the other where the ionisation energy loss equals the radiative energy loss at high energy,

$$\left(\frac{dE}{dx}\right)^{ion}_{E\to\infty} = \left(\frac{dE}{dx}\right)^{rad}_{E\to\infty} = \frac{E}{X_0}.$$

The latter is the one recommended by the PDG [505] and we follow that recommendation.

The critical energy can be parameterised as $710\ \mathrm{MeV}/(Z+0.92)$ (for gases) or $610\ \mathrm{MeV}/(Z+1.24)$ (for solids) to within a few %. For muons, the lightest charged particle with a mass above the electron, the critical energy is of the order of TeV.

The bremsstrahlung radiation spectrum is extending over the full range up to the energy of the incoming particle. It peaks for small energies. The probability for a large energy transfer is driven by screening effects, with a

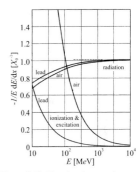

Figure 2.32: Fractional energy loss to ionisation and excitation, and fractional energy loss by radiation for electrons per radiation length of air and of lead [433].

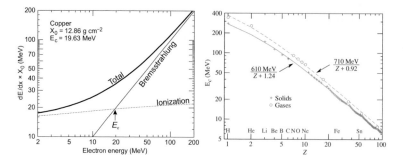

Figure 2.33: Left: Electron energy loss due to bremsstrahlung compared to energy loss due to ionisation (modified from [379]). Right: Critical energy for materials with different Z [505].

higher probability for the emission of a high-energy photon for incoming particles with higher energy.

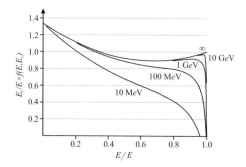

Figure 2.34: Differential radiation probability per radiation length in lead for electrons of various energies (modified from [432]). The vertical scale is defined by

$$\left(\frac{dE}{dx}\right)_{rad} = -\int_0^E \frac{E_\gamma}{E} f(E, E_\gamma)\, dE_\gamma.$$

Classically, the radiation in the radiation zone is a dipole radiation in the frame of the electron, with the maxima perpendicular to the acceleration. Lorentz transformation into the lab frame gives a narrow, but not sharp, cone of radiation with a cone angle of $\theta = 1/\gamma$.

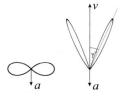

Figure 2.35: Bremsstrahlung emission direction in the electron frame (left), and in the lab frame (right).

This is again an example for the head-light effect (see exercise 4).

2.12 HADRONIC INTERACTIONS

While the processes involved in electromagnetic interactions are well understood, hadronic interactions are significantly more challenging, as they involve complex nuclear physics and low momentum transfers, so that perturbative QCD is not applicable.

Before we consider interactions of hadrons with nuclei, we will first consider the simpler case of hadronic interactions on protons. The total cross-section is found to change slowly with the energy in the centre-of-mass system.

There is a significant fraction of elastic scattering. The inelastic cross-section is dominated by processes with low momentum transfer and therefore cannot be predicted using perturbative QCD and we rely on simple phenomenological models. The most crucial empirical observation is that the secondary particles have low transverse momentum p_T (relative to the direction of the incident hadron), with a mean value of $\langle p_T \rangle \sim 400$ MeV/c.

In the longitudinal direction (parallel to the incident hadron) the secondary particles are distributed approximately uniformly in rapidity

In the lab frame elastic scattering results in energy transfer from the incident proton to the target proton and therefore does result in a hadronic secondary.

The mean transverse momentum $\langle p_T \rangle$ increases with the centre-of-mass energy ($\sqrt{s}$) and can empirically be parameterised as [195]

$$\langle p_T/(GeV/c) \rangle = 0.413 -$$
$$-0.0171 \ln(s/GeV^2) +$$
$$+0.0143 \ln^2(s/GeV^2).$$

$$y = \frac{1}{2} \ln\left(\frac{E + p_{Lc}}{E - p_{Lc}}\right), \tag{2.54}$$

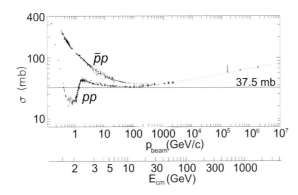

where E is the energy and p_L is the momentum component in the direction of the incident hadron. It can be shown that at high energies (compared to the mass) the rapidity approaches the pseudorapidity defined by

See exercise 9.

$$\eta = -\ln\tan(\theta/2), \qquad (2.55)$$

where θ is the angle to the incident particle.

The rapidity increases by a constant value in Lorentz boosting from the centre-of-mass to the lab frame. Therefore, in the lab frame the secondary particles produced in the hadronic interaction will tend to have large values of η, i.e. they will be at small values of θ. This implies that the secondary particles tend to be emitted in a narrow 'jet' of particles in the direction of the incident hadron.

The interactions of hadrons with nuclei have many additional features. The incident hadrons can interact with the entire nucleus ('coherent scattering'), but the most important process is still scattering off an individual nucleon. This will result in secondary hadrons in a very similar way to that for a proton target. The secondary hadrons can interact with other nucleons in the same nucleus and create an intra-nuclear cascade. Some of the secondary nucleons or created hadrons can also escape the nucleus. The nucleons that are captured will undergo further collisions such that the nucleus ends up in thermodynamic equilibrium in a highly-excited state. This excited nucleus will decay by emitting nucleons or light nuclei (e.g. deuterium, tritium or helium), a process that is called 'spallation'. Some of the incident energy ends up in nuclear recoil.

This is the compound nucleus model.

Spallation is used to create very intense beams of neutrons in spallation sources, such as ISIS at the Rutherford Appleton Laboratory, U.K.

At the end of the spallation process, the remaining excitation energy is insufficient for more nucleons to escape and the nucleus fully de-excites by emitting γs with an energy of ∼MeV [342].

Energy is conserved in these processes but the energy used to release nucleons from the nuclei is effectively lost for detection. An additional part of the energy that escapes detection is due to neutrinos from the weak decays of pions and kaons (e.g. $\pi \to \mu\nu_\mu$).

In very heavy nuclei like uranium, low energy neutrons can induce fission, which results in the emission of more low energy neutrons.

Hadronic interactions are primarily important for the detection of neutral hadrons, as they do not directly lose energy to ionisation. It is rather the ionisation generated by the secondaries created in the hadronic interactions, or the nuclear recoil, which we can use for detection. The absence of ionisation loss also accounts for the high penetration capabilities of neutral hadrons. The most important neutral hadron, of course, is the neutron.

This is a small effect in dense materials used for hadron calorimeters at accelerators but is a very significant effect when studying the hadronic showers in the atmosphere induced by cosmic rays.

LENGTH SCALE OF HADRONIC INTERACTIONS

The 'nuclear interaction length' is defined as the mean free path of a hadronic particle before undergoing an inelastic collision. It is given by

$$\lambda_{int} = \frac{1}{n\sigma_{in}} \simeq \frac{m_a}{N_A \rho \sigma_{in}}, \tag{2.56}$$

where σ_{in} is the inelastic cross-section, n is the volume number density of the nuclei, m_a their atomic mass (in g/mol), ρ the mass density and N_A is Avogadro's number. As discussed in section 2.12 the typical hadron-hadron cross-sections only vary logarithmically with energy.

We can make an approximate estimate of λ_{int} based on the scaling of the radius of nuclei with atomic number A and assuming that the cross-section is given by the geometrical cross-section, $\sigma = \pi R^2$, with the nuclear radius $R = R_0 A^{1/3}$ and $R_0 = 1.2$ fm.

The values of the interaction lengths can be calculated more accurately. The nuclear interaction length is conventionally defined by interactions of neutrons with a momentum of 200 GeV/c. At high energy the nuclear interaction length changes very little with momentum. The difference between neutron and proton interaction lengths is negligible in this energy range. However, $\pi^\pm$ have significantly smaller nuclear interaction cross-sections, and thus the interaction length for these particles are longer.

As we will discuss in chapter 11, the radiation length X_0 and the nuclear interaction length λ_{int} define the dimensions of high-energy electromagnetic and hadronic calorimeters, respectively. These two length scales have a different dependence on the size of the nuclei in the target material. While ρX_0 with the mass density ρ scales with A/Z^2 (eq. (2.17)), i.e. decreases for larger nuclei, $\rho \lambda_{int}$ scales with $A^{1/3}$ and thus increases for larger nuclei. The two are comparable for small absorber nuclei, but for large absorber nuclei, nuclear interaction lengths are typically an order of magnitude longer than radiation lengths. In chapter 12 we will discuss how we can use this difference to our advantage for electron/hadron separation. Exercise 10 compares these measured values with simple estimates based on the atomic number.

There is also the 'nuclear collision length', which is smaller than the nuclear interaction length because it also includes elastic and quasi-elastic (diffractive) reactions. Here we focus on the interaction length, as it is the relevant scale length for creating secondaries that result in a signal in a detector.

Another common estimate for the interaction length uses the black disk limit $\sigma(hN) = \sigma(hp)A^{2/3}$, where $\sigma(hN)$ and $\sigma(hp)$ are the cross-sections for the strong interaction of an incoming hadron with a nucleus and a proton, respectively, and the observation that $\sigma(hp)$ is about 37.5 mb (see Figure 2.36), so that in the energy range 1 to 1000 GeV

$$\lambda_{int} \simeq \frac{35 \text{ cm}}{\rho/(\text{g/cm}^3)} A^{1/3}, \tag{2.57}$$

with the mass density ρ. The error associated with these approximation can be estimated from Figure 2.36.

Table 2.2: Comparison of interaction lengths λ_{int} with radiation lengths X_0 for common solid absorber materials [379].

Element	X_0 [cm]	λ_{int} [cm]
Fe	1.76	16.8
Cu	1.44	15.3
Pb	0.56	17.6
U	0.32	11.0

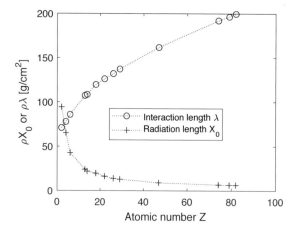

Figure 2.37: Nuclear interaction length $\rho\lambda$ and radiation length ρX_0 for different elements (data from [271]). To get to a length in centimetres, the value in the plot needs to be divided by the mass density ρ in g/cm^3.

2.13 INTERACTION OF NEUTRINOS WITH MATTER

Neutrinos (and anti-neutrinos) can only interact with matter by weak interactions. They can do that either via charged current or neutral current interactions. From dimensional analysis the cross-sections will vary with centre-of-mass energy ($\sqrt{s}$) as $\sigma \approx sG_F^2/(\hbar c)^4$, where G_F is the Fermi coupling constant.

In the lab frame, if the energy of the neutrino is E and the target mass is m, then $s = 2mc^2E$. We can then see that the neutrino cross-section for νe scattering will be three orders of magnitude smaller than for scattering of a neutrino off a nucleon. The consequences of these very small cross-sections are explored in exercise 11.

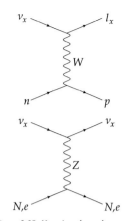

Figure 2.38: Neutrino charged current (top) and neutral current (bottom) reactions.

This will turn out to be invaluable for detecting low energy $\bar{\nu}_e$.

For charged current neutrino interactions we need to consider different types of processes as the momentum transfer (Q^2) increases:

- At very low values of Q^2 we have quasi-elastic scattering in which the target nucleon remains intact and no additional particles are created, e.g. $\bar{\nu}_e p \to ne^+$;

- At higher energy we can have coherent production of mesons, but the target nucleon remains intact, e.g. $\nu_e n \to e^- p\pi^0$;

- At energies $E_\nu \sim 1$ GeV we can have nuclear resonances, e.g. the Δ in reactions like $\bar{\nu}_\mu n \to \mu^+ \Delta^-$, which subsequently decays, e.g. $\Delta^- \to n\pi^-$;

- For higher energies we enter the Deep Inelastic Scattering (DIS) region. In DIS the neutrinos interact with a single quark in the nucleus, allowing for QCD theoretical predictions.

In addition, the struck nucleon can have final state interactions with other nucleons in the nucleus in a similar way as we discussed for hadronic interactions.

The one difference for neutral currents is that at low energy we can have elastic scattering instead of quasi-elastic scattering, e.g. $\nu_\mu e \to \nu_\mu e$.

In all instances it is the charged particles (or photons from π^0 decay), produced or liberated in the primary interaction of the neutrino, that interact with the detector material in a way described previously in this chapter and thus create a signal.

Key lessons from this chapter

- There are three dominant ways in which photons lose their energy in matter, photo-electric effect (dominating at low energies), Compton scattering (dominating around 1 MeV), and pair production (dominating at energies >1 MeV). The absorption cross-section for the photoelectric effect does increase strongly with Z, while the increase is more moderate for pair production, and much smaller for the Compton effect.

- Charged particles lose energy in electromagnetic interactions with the electrons in matter. The mean energy loss can be described by the Bethe equation. It is charac-terised by a drop $\propto \beta^{-2}$ at low energies, a minimum at $\beta\gamma \approx 4$ (which is referred to as a minimum-ionising particle), and a logarithmic rise at higher energies. This increase is larger for less dense material and the measured energy depends on the thickness of the material.

- The energy transfer from a charged particle to the material is happening through a series of small-energy transfers. The distance between collisions and the energy trans-ferred in each collision have stochastic distributions, and thus the energy deposited in a finite slab of material will display a straggling distribution with finite width and a long tail to high energies. The energy transfer can result in ionisation and/or excita-tion of the absorber material.

  - Ionisation electrons liberated in very high energy energy transfers ($>$keV) are called δ-electrons.

- Other forms of charged particle energy loss are Cherenkov and transition radiation. They only constitute a small fraction of the charged particle energy loss, but under certain conditions they can result in the production of long range photons that can be used for detection. Also, the properties of this radiation does depend on kinematic properties (β or γ) of the incoming particle, which thus becomes accessible for mea-surement.

- The most powerful approach to calculation of charged particle energy loss are dielec-tric models of the energy transfer like the PAI model and its extensions.

- Electromagnetic interactions with the nuclei in the material will result in multiple scattering of the particle, which does not lead to energy loss in the material, but a deflection of the particle trajectory.

- Light charged particles (primarily electrons) will also lose energy to bremsstrahlung in interactions with the nuclei.

- Bremsstrahlung is a closely related process to pair production and thus they can be described by the same length scale variable, the radiation length X_0. The radiation length scales with ρ^{-1} and Z^2/A (for the dominating interactions with the nuclei in the matter). The radiation length is a property of the absorber material.

  - While not a radiative process, the similarity of the interaction with the nuclei in the material also makes the radiation length a useful length parameter to de-scribe multiple scattering effects.

- Hadronic interactions are considerably more complex, but we also assign a character-istic length to hadronic interactions, the nuclear interaction length λ_{int}, which is again a property of the material. For high Z the interaction length is significantly longer than the radiation length.

- Neutrinos interact with the detector through weak interaction processes, and the actual signal in the detector is provided by the charged particles created or liberated in this interaction.

EXERCISES

1. *Energy dependence of the photoelectric effect.*

   The rate of scattering of a photon off an atom can be found from Fermi's Golden Rule

   $$\Gamma = \frac{2\pi}{\hbar} \left| \langle f | H' | i \rangle \right|^2 \rho(E),$$

   with the interaction described by a perturbation Hamiltonian H' and the density of states $\rho(E)$. The photon can be described by the vector potential $\vec{A} = A_0 \vec{\varepsilon} \exp(i\vec{k} \cdot \vec{r})$, and the interaction Hamiltonian $H' = e\vec{A} \cdot \vec{p}/m$.

   a) Evaluate the matrix element for the interaction with a hydrogen atom in the ground state ($|i\rangle = (\pi a_0^3)^{-1/2} \exp(-r/a_0)$, with the Bohr radius a_0) and the assumption that the final state can be described by a free electron wave function $|f\rangle = \exp(i\vec{p} \cdot \vec{r}/\hbar)$. (Use $\vec{q} = (\hbar\vec{k} - \vec{p})/\hbar$ to describe the momentum transfer to the atom.)

   b) For high energy photons (well above the binding energy of the atom) the electron momentum will equal the momentum transfer. What will be the angular distribution of the electron emission? What approximation can you make to the result from part a)?

   c) Show that the density of states is proportional to p.

   d) Find the differential cross-section $d\sigma/d\Omega = \Gamma/(\hbar\omega A_0^2)$ and the cross-section σ and demonstrate that the latter in this approximation is proportional to q^{-7}, or $E_\gamma^{-7/2}$.

2. *Compton scattering.*

   Verify eq. (2.10). Hint: use 4-momentum conservation to "isolate and square" the 4-momentum of the scattered electron.

3. *Weizsäcker-Williams model* (see also [498, 290]).

   a) Show that the field of a point charge moving in the x_1 direction passing a system S at a distance b from the direction of motion at the location of the system is given by

   $$E_2(t) = \frac{q\gamma b}{(b^2 + \gamma^2 v^2 t^2)^{3/2}},$$
   $$B_3(t) = \beta E_2(t),$$
   $$E_1(t) = -\frac{q\gamma v t}{(b^2 + \gamma^2 v^2 t^2)^{3/2}},$$

   with other components vanishing.

b) For $\beta \simeq 1$, $E_2(t)$ and $B_3(t)$ are equivalent to a pulse of plane-polarised radiation in the direction of the motion. $E_1(t)$ is technically lacking a B field to form a similar transverse pulse. However, if the motion is non-relativistic in the frame of S, one can add the missing B field without changing the physics, as the particles in the system only respond to electric forces.

In the Weizsäcker-Williams model the particle is seen as accompanied by a cloud of virtual photons. The frequency spectrum of these virtual quanta for the fields in part a) for a given impact parameter b is given by

$$\frac{dI(\omega, b)}{d\omega} = \frac{c}{2\pi} |E(\omega)|^2 \,,$$

where $E(\omega)$ is the Fourier transform of the transverse field in the pulse.

Find the Fourier transform of the electric fields and hence the frequency spectrum of the pulses. (You will need that $\int_{-\infty}^{\infty} \exp(iax)/(1+x^2)^{3/2}\, dx = 2aK_1(a)$ and $\int_{-\infty}^{\infty} x\exp(iax)/(1+x^2)^{3/2}\, dx = 2iaK_0(a)$, where the K_i are modified Bessel functions.)

c) To find the frequency spectrum of the radiation pulse for the interaction with matter we have to integrate this over all impact parameters from a minimum impact parameter b_{min}. From the uncertainty principle $b_{min} = \hbar/Q_{max}$. Show that for bremsstrahlung this is given by $b_{min} = \hbar/(2Mv)$.

d) Find $dI(\omega)/d\omega = \int_{b_{min}}^{\infty} dI(\omega, b)/d\omega$. What are the limits for this spectrum for low and high frequencies?

4. *The 'searchlight' (or 'headlight') effect.*

a) As derived in exercise 3 part d), the high-frequency tail for the spectrum for $\beta \to 1$ is given by

$$\frac{dn(\omega)}{d\omega} = \frac{1}{\hbar\omega} \frac{dI(\omega)}{d\omega} \propto \frac{\alpha}{\omega} e^{-\frac{2\omega b_{min}}{\gamma c}} \,.$$

b) The critical frequency ω_c describes the end point of the spectrum. Show that this frequency can be described by $\omega_c \simeq \gamma c/b_{min}$. If the radiation is seen as a beam with waist b_{min}, argue that the opening angle is given by γ^{-1}.

5. *Derivation of the Thomas-Reiche-Kuhn (TRK) sum rule.*

Start from the definition of the dipole oscillator strength, eq. (2.4). Show that this can be written as

$$f_n = \frac{i}{3\hbar} [\langle \Psi_0 | \vec{p} | \Psi_n \rangle \langle \Psi_n | \vec{x} | \Psi_0 \rangle - \langle \Psi_0 | \vec{x} | \Psi_n \rangle \langle \Psi_n | \vec{p} | \Psi_0 \rangle] \,,$$

then use the closure relation $\sum_n |\Psi_n\rangle\langle\Psi_n| = 1$ for the hydrogenic wave functions to show eq. (2.7).

6. *Derivation of the Bethe equation for energy loss using non-relativistic quantum mechanics (NRQM).*

If you are already familiar with the NRQM derivation of Rutherford scattering start at b).

a) Consider the scattering of a charged particle with mass M on an electron with mass m_e for $M \gg m_e$. The differential scattering cross section is given by

$$\frac{d\sigma}{d\Omega} = \frac{m_e^2}{4\pi^2} \left| \frac{V(q)}{\hbar^2} \right|^2 \,,$$

where the matrix element for Coulomb scattering is

$$V(q) = \int d^3 r \exp(-i\vec{q}\cdot\vec{r}) \frac{e^2}{4\pi\varepsilon_0 r},$$

where $\vec{q} = \vec{p}_f - \vec{p}_i$ and $\vec{p}_f$ ($\vec{p}_i$) is the final (initial) momentum. Show that the differential cross-section is

$$\frac{d\sigma}{d\Omega} = \left(\frac{e^2}{4p^2 \sin^2(\theta/2)}\right)^2,$$

where θ is the scattering angle and p the momentum of the incoming particle.

b) Let p be the magnitude of the 3-momentum of the incident charged particle in the rest frame of the electron before the collision and v be the speed of the incident charged particle. Show that in the frame at which the electron was initially at rest, the kinetic energy of the scattered electron is $T = (p^2/m)(1 - \cos\theta)$. Hence show that

$$\frac{d\sigma}{dT} = \frac{2\pi e^2}{m_e v^2 T}.$$

c) The rate of energy loss per unit length in a medium with atomic number Z is then given by

$$\frac{dE}{dx} = NZ \int_{T_{\min}}^{T_{\max}} \frac{d\sigma}{dT},$$

where N is the number of atoms per unit volume. We assume that $T_{\min}$ is the ionisation energy I. Show that $T_{\max} \approx 2\beta^2 \gamma^2 m_e c^2$. Hence show that the mass stopping power is

$$\frac{1}{\rho}\frac{dE}{dx} = \frac{KZ}{m_A \beta^2} \ln\left(\frac{2\gamma^2 \beta^2 m_e c^2}{I}\right),$$

where $K = 4\pi N_A r_e^2 m_e c^2$, $r_e = e^2/(4\pi\varepsilon_0 m_e c^2)$, m_A is the atomic molar mass of the absorber and N_A is Avagadro's number.

7. *Calculation of the Thomson cross-section.*

   a) Consider a transverse electromagnetic wave in vacuum with the electric field propagating in the z direction, $\vec{E} = E_0 \exp[i(kz - \omega t)]\hat{x}$. Determine the magnetic field $\vec{B}$.

   b) Calculate the Poynting vector.

   c) Calculate the acceleration of an electron in this field. Use Larmor's formula to determine the emitted radiation field. Determine the Poynting vector for the radiated field.

   d) Combining the results in part b) and c) determine the differential cross-section for Thomson scattering and finally integrate this to get the total cross-section. Check your answer with eq. (2.9).

8. *Cherenkov light in the eye.*

   Some people claim that muons from cosmic radiation can be observed through the Cherenkov radiation in the human eye. Comment on the plausibility of this statement. (The human eye will trigger a conscious response to bursts of light with about 10 photons over 100 ms.)

   After you have found your answer, look at [364].

9. *Rapidity and pseudorapidity.*

   Show that for energies E such that $E \gg m$, where m is the mass, the rapidity y is equal to the pseudorapidity η.

10. *Nuclear interaction lengths.*

    Estimate the hadronic interaction lengths using eq. (2.57) and compare with the values given in Table 2.2.

11. *Neutrino interaction cross-section.*

    Consider a neutrino experiment with 1 MeV neutrinos. Estimate the cross-section on protons. The target consists of a cubic box of water with sides of 1 m. Approximately how many neutrinos will have to hit the target in order to have 1000 interactions on a free proton (i.e. a hydrogen atom)? Clearly, neutrino experiments will need a combination of intense sources and large detectors.

    $[G_F/(\hbar c)^3 = 1.1 \times 10^{-5}$ GeV^{-2}, the proton mass $m_p = 1.7 \times 10^{-19}$ kg $= 0.939$ GeV/c^2, and you can use $\hbar c = 0.2$ GeVfm.]

3 Electronic signals and noise

In chapter 2 we considered the different ways in which high energy particles deposit their energy in matter. Now we will consider how to use the results of these interactions to create a measurable signal in a detector.

One approach is to use the ionisation created in the detector material. We will consider the alternative approach of detecting excitations in chapter 5.

As we have seen, the typical energy loss for a minimum-ionising charged particle is about 2 MeVcm^2/g, or typically a few 100 electron/ion pairs per cm of gas or a few 10000 electron/hole pairs in a 500 μm thick semiconductor. If these charges flow over a time of a few 100 ns this results in currents of nA and below. It is the challenge of electronic particle detectors to make such a small signal detectable among the electronic noise that is intrinsic to any electronic readout system.

Whether the signal will be visible above the noise will depend not on the absolute magnitude of the signal (S) or the noise (N) but the signal-to-noise ratio (S/N). Clearly, if $S/N \ll 1$ it will be impossible to separate the signal from the background and conversely, if $S/N \gg 1$ then it will be very easy to separate the signal from background. The required value of S/N will depend on the requirements of a particular system but we can get a first feeling for the required value in exercise 1.

An alternative parameter describing the noise performance of a charge-processing readout system is the equivalent noise charge (ENC). The ENC is defined as the charge at the input which would give the same magnitude charge signal as the noise. The ENC is usually give in number of electrons but it is sometimes quoted in units of fC.

To understand the performance of electronic particle detectors it is therefore necessary to consider the process of signal generation and noise theory. In this chapter, we will first discuss how electronic signals are generated and how they can be calculated, followed by a short introduction into noise theory. We will then discuss how electronic filtering can be employed to optimise S/N. In order to understand the actual values of S/N that can be achieved, we will need to consider the noise in transistors. We will then be able to combine the results to discuss overall system optimisation.

3.1 ELECTRICAL SIGNAL GENERATION

When electric charges are close to grounded conductors, they induce an electric charge on the surface of the conductor. The magnitude of the induced charge depends on the distance to the conductor. When the charge moves, the induced charge densities on the conductor will change, causing a current to flow.

We can use the Ramo-Shockley theorem [455, 418] to calculate the induced current. To derive this theorem we consider a system of several grounded electrodes and a mobile charge q. Between the conductors and

This ionisation charge could be electrons and positive ions in a liquid or gas or electrons and holes in a semiconductor.

As will be discussed later, this charge can be amplified inside the detector using very large electric fields, but not usually by an amount that would change our conclusion here.

Sometimes also the acronym SNR is used.

It is the charge signal which would result in an S/N of 1.

This is demonstrated for a very simple case in exercise 3.

This derivation follows very closely that given in [464].

DOI: 10.1201/9781003287674-3

the mobile charge there is no charge. Therefore, the potential V satisfies Laplace's equation there, $\nabla^2 V = 0$.

Next, we will consider a second configuration, where we remove the mobile charge and put 1 V on one conductor (C_1) resulting in a charge Q_1 on this conductor. Again, between the conductors there is no charge, so the new potential V_1 also satisfies Laplace's equation there, $\nabla^2 V_1 = 0$.

Consequently, the superposition $V_1 \nabla^2 V - V \nabla^2 V_1$ also equals 0 between the conductors. Applying Green's second identity to the volume between the conductors and a sphere surrounding the moving charge gives

$$\int \left(V_1 \vec{\nabla}^2 V - V \vec{\nabla}^2 V_1 \right) dv = \oint \left(V_1 \vec{\nabla} V - V \vec{\nabla} V_1 \right) \cdot d\vec{s} = 0,$$

where the closed surface integral covers the surfaces of the conductors and the sphere around the moving charge.

We can split the right hand side integral into three parts:

1. The integral over all conductor surfaces except C_1 vanishes because $V = V_1 = 0$ on these conductors.

2. On the surface of C_1, $V_1(C_1) = 1$ V, and $V = 0$ and using Gauss's law we get a contribution $V_1(C_1) \times \oint \vec{\nabla} V \cdot d\vec{s} = Q_1 V_1(C_1)/\varepsilon_0$.

3. On the surface of the sphere the second term in the integral vanishes because there is no charge enclosed by the sphere for V_1. Applying Gauss's law to the first term and reducing the radius of the sphere to 0 gives $\oint_{\text{sphere}} V_1 \vec{\nabla} V \cdot d\vec{s} \to q V_1(\vec{r}_q)/\varepsilon_0$ where $\vec{r}_q$ is the position of the mobile charge.

Combining these results, we obtain for the charge induced on the conductor C_1

$$Q_1 = q \frac{V_1(\vec{r}_q)}{V_1(C_1)} = q V_w(\vec{r}_q),$$

where we have introduced a weighting potential V_w, with dimensions of inverse length.

It is impractical to measure a static charge, but if the charge moves with a velocity $\vec{v}$, the induced current on C_1 is

$$i_1 = \frac{dQ_1}{dt} = q \frac{dV_w}{dt} = q \vec{\nabla} V_w \cdot \vec{v}(t) = -q \vec{E}_w \cdot \vec{v}(t), \tag{3.1}$$

with the weighting field $\vec{E}_w = -\vec{\nabla} V_w$. This is the Ramo-Shockley theorem.

The Ramo-Shockley theorem is derived under the assumption that magnetic effects are negligible and the electric field propagates instantaneously. In addition we assume that the charges flow freely in a nonconducting medium such as a gas.

The weighting field has dimensions of inverse length. It generally has no relation to the actual electrical field. It depends only on the geometry of the electrodes and determines how the motion of the charge couples to a specific electrode to induce a signal. It is obtained by grounding all electrodes in the system apart from the one on which the induced current is measured, and placing that electrode at unit potential. Only for a system of two conductors will the weighting field have the same form as the real electric field.

Numerically, $V_w(\vec{r}_q) = V_1(\vec{r}_q)$, but without the unit, as we have defined V_1 on the surface of the conductor to be 1 V.

If only the total charge induced on C_1 is required, it can be found from

$$Q_1 = \int i_i \, dt = -q \int_{x_i}^{x_f} \vec{E}_w \cdot \vec{x} =$$

$$= q \left[V_w(x_i) - V_w(x_f) \right].$$

where x_i and x_f are the starting and the final positions of the moving charge.

The effects of charges flowing in a conducting medium are examined in [423, 424] and a derivation which allows for relativistic effects is given in [426].

To get an induced current, we need to make the free charges move. In electronic particle detectors we use electrical drift fields in the detector to, first, separate the ionisation charges created by the interaction of the incoming particle with the detector matter, and second, to create the motion of the charges required to induce a current on the readout electrodes. The current is induced from the moment the ionisation charge is created and starts to move, for as long as the charge is moving, and it ends when the movement of the charge stops (typically when it reaches the oppositely charged electrode).

In silicon detectors where speed is not a concern (for example some mobile phone cameras) diffusion of charge carriers can be sufficient.

We will illustrate the Ramo-Shockley theorem here for a simple parallel plate capacitor and a single electron-ion pair created in its gap.

Sometimes in the literature the term 'charge collection' is used, slightly carelessly. It needs to be understood that the process of collection happens during the approach of the charge towards the electrode, and it ends with the arrival of the charge on the electrode.

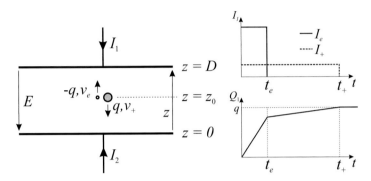

Figure 3.1: Ionisation in a parallel plate capacitor. It is assumed that the capacitor has a width and length that are much greater than the thickness so we can ignore the effects of fringe fields.

For this geometry the electric field is $E = -V_0/D$. The weighting field is $\vec{E}_w^1 = -(1/D)\hat{z}$ for the upper electrode, while it is $\vec{E}_w^2 = (1/D)\hat{z}$ for the lower electrode. For simplicity we will assume here that the drift velocity for the charges is proportional to the drift field ($v = \mu E$, with the mobility μ) and thus constant in this geometry. Therefore the induced currents on the two electrodes are

We will study the drift of charge carriers in gases in section 4.1, in liquids in section 4.2 and in semiconductors in section 4.3 in more detail.

$$I_1(t) = \underbrace{(-q)\left(-\frac{1}{D}\right)v_e}_{\text{from electron}} + \underbrace{q\left(-\frac{1}{D}\right)(-v_+)}_{\text{from ion}} = \frac{q}{D}(v_e + v_+),$$

$$I_2(t) = -I_1(t).$$

$Q = \int_0^t I\,dt$, and thus the total charge induced on the upper plate is

$$Q_1^{\text{tot}} = \int_0^\infty I_1\,dt = \frac{q}{D}(v_e t_e + v_+ t_+) = \frac{q}{D}\left(v_e\frac{D-z_0}{v_e} + v_+\frac{z_0}{v_+}\right) = q.$$

This discussion illustrates the use of the Shockley-Ramo theorem for the simplest possible case. We will use it for more complicated geometries in later chapters.

3.2   ELECTRONIC NOISE

Electronic noise is a fluctuation of the output of the electronic readout chain of a detector that is not caused by the input signal of interest. All components in the readout chain contribute to noise.

The readout chain comprises the detector, and subsequent analogue and digital electronics.

Electrical current describes the motion of charge carriers. In a metallic conductor these are typically electrons, but in other media other charge carriers can contribute as well, for example, holes in semiconductors. If the charges move between two electrodes separated by a distance l the current is given by

$$i = \sum_{i=1}^{N} \frac{e v_i}{l},$$ (3.2)

where we are summing over the charge carriers, which have individual speeds v_i.

   In this simple equation there will be two possible sources of fluctuations of the current. One is the fluctuation in the number of charge carriers and the other are the fluctuations in the speeds. Assuming these two effects are uncorrelated, we can write the fluctuation of the current (i.e. the noise) as

$$\langle \delta i \rangle^2 = \left(\frac{\langle \delta N \rangle e v}{l} \right)^2 + \left(\frac{N e \langle \delta v \rangle}{l} \right)^2.$$ (3.3)

Fluctuations in the number of charge carriers give rise to 'shot noise'. Fluctuations in the speeds can arise from thermal motion and hence this type of fluctuation is called 'thermal noise'. We call these types of noise 'random noise' as the origin of these fluctuations is stochastic. A third contribution to random noise, which is not covered by the simple picture of eq. (3.2), is '$1/f$' noise.

Thermal noise is also sometimes referred to as Johnson noise, after its discoverer.

   If we compare eq. (3.2) with the Ramo-Shockley theorem, eq. (3.1) it is obvious that similar fluctuations will be created in the detector itself. In addition, stochastic processes in the generation of the charge from the detector, some of which we have already discussed in chapter 2 (e.g. straggling distributions), can create very significant signal fluctuations.

   The above are the fundamental noise sources in our detectors and associated amplifiers but there can in addition be noise from external sources. Finally, in modern particle detectors the output of the readout chain is usually digitised, which introduces additional deviations from the ideal signal called 'digitisation noise'.

Because of the different origin of this noise (stochastic fluctuation in the ionisation charge generation) it is practical to treat this separately ('signal noise').

   In the following we will study each of these contributions in more detail.

Shot Noise

Shot noise arises from the random fluctuations in the emission of charge carriers. Shot noise is usually small, but it can be a major noise source for small currents, low temperatures and high frequencies.

A current of 1 A comprises 6.25×10^{18} electrons per second. Even fluctuations of several billion electrons per second will not have a large effect.

   If we consider the flow of electrons between two electrodes, for example in a pn junction diode (This can be a silicon detector, or a semiconductor component in an electric circuit), the emission of charge carriers across the junction will be uncorrelated. Generally, each emission event will create a signal $h(t)$ with some finite time dependence. (For example $h(t) = (U(t)/C) \exp(-t/\tau)$, with $U(t)$ a unit step function, and τ a time constant given by complex impedances in the circuit.) and the combined noise signal is given by the superposition of these. To calculate the noise variance we use Campbell's theorem [180], which states that this variance is given by the sum of the mean square contributions of all impulses preceding in time.

The stochastic nature of the charge movement is important. For example, in a regular (i.e. 'ohmic') resistor, the movement of individual electrons is correlated, and hence shot noise is negligible.

We are following here the derivation given in [413].

We denote the contribution at time t_0 from an impulse occurring at time t_i as $V_i(t_0) = q_e h(t_0 - t_i) = q_e h(u)$, and thus

$$\sigma_v^2 = \langle v^2(t_0) \rangle - \langle v(t_0) \rangle^2 = q_e^2 \sum_i h^2(t_0 - t_i) \rightarrow \langle n \rangle q_e^2 \int_{-\infty}^{\infty} h^2(u)\, du,$$

where n is the rate of impulses. The transition to continuous variables in the last part is justified in the case of high rates, $\langle n \rangle \tau \gg 1$, which is the case for macroscopic currents.

In the frequency domain the system response is given by the transfer function $H(\omega)$, which is the Fourier transform of $h(t)$. We can now use Parseval's theorem,

$$\int_{-\infty}^{\infty} h^2(t)\, dt = \frac{1}{2\pi} \int_{-\infty}^{\infty} |H(\omega)|^2\, d\omega = \frac{1}{\pi} \int_0^{\infty} |H(\omega)|^2\, d\omega,$$

to obtain

$$\sigma_v^2 = 2\langle n \rangle q_e^2 \int_0^{\infty} |H(f)|^2\, df.$$

If we assume that this variance is the result of an input spectral density i_n^2, which can be associated with a variance

$$\sigma_v^2 = \int_0^{\infty} i_n^2 |H(f)|^2\, df,$$

we find by comparison for the spectral density of the shot noise

$$i_n^2 = 2\langle n \rangle q_e^2 = 2 I q_e, \tag{3.4}$$

using $I = \langle n \rangle q_e$.

The spectral density for shot noise is independent of the frequency. Numerically, the power density for shot noise dP/df has a magnitude of 3.2×10^{-19} W/AHz. As the noise power scales with I, the noise current i_n scales with $\sqrt{I}$, and therefore the relative noise current i_n/I is proportional to $1/\sqrt{I}$, and increases for smaller currents.

Thermal Noise

Thermal noise or Johnson-Nyquist noise is the result of variations of the velocities of the charge carriers and can be seen as the effect of Brownian motion of charge carriers.

In a conductor of length l the average current due to the movement of one electron is given by $I(t) = q\bar{v}(t)/l$. The autocorrelation function for the current is then given by

$$\langle I(t)I(t+t') \rangle = \frac{q^2 \left\langle \overline{v(t)}\, \overline{v(t+t')} \right\rangle}{l^2}.$$

We will assume that electrons do not interact with each other, but will scatter with scattering centres in the band gap, with a probability given by $\exp(-t/\tau)$. The velocity will then de-correlate from its previous value at a similar rate,

$$\left\langle \overline{v(t)}\, \overline{v(t+t')} \right\rangle = \overline{v^2} e^{-|t|/\tau}.$$

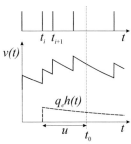

Figure 3.2: Shot noise in the time domain.

It is common to denote the spectral densities

$$e_n^2 = dv_{noise}^2/df,$$
$$i_n^2 = di_{noise}^2/df,$$

and we will adopt this convention, even though it leads to ambiguity for the current.
In calculations, e_n and i_n can be used as voltages and currents at a given frequency.

Noise with a uniform power spectral density is called 'white' noise.

For a comprehensive discussion of thermal noise, including historical context, and a discussion of thermal noise at high frequencies and low temperatures, see [7].

The derivation here follows [74].

Next, we will employ the equipartition theorem, which states that in equilibrium at a temperature T the energy stored in a degree of freedom is on average $(1/2)k_B T$, hence for the kinetic energies of the moving charge carriers

We only are interested in the longitudinal degree of freedom here.

$$\frac{\overline{mv^2}}{2} = \frac{1}{2}k_B T.$$

Putting all this together we obtain

$$\langle I(t)I(t+t')\rangle = \frac{q^2 k_B T}{ml^2}e^{-|t|/\tau}.$$

For a continuous flow of a total number $N = nAl$ electrons, where n is the electron density, and A the cross-section of the conducting channel, the autocorrelation function becomes

$$\langle I(t)I(t+t')\rangle = \frac{Nq^2 k_B T}{ml^2}e^{-|t|/\tau} = \frac{nAq^2 k_B T}{ml}e^{-|t|/\tau}.$$

Next, we will make use of the Wiener-Khinchin theorem [494, 305], which states that the power spectral density is given by the Fourier transform of the autocorrelation function. Here for the current this can be written as

$$i_n^2 = 2\mathscr{F}\left\{\langle I(t)I(t+t')\rangle\right\}.$$

In our case

$$i_n^2 = \frac{2nAq^2 k_B T}{ml}\mathscr{F}\left\{e^{-|t|/\tau}\right\} = \frac{2nAq^2 k_B T}{ml}\frac{2\tau}{1+\omega^2\tau^2}.$$

In a typical conductor, $\tau \lesssim 1$ ps, whereas ω of interest is usually $\lesssim 1$ GHz. Hence $\omega^2\tau^2 \ll 1$ and we can ignore this term.

Finally, we use the Drude conductivity for a conductor, $\sigma = nq^2\tau/m$ with the charge carrier density n, and replace $\sigma A/l = 1/R$ to find

An equivalent argument is that the current signals (and thus the autocorrelation function) are very short, hence can be described by δ-pulses, and the Fourier transform of a δ-function is uniform in frequency.

$$i_n^2 = \frac{4k_B T}{R}. \tag{3.5}$$

With $P = U^2/R = I^2 R$ the noise current power density can also be expressed as a noise power density $dP/df = 4k_B T$ or a noise voltage power density $e_n^2 = 4k_B TR$. The power spectra for thermal noise are also uniform in frequency and so are another example of white noise. At room temperature $dP/df \simeq 1.7 \times 10^{-20}$ W/Hz.

Thermal noise can be reduced by lowering the temperature. Thermally induced signal fluctuation of the signals from the detector will also decrease for lower temperatures. Both are reasons, why ultra-low noise detectors operate at cryogenic, sub-K temperatures. The most advanced technologies are on the brink to become sensitive to quantum effects in the sensing and readout (see for example [501]).

1/f noise

$1/f$ noise or 'flicker noise' comprises a number of noise mechanisms that are not well understood, but display a similar spectral characteristic in that the power spectral density falls as $1/f^\alpha$ with $0.5 < \alpha < 2$. In electronics it is often caused by trapping of charge carriers or shows up as a resistance fluctuation.

Such a spectrum is also referred to as 'pink' spectrum.

If the mean time constant for a trapping process is τ, the resulting noise spectrum will have the form (see exercise 6)

This type of noise is also called 'random telegraph noise' (RTN).

$$i_n^2 = \frac{A\omega^2}{1+(\omega\tau)^2}. \tag{3.6}$$

A combination of a few processes with different values of τ can lead to an approximate $1/f$ spectrum (see exercise 6).

The variance of the noise voltage due to $1/f$ noise does not scale with the bandwidth, but with the ratio of the high and low cut-off frequencies,

$$\sigma_v^2 \propto \int_{f_{lo}}^{f_{hi}} \frac{df}{f} = \ln\left(\frac{f_{hi}}{f_{lo}}\right).$$

This has the consequence that if the spectral response of the readout is limited by a simple bandpass filter (for which the ratio of high and low cut-off frequency is usually constant) the contribution from $1/f$ noise is usually independent of the centre frequency of the bandpass filter (which corresponds to the inverse of the time constant of the measurement).

Unlike thermal or shot noise, $1/f$ noise is device technology dependent. For example, it is the major noise mechanism in MOSFETs at low frequency, where it is caused by defects in the surface layer of the conductive channel.

A MOSFET is a widely used type of field effect transistor. The acronym stands for Metal Oxide Semiconductor Field Effect Transistor.

Interference noise

The previous sections discussed random noise which originates in the physical process of charge conduction. As such, these types of noise are often referred to as irreducible. In addition, there can be large unwanted noise contributions caused by external sources that usually can be minimised by proper design of the electronic system.

Typical instances of interference noise are

Because of the correlation with external sources, this is appropriately called 'interference', to distinguish it from the random noise discussed so far. emf or electromotive force is the electrical action produced by a non-electrical source (e.g. an external elecromagnetic field). Effectively, it is an induced voltage in the circuit.

- External noise sources that couple an emf into your electronics by capacitive or inductive coupling;

- Shared current paths on a common ground bus ('ground loops') that introduce shifts of the ground potential in a circuit. Ground loops are also vulnerable to stray AC fields, which will create an emf due to Faraday's law;

- Ripple on DC supply lines, often when the DC voltage has been generated from AC sources or by switching.

Often, external sources generate noise at specific frequencies and the generated noise spectrum is non-uniform with limited bandwidth, and maxima at these frequencies (for example interference from power lines at line frequency). In multi-channel systems interference noise is often coherent.

Coherent noise in a set of channels belonging to a section of the readout (e.g. a chip or a module) is also sometimes called 'common-mode' noise.

The good news is that interference noise can be prevented if the action of the external noise source on the system of interest is interrupted. Measures to prevent interference noise are

- Shielding (Faraday cage): Shielding reduces external noise in two ways. First, it will reflect the incident electromagnetic wave. The reflected electric field is $E_{refl} = E_0(1 - Z_{shield}/Z_0)$, with $Z_0 = \mu_0/\varepsilon_0 = 377\,\Omega$, the impedance of free space. The impedance of the shield Z_{shield} is typically much lower than this, so most of the incoming wave is reflected. Secondly, shielding will attenuate the penetrating electric field. The absorbed wave gives rise to a local current, which creates a field that counteracts the primary excitation. The net current decreases exponentially as the wave penetrates deeper into the medium, $i(x) = i_0 e^{-x/\delta}$, where δ is the 'skin depth', <u>which</u>, for frequencies much below $(\rho\varepsilon)^{-1}$, is given by $\delta = \sqrt{\rho/(\pi f \mu)}$ (In aluminium at $f = 1$ MHz the skin depth is about 85 μm). Thus, in principle, shields can provide a very power- ful means to reduce interference, as long as holes in the shields with a size of $\lambda = c/f$ and larger, where f is the frequency of interest, are avoided. It is therefore important to shield all parts, including connector housings, and loss of conductivity across seams in the shield due to corrosion or insulating surface finish must be avoided.

Here ρ is the resistivity, $\varepsilon = \varepsilon_r \varepsilon_0$ is the permittivity, and $\mu = \mu_r \mu_0$ is the permeability of the conductor.

In a colliding beam experiment it is essential to provide good shielding from the large currents caused by the beams.

- Break ground loops: Avoiding ground loops has two benefits: it re- duces the risk of shared current paths and it removes possible in- ductive loops. The best topology for ground connections is a star configuration, where ground is distributed from one, well-grounded, central point.

- Minimise the size of inductive loops: Route signal and return lines closely together. Twisted wires will further improve inductive pick- up as the field direction is reversed in consecutive twists.

- Low-impedance ground connections: High conductivity ground con- nections will reduce voltage drops. Be careful about inductances, which can introduce serious impedances at frequencies of interest.

- Ferrite sleeves or beads: These are made of materials that can be magnetised, but are not conductive. In close proximity to a conduc- tor, they will create a high impedance for high frequency signals, and thus create a low-pass filter. They are therefore useful for wires from power supplies for which we want to filter out high frequency noise over a broad frequency range.

Typical cut-off frequencies for ferrites are about 10 MHz.

- Use differential signals and receivers with high common mode re- jection ratio (CMRR). Many external types of noise will generate similar currents/voltages in the two conductors, which then cancel in such a receiver.

- Manage return current paths: Keep digital and analogue return paths separate. This is to minimise cross-talk from the large digital sig- nals into the small analogue signals. The most sensitive part of the circuit is typically the connection of the sensor/detector to the first amplifier (preamplifier).

In the subsequent discussions in this chapter we assume that interfer- ence noise has been minimised by proper design.

Digitisation noise

It is standard practice to convert the output of modern particle detectors into digitised data. Major benefits of this format are the ease of high- volume storage, good replicability, and accessibility for processing in computer-based offline analysis. However, these benefits come at the cost of a reduction of information contained in the analogue output signal from the detector. The key challenge in designing the signal processing for a particle detector system is that this data reduction does not unduly reduce the ability of the detector to measure the properties of interest.

A common feature of different digitisation processes is the mapping of a continuous variable onto a digital scale, which increments in discrete steps. This means that the output of the digitisation will differ from the analogue value, except in the centre of the interval. The distribution describing the error has a square distribution and the average error can be quantified as the RMS of that distribution.

This is an error that does not follow a Gaussian distribution. This needs to be taken in consideration if the error needs to be combined with other error contributions.

$$\mathrm{RMS}_t = \sqrt{\langle t^2 \rangle - \langle t \rangle^2} =$$

$$= \sqrt{\frac{1}{T}\int_{-T/2}^{T/2} t^2 \, \mathrm{d}t - \left(\frac{1}{T}\int_{-T/2}^{T/2} t \, \mathrm{d}t\right)^2} = \frac{T}{\sqrt{12}} \simeq 0.29T, \quad (3.7)$$

where T is one digitisation interval.

It is obvious that a small digitisation step size reduces the digitisation error. However, a small digitisation step size requires a larger number of bits to cover the range to the largest signal to be digitised. The ratio between the largest and the smallest signal that can be digitised is called the 'dynamic range' (DNR) and can be given in number of bits or in dB. The smallest measurable signal is 1 Least Significant Bit (LSB). In a linear system the DNR is then given by the number of bits available for the digitisation.

Often, a signal is not only digitised into a single number (like for example the peak voltage, or the charge in the signal), but it is sampled at regular intervals to obtain a digitised image of the signal in the time domain. Sampling again introduces a loss of information, this time the knowledge of the signal in between the sampled times. Consequently, signals with a frequency above half the sampling frequency will appear to have an unphysical frequency ('aliasing').

The frequency distribution of the amplitude digitisation noise is generally not trivial. If the signal is uncorrelated with the sampling, it will have a constant spectral density between $f = 0$ and $f = f_N$. However, if the signal is periodic, the quantisation noise can become periodic, too. Dithering is a method to effectively randomise the phase, and equalise the spectral density.

The obvious way to reduce digitisation errors is to increase the DNR and, in sampling readout systems, to increase the sampling frequency. However, there are often practical limits (size of data storage, data throughput rates, speed, cost, etc.) in this optimisation.

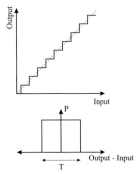

Figure 3.3: Ideal digitisation characteristics (top) and digitisation error distribution (bottom).

In dB,
$$\mathrm{DNR} = 20 \log_{10} \times$$
$$\times (2^{n_{\mathrm{bits}}} - 1) \, \mathrm{dB} \simeq 6.02 n_{\mathrm{bits}} \, \mathrm{dB}.$$

We call $f_N = f_s/2$, where f_s is the sampling frequency, the Nyquist frequency. The reconstructed frequency for a signal frequency $f_0 > f_N$ is $f_0' = f_N - (f_0 - f_N) = f_s - f_0$.

In dithering a small amount of random noise is added to the signal.

3.3 AMPLIFICATION

Due to the small amount of charge generated in a particle detector almost all electronic processing of particle detector signals requires electronic amplification of the signal. The detector usually provides us with a signal made up of charges, which, if their arrival is staggered in time, can be described as a current. However, we prefer as output of the analogue signal

processing a voltage signal. We therefore connect our particle detectors to
an electronic amplifier, which amplifies input charge or current to a voltage output.

An ideal voltage generator has negligible output impedance.

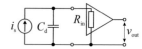

Figure 3.4: Capacitive source and amplifier.

As we have seen, we typically require electric fields to collect the ionisation charges created by the interaction of the incoming particle in the
detector material. We can therefore describe the detector as a capacitive
source, made up of an ideal current source i_s, with a parallel detector capacitance C_d. Here, we will assume that the current is constant, and flowing during the charge collection time t_c. If we connect this source to an
amplifier with input resistance R_{in}, we can distinguish two limiting cases.
If $R_{in}C_d \gg t_c$, the charge on the detector discharges slowly through the amplifier input resistance, and $v_{out} = V_0 \exp(-t/(R_{in}C_d))$, with $V_0 = Q_s/C_d$
and $Q_s = \int i_s \, dt$. In this case the amplifier input is given by the voltage created by the charge across the detector capacitance. On the other hand, if
$R_{in}C_d \ll t_c$, then the capacitor discharges quickly, and the output voltage is
proportional to i_s.

Today by far the most widely used electronic amplification devices
are transistors. These are three-terminal components, where a charge (or
a current) on the centre terminal controls the flow of current between the
outer terminals. Two major types exist. Bipolar junction transistors (BJTs)
consist of two back-to-back diodes that are connected to a common thin
region called the base. The emitter-base junction is forward biased and
the collector-base junction is reverse biased. A current injected into the
base, the terminal connecting to the doped area between the two junctions,
allows for a current to flow through the diode junctions, even though one
of them is in reverse bias. BJTs are designed so that a small base current
controls a large current between the collector and the emitter.

An alternative description that is often used is that a small change in base-emitter voltage causes a large change in current, i.e. the device has a large transconductance, g_m.

The second type of transistors are field effect transistors (FETs), in
which the width of a conducting channel between the outer terminals (here
called the drain and source) is controlled by a charge close by, which is
brought to the transistor on the centre terminal (here the gate). In an FET
the gate is insulated from the drain-source current flow, so that very little
or no current is flowing into it. There are two types of FETs: In junction
FETs (JFETs) the gate isolation is achieved by a reverse-biased diode
junction, whereas in metal oxide semiconductor FETs (MOSFETs) a
thin isolator, typically SiO_2, isolates the gate. Complementary MOS, or
CMOS, is a widely available industrial fabrication process that uses complementary and symmetrical pairs of p-type and n-type MOSFETs for
logic functions, but is increasingly used for analogue applications due to
the small feature size available and its affordability.

The MOSFET is the most copious device produced in history: An estimated 10^{22} MOSFET transistors have been produced in the past 60 years, mostly in digital ICs.

CMOS gates have the very important advantage compared to BJTs that the power consumption is very low when the gate is not switching.

While these devices provide the amplification we need, a number
of performance shortcomings (non-linearity, non-ideal input and output
impedance, the need to provide a very stable bias voltage to achieve a defined gain, and a current instead of a voltage output) make these devices
poor stand-alone amplifiers. Practically, we therefore embed them in more
complex amplification circuits, of which the most powerful is the feedback amplifier in which the gain is stabilised by feeding back part of the
output signal to the input. As the fraction of the signal fed back is usually
controlled by passive components, it is these components that control the
gain, and not the more complex transistor components.

For the amplification of particle detector signals the most useful feed-back amplifier circuit is the charge-sensitive feedback amplifier.

Also known as the 'charge-integrating amplifier'.

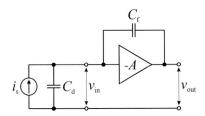

Figure 3.5: Conceptual charge-sensitive feedback amplifier.
A simple implementation uses op-amps, but similar circuits requiring fewer transistors exist (e.g. the complementary cascode, see for example [414]).

We assume an ideal inverting amplifier with infinite input impedance and a large 'open-loop' gain $-A$, so that $v_{out} = -Av_{in}$ and consequently the voltage across C_f is $v_f = (A+1)v_{in}$. Hence, the charge induced on C_f is $Q_f = C_f v_f = C_f(A+1)v_{in}$. Because of the infinite amplifier input impedance $Q_f = Q_{in}$, and thus the input capacitance of the circuit $C_{in} = Q_{in}/v_{in} = C_f(A+1)$.

The charge gain of the amplifier is the output voltage for a given input charge, and is given by

$$A_Q = \frac{v_{out}}{Q_{in}} = \frac{Av_{in}}{C_{in}v_{in}} = \frac{A}{A+1}\frac{1}{C_f} \simeq \frac{1}{C_f}, \qquad (3.8)$$

where the last approximation holds for $A \gg 1$.

The signal charge Q_s is divided by the capacitive divider C_d and C_{in}, so that the fraction of the charge supplied to the amplifier is

$$\frac{Q_{in}}{Q_s} = \frac{C_{in}}{C_d + C_{in}}.$$

To effectively couple the charge into the amplifier C_{in} should be much larger than C_d. In that case the input charge to the amplifier is independent of C_d.

For practical purposes a discharge path for the feedback capacitor C_f is usually required. This can be achieved by a parallel resistor R_f or switches.

R_f will discharge the capacitor with a time constant of $\tau = R_f C_f$.

CASCODE AMPLIFIERS

To achieve the desired signal gain before further signal processing multiple gain stages are needed. In order to minimise the effects of noise in the second or subsequent stages, we require a large gain in the first stage (usually called the 'pre-amplifier').

The gain of a single transistor common source FET amplifier is given by (see exercise 4)

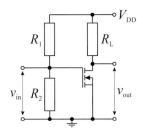

$$A_v = \frac{g_m R_L}{1 + \frac{R_L}{r_o}}, \qquad (3.9)$$

where r_o is the small signal output resistance of the transistor, $r_o = dV_d/dI_{ds}$. Typically, r_o is between 100 kΩ and 1 MΩ, and generally $r_o \gg R_L$. The output impedance of the common source amplifier is $r_o || R_L \simeq R_L$. To increase the gain we need to maximise the transconductance g_m, and r_o.

Figure 3.6: Circuit diagram for a single FET common source amplifier. The voltage divider R_1/R_2 sets the gate voltage. This is called a common source amplifier, as the source is connected to both input and output.

We will see in section 3.6 that we want a large value of g_m to optimise the S/N.

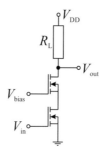

Figure 3.7: Circuit diagram for a cascode FET amplifier. The circuit can be further improved by replacing the load resistor R_L with a current source.

An improved amplifier circuit to achieve a large output impedance r_o is the 'cascode' configuration, comprising two FETs. A cascode consists of a grounded common source FET amplifier (the lower FET in Figure 3.7), which is loaded with a common gate amplifier (the upper FET in Figure 3.7).

The output impedance of the cascode is given by (see [464] and exercise 4)

$$r_o = r_{o,1} + r_{o,2} + r_{o,1}\, r_{o,2}\, g_{m,2}, \tag{3.10}$$

For a voltage output the output impedance is ideally small. This can be achieved using a subsequent buffer amplifier, for example an emitter (BJT) or source (FET) follower.

where $r_{o,1}$ and $r_{o,2}$ are the output resistances of the two transistors, respectively, and $g_{m,2}$ is the transconductance of the second transistor. Therefore the cascode circuit can have the same input impedance as the single transistor but a significantly higher output impedance and thus increase the gain.

Another benefit of the cascode is that it has a smaller input capacitance as it avoids the Miller effect. Thus it has a high bandwidth (see for example [289]).

3.4 NOISE IN DETECTOR AND AMPLIFIER SYSTEM

We can now combine the results from the previous sections to consider the noise of the combined system of a detector and electronic amplifiers.

First, we will show that the S/N in an amplifier chain is dominated by the noise from the first amplifier in the chain. If you have a chain of amplifiers, each with a gain A_i and an associated input noise N_i, amplifying a signal S, then in the first stage both the signal and the first stage input noise get amplified by the same gain and the output signal-to-noise ratio at the output of the first stage is S/N_1. After the second stage

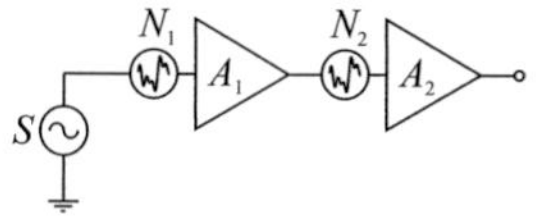

Figure 3.8: Noise in an amplifier chain.

$$\frac{S}{N_2} = \sqrt{\frac{(SA_1 A_s)^2}{(N_1 A_1 A_2)^2 + (N_2 A_2)^2}} = \frac{S}{N_1}\sqrt{\frac{1}{1 + \left(\frac{N_2}{N_1 A_1}\right)^2}}.$$

If the gain of the first stage is reasonably high, the square root will tend to 1 and the signal-to-noise ratio after the second stage is the same as for the first stage. In an amplifier chain the signal-to-noise ratio is therefore dominated by the first stage (preamplifier), and we will in the following concentrate on this.

A series resistor between the detector and the preamplifier will generate thermal noise as discussed in section 3.2. This will appear as a voltage source in series with the preamplifier and is therefore called series noise. Similarly, internal noise in the preamplifier, like for example from the input transistor in the preamplifier, will contribute to the series noise.

Let the total series noise have a spectral density e_n. We will now consider the combined circuit of the detector and a charge integrating amplifier as introduced in section 3.3. We model the amplifier as ideal (i.e. no noise) because we have included the amplifier noise in e_n.

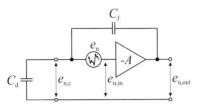

Figure 3.9: Detector and preamplifier circuit model with series noise.

Taking the preamplifier to be an ideal op-amp, no current flows into the inverting input, so that the impedances of the two capacitors constitute a voltage divider for $e_{n,out}$. Using the capacitor impedance $Z_C = (\omega C)^{-1}$,

$$\frac{e_{n,out}}{e_{n,c}} = \frac{(\omega C_d)^{-1} + (\omega C_f)^{-1}}{(\omega C_d)^{-1}} = 1 + \frac{C_d}{C_f}.$$

$e_{n,c} = e_n + e_{n,in}$ and $e_{n,out} = -Ae_{n,in}$, and thus

$$e_{n,c} = e_n \left(1 + \frac{1}{A} + \frac{1}{A}\frac{C_d}{C_f}\right)^{-1} \simeq e_n,$$

where the approximation is valid for $A \gg 1$. In other words, the noise on the voltage across the capacitor is given by the series noise.

Using the charge gain A_Q of the feedback circuit (eq. (3.8)), the equivalent input noise charge is

That result also was valid for $A \gg 1$.

$$Q_{n,in} = \frac{e_{n,out}\sqrt{\Delta f}}{A_Q} = e_{n,out}\sqrt{\Delta f}C_f \simeq e_n\sqrt{\Delta f}(C_d + C_f), \qquad (3.11)$$

where Δf is the bandwidth of the readout system, over which the noise is integrated. It is obvious that to reduce the noise charge C_f should be small, and ideally we can ignore it compared to C_d. In that case the equivalent noise charge is proportional to the detector capacitance. As seen in the previous section the signal charge is independent of C_d, and thus the charge S/N is proportional to $1/C_d$. The exact result can be different for other preamplifier configurations but the resulting signal to noise always shows this dependency.

Note that the capacitor itself is noiseless, the result is a consequence of the capacitive divider circuit. This clearly gives one motivation for using small areas for the sensitive elements in particle detectors, in particular semiconductor detectors. Other compelling reasons are the desire to improve spatial precision and reduce the rate of double hits in one element.

But the series noise from a series resistor and from the amplifier are not the only noise sources for the readout of a particle detector. We can add several other noise sources.

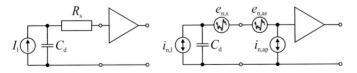

Figure 3.10: Detector and preamplifier circuit (left) and a more complete noise model (right).

First, we can add 'parallel noise' from current sources that are parallel to the detector capacitance. Two major sources for this type of noise can be present: amplifier parallel noise (with a current power density $i_{n,ap}^2$) and, in particular in semiconductors, shot noise due to sensor leakage current

This is shot noise with a power spectrum as described by eq. (3.4).

I_{leak}. We assume that the input impedance of the (ideal) amplifier and any other parallel resistances are infinite, and thus the only impedance through which the current can flow is the detector capacitance, generating a voltage $i_n/(\omega C_d)$. Hence, the noise voltage power density at the amplifier input due to this current noise is

$$e_{n,in}^2 = \frac{i_n^2}{(\omega C_d)^2} = \frac{2q_e I}{(\omega C_d)^2}.$$

Finally, we can also add a contribution from $1/f$ noise in the input transistor of the amplifier, $e_{n,1/f}^2 = A_{1/f}/f$, where $A_{1/f}$ is an appropriate proportionality constant.

We can assume that the different noise sources are uncorrelated, so we can add them in quadrature to get the noise spectral density at the preamplifier input

$$e_{n,in}^2 = e_{n,s}^2 + e_{n,as}^2 + e_{n,1/f}^2 + e_{n,l}^2 + e_{n,ap}^2 =$$
$$= \underbrace{4k_B T R_s + e_{n,as}^2}_{e_{n,s}^2} + 2\pi A_{1/f}\frac{1}{\omega} + \underbrace{\left(2q_e I_l + i_{n,ap}^2\right)}_{i_{n,p}^2}\frac{1}{C_d^2 \omega^2}, \quad (3.12)$$

where we have collected different contributions according to their dependency on the frequency.

Additional components in a real detector circuit might need to be added, but by now it should be clear how this can be done. C_d, for example, will include other parallel capacitances, etc.

It is interesting to note that the capacitance C_d converts the white noise spectrum of the parallel shot noise into an effective $1/f^2$ spectrum.

Remember that for the charge integrating feedback amplifier with the usual assumptions ($A \gg 1$ and $C_d \gg C_f$), $e_{n,out}^2 = (C_d/C_f)^2 e_{n,in}^2$.

3.5 PULSE SHAPING AND SHAPING TIME OPTIMISATION

We have seen that the spectral density of random noise is difficult to reduce. However, the total noise is given by the product of the spectral density and the bandwidth, and hence the most efficient way to reduce the effect of random noise is to reduce the bandwidth of the readout system.

The signal processing tool to limit the bandwidth at the low and the high frequency end is a bandpass filter. Many implementations of analogue bandpass filters exist, but for the discussion here we will focus on the conceptually simplest implementation, the buffered RC-CR filter, which comprises a first order low-pass and a first order high-pass filter, each with a cut-off frequency $\omega_c = (RC)^{-1}$, separated by a unity gain buffer amplifier. The purpose of the buffer amplifier, which we assume has an infinite input impedance, is to prevent the output of the first stage 'loading' the second stage. While in principle we could use different values of R and C in the two stages, the choice of the same values simplifies the calculations, and a more general analysis shows that this choice is also optimal for maximising the S/N.

Obviously this needs to be done in a way that does not reduce the signal spectrum more than the noise spectrum.

More complex bandpass filters can achieve slightly better S/N, but the basic discussion is the same.

Often the low-pass and the high-pass filter are labelled as integrator and differentiator, although this is strictly speaking not correct, as these circuits would only have the frequency characteristics of an actual integrator or differentiator if $R \to 0$.

Figure 3.11: Simple RC-CR bandpass filter with unity gain buffer amplifier.

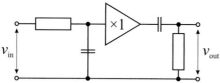

A useful tool to analyse analogue filters is the Laplace transform. To find the response of a filter in the time domain, we need to find the transfer function of the filter, which is defined by the ratio of output to input voltage as a function of complex frequency s. We then multiply it with the Laplace transform of the signal, and then calculate the inverse transform to obtain the result in the time domain.

The transfer function for this *RC-CR* filter is given by the product of the transfer functions for the two parts, $H_{RC}(s) \times H_{CR}(s)$. These transfer functions can be found using the voltage divider rule,

$$H(s) = H_{RC}(s) \times H_{CR}(s) =$$
$$= \frac{(sC)^{-1}}{R + (sC)^{-1}} \times \frac{R}{R + (sC)^{-1}} = \frac{sRC}{(1 + sRC)^2}. \quad (3.13)$$

The Laplace transform is a generalisation of the Fourier transform. It is a transformation into a complex frequency space

$$F(s) = \mathscr{L}\{f(t)\} = \int_0^\infty f(t)e^{-st}\, dt,$$

with $s = \alpha + i\omega$. α is sometimes called the Neper frequency and is a measure of how fast the transient response of the circuit will die away after the stimulus has been removed. It is related to the attenuation of a damped RLC circuit. ω is the standard angular frequency.

Next, we need the Laplace transform of the signal. As we have seen, the exact shape of the signal depends on the charge collection in the detector, and the time constant of the preamplifier. However, it is common to consider the response of the filter to a step function. The Laplace transform for a step function is of the form $1/s$. If the total charge of the pulse is q we can model the voltage output of the preamplifier as $q/(C_f s)$.

Combining this input function with the transfer function we can write down the output for the signal pulse as

$$v_{\text{out}}^{\text{signal}}(s) = \frac{qRC}{C_f(1 + sRC)^2}. \quad (3.14)$$

We now use the inverse Laplace transform to determine the output pulse in the time domain. This is discussed in exercise 2. The result is

$$v_{\text{out}}^{\text{signal}}(t) = \frac{qt}{C_f \tau} e^{-t/\tau}, \quad (3.15)$$

where the time constant $\tau = RC$.

We can differentiate this equation to see that the maximum of the pulse occurs at a time $t_{\max} = \tau$ and therefore the maximum signal amplitude (in terms of charge) after the shaper is $q\exp(-1) \simeq 0.368q$.

It is the modification of the signal shape in the time domain that leads to the designation 'shaper' for such a circuit in detector signal processing. In addition to limiting the bandwidth a shaper can also be useful for baseline restoration (i.e. resetting the output to zero after a suitable time).

After studying the effect of the shaper on the signal we will now study its effect on the noise, which is given in eq. (3.12). We obtain the total RMS noise after the shaper from

$$V_{\text{RMS}}^2 = \int_0^\infty e_{n,\text{shaper}}^2\, d\omega = \int_0^\infty e_{n,\text{in}}^2 \left(\frac{C_d}{C_f}\right)^2 |H(\omega)|^2\, d\omega,$$

where we are using the transfer function in the frequency domain,

$$H(\omega) = \frac{i\omega\tau}{[1 + (i\omega\tau)^2]^2}.$$

From comparison with eq. (3.12) we obtain

$$V_{\text{RMS}}^2 = \frac{\pi}{4C_f^2} \left(e_{n,s}^2 \frac{C_d^2}{\tau} + 4A_{1/f}C_d^2 + i_{n,p}^2 \tau\right). \quad (3.16)$$

Such a situation is not uncommon, for example, the current pulse in a silicon detector is very fast (typically lasting $\mathcal{O}(10\,\text{ns})$) and if we integrate this with a charge sensitive amplifier with a large time constant $R_{in}C_d$, the input to the RC-CR filter will be a step function.

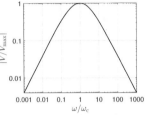

Figure 3.12: Transfer function of the buffered RC-CR filter in the frequency domain.

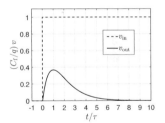

Figure 3.13: Response of buffered RC-CR filter to a step function input in the time domain.

Noise is a continuous phenomenon and thus we can drop the aperiodic time-dependent effects of the shaper.

The required integrals are

$$\int_0^\infty \frac{\omega^2\tau^2}{[1+(\omega\tau)^2]^2}\, d\omega = \frac{\pi}{4\tau},$$
$$\int_0^\infty \frac{\omega^2\tau^2}{[1+(\omega\tau)^2]^2}\, \frac{d\omega}{\omega} = \frac{1}{2},$$
$$\int_0^\infty \frac{\omega^2\tau^2}{[1+(\omega\tau)^2]^2}\, \frac{d\omega}{\omega^2} = \frac{\pi\tau}{4}.$$

We see that the contributions from the different noise sources have different dependencies on the shaping time constant τ. The parallel noise contribution increases with shaping time. This is what we would expect because as the shaping time increases more charge from the leakage current accumulates, and we expect the Poisson fluctuations to increase linearly with $\sqrt{\tau}$. The series noise contributions scale with $1/\tau$ because the series noise is uniform at all frequencies, and if τ decreases the absolute bandwidth increases and the noise is integrated over this larger bandwidth. Finally, the $1/f$ noise contribution does not depend on the shaping time, as we have seen before (see section 3.2). We can also see that the series and the $1/f$ noise contributions increase for larger detector capacitance.

We can combine the results so far to find the equivalent noise charge (ENC). Assuming that we sample the signal from one electron then after the shaper the signal peak is $q_e/(eC_f)$. Combining this with the noise at the output of the shaper (eq. (3.16)) we find that the squared value of the ENC is given by

$$(\text{ENC})^2 = \frac{e^2\pi}{4q_e^2}\left(e_{n,s}^2\frac{C_d^2}{\tau} + 4A_{1/f}C_d^2 + i_{n,p}^2\tau\right). \qquad (3.17)$$

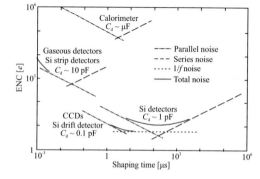 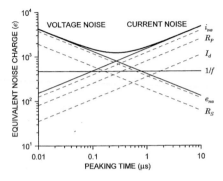

Figure 3.14: ENC as a function of the shaping time. Left: For different detector systems (after [413]). Right: For a typical silicon detector ($I_d = 10\ nA$, $i_{n,p} = 0.2\ pA/\sqrt{Hz}$, $e_{n,as} = 5\ nA/\sqrt{Hz}$, $A_f = 10^{-11}\ V^2$, $R_s = 400\ \Omega$, $C_d = 10\ pF$) [463].

If there are no other system constraints then we can find the optimal shaping time, τ_{min}, by differentiating eq. (3.17) with respect to τ (exercise 5). The result of this calculations is that the optimal shaping time is achieved when the series and parallel noise contributions are equal.

Other considerations, like for example the need for fast baseline restoration in high-rate experiments, can affect this optimisation.

3.6 TRANSISTOR NOISE

So far our discussion of noise optimisation was based on an ideal amplifier, for which we described the amplifier noise performance using generic series and parallel noise contributions. We will now investigate in slightly more depth the noise contributions from real transistors. The noise sources in BJT and FET amplifiers are very different, so we will discuss them separately.

Bipolar junction transistors

The base current I_b in BJTs in general is comparatively large. This results in shot noise due to the base current with a spectral power density (using eq. (3.4))

$$i_{n,b}^2 = 2q_eI_b. \qquad (3.18)$$

This noise current is in series with the noise from the detector leakage current so the two noise contributions add up for the parallel current noise.

The shot noise due to the collector current can be referred back to an equivalent series voltage noise density at the input, $e_{n,in}$, using the standard bipolar junction transistor relation $i_c = g_m v_b$. It is given by

$$e_{n,in}^2 = \left(\frac{k_B T}{q_e I_c}\right)^2 2 q_e I_c = \frac{2(k_B T)^2}{q_e I_c}. \tag{3.19}$$

The noise optimisation for bipolar junction transistors now has a different outcome. If we increase I_c we increase g_m and therefore decrease the voltage noise (eq. (3.19)). However, this will imply an increase in the base current I_b, and hence an increase in the current noise (eq. (3.18)). The result is that we can maintain the same ENC as the shaping time τ is changed, provided we adjust I_c.

In the Ebers-Moll model of a BJT the transconductance is given by

$$g_m = \frac{dI_c}{dV_b} = \frac{q_e}{k_B T} I_c,$$

where the derivative and I_c are evaluated at the working point. At room temperature for a BJT a good approximation is $g_m \simeq I_c/(25\ mV)$.

MOSFETs

For MOSFETs the gate leakage currents are generally negligible so there is no contribution to the current noise. The conducting channel of the MOSFET acts like a resistor and therefore there will be thermal noise. The MOSFET channel is not uniform over it's length and the calculation of the thermal noise is more complicated, but it will still act like a resistance $R \propto 1/g_m$. Therefore the spectral density of the resulting noise current can be written as

For JFETs the leakage currents are not negligible.

$$i_n^2 = \gamma_n g_m 4 k_B T, \tag{3.20}$$

where γ_n is a constant of the order of unity.

It is convenient to refer this current noise power density to an equivalent voltage noise power density at the input of the transistor

The value of $\gamma_n = 2/3$ is often used.

$$e_{n,in}^2 = \frac{\gamma_n 4 k_B T}{g_m}. \tag{3.21}$$

For MOSFETs the variation of drain current with gate voltage is given approximately by

$$I_d = k(V_{gs} - V_T)^2, \tag{3.22}$$

where k is a constant which depends on the device geometry and the capacitance, and V_T is the threshold voltage. Therefore the transconductance for a MOSFET is

This assumes that the MOSFET is operated in saturation ($V_{gs} > V_T$). If we are trying to optimise the S/N at low power then it is advantageous to operate in a different mode.

A very simple justification of this 'square-law' form is given in [289]. An excellent full first principles physics derivation is given in [406]. This text clearly explains the approximations required to derive this result and also includes a full derivation of the variation of I_d with V_{gs} when these simplifying approximations are relaxed.

$$g_m = \frac{dI_d}{dV_{gs}} = 2k(V_{gs} - V_T) = 2\sqrt{k I_d}. \tag{3.23}$$

Substituting this into eq. (3.21) we find

$$e_n^2 = \gamma_n \frac{2 k_B T}{\sqrt{k I_d}}. \tag{3.24}$$

3.7   NOISE SYSTEM OPTIMISATION

Combining the results from the generic noise analysis and the noise in MOSFET and bipolar junction transistors we can now consider the overall system optimisation. This optimisation is different for the two transistor technologies. For BJTs, we found that we can decrease the voltage (series) noise by increasing the collector current (eq. (3.19)), but this increases the current (parallel) noise (eq. (3.18)). Therefore the optimal noise using a bipolar transistor can be found for any shaping time by adjusting the collector current. However, in practice this might result in an unacceptably large current and hence power consumption.

The noise minimisation analysis is different for CMOS transistors because the current noise only arises from the detector shot noise. This means that the simple optimisation of eq. (3.17) considered in exercise 5 is valid. From eq. (3.24) we can see that improved noise is obtained by operating at high currents, I_d. Operating at higher currents obviously implies higher power consumption.

Increased power consumption might not be an issue for a single channel detector, but it is critical for large tracking detectors used in collider experiments. If the power consumption is larger, more metal (usually copper) conductors will be needed to transfer the current from the power supplies to the on-detector electronics. In addition, an increase in electrical power will require more substantial cooling circuits to remove the heat. All this will introduce more passive material in the detector, increasing multiple scattering for charged particles and hence degrading the resolution for momentum and impact parameters. The additional material will also cause more electron bremsstrahlung impairing the measurement of electrons in the electromagnetic calorimeters.

In section 3.5 we saw that we can select the optimal shaping time τ_{min} to obtain the minimum ENC. Such an optimisation would be acceptable for a low-rate detector but not for a detector in a high-rate environment. For example, at the LHC, we have collisions every 25 ns. The extra collisions that occur in one bunch crossing on top of the interesting triggered events create background particles ('pile-up'). Therefore to minimise pile-up at the LHC we need to keep the shaping time less than about 25 ns.

All these constraints need to be very carefully considered in the overall detector optimisation. Faster pulse shaping to minimise pile-up leads to an increase in noise. For CMOS transistors, the noise can be reduced by increasing I_d, but this increases the power consumption and requires an increase in passive material. As common in detector physics there is no simple optimisation and the physics requirements for particular detectors will lead to different solutions.

For the LHC phase 1 silicon detectors, there was an advantage in terms of noise at low power using bipolar junction transistors for the front-end amplifiers. On the other hand, the greater density available with CMOS technology was required for the digital circuitry. Therefore mixed technology BiCMOS processes were used in some applications. With the improvement in CMOS technology this advantage is less compelling and pure CMOS technology is being used for the ATLAS and CMS trackers for the HL-LHC upgrade.

3.8   DIGITISATION

The original output of any particle detector will be an analogue signal generated by the passing particle, possibly amplified internally and by a subsequent pre-amplifier. For processing of this data, relevant features need to be extracted and converted into digital information. These features can be the presence of a signal (analogue-to-binary conversion), the measurement of the arrival time of the signal, the magnitude of a signal (analogue-to-digital conversion), or its complete time structure (sampling).

ANALOGUE-TO-BINARY CONVERSION

In the simplest case the information required from a particle detector is a binary hit/no-hit decision, for example, if the information is to be used in a coincidence in an experiment trigger (see chapter 13). The binary signal is obtained as the result of the comparison with a threshold.

In general, the output of the detector will also carry noise and non-physics signals, for example from thermal excitations in the detector. These will also fire the comparator if they are above threshold. The first task will therefore be to select a suitable digitisation threshold that max-imises the number of detected real signals, while rejecting most of the fake ones. Both signal and backgrounds will increase as the threshold is low-ered. For a working detector the efficiency for the signals will be high for thresholds, where the rate of accepted background events is still small. As the rate of real signals is defined by external factors the accepted rate is constant once the comparison is fully efficient, and will only increase again when significant amounts of background are accepted. The graph of rate of accepted hits shows a 'plateau', and the threshold is usually set within the plateau. A better way to set the threshold is if we can identify signal and noise events (for example using other detectors). The threshold can then be optimised to obtain a high efficiency with low noise rates.

Electronically, this can be done with specialised 'comparator' integrated circuits. A slightly more robust com-parator circuit is the Schmitt trigger (see for example [289]). In time-critical applications constant fraction discrim-inators (see below) can be used for the digitisation.

Sometimes the term 'discriminator' is used instead of comparator.

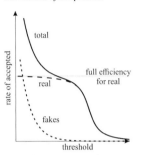

Figure 3.15: Schematic plateau curve.

ELECTRONIC TIME MEASUREMENTS

While in many cases it is the magnitude of an electronic signal that is re-quired from a detector, there are instances where we want to know the time interval between a start and a stop signal. The electronic device for this purpose is a time-to-digital converter (TDC). First, the start and stop signals are converted into a binary signal using a comparator. In the sim-plest case the TDC then counts the number of cycles of a free-running clock which occur between the leading edges of the binary start and stop signals.

If the times to be measured are too short to count the periods of a clock, ramp TDCs can be used. In these devices a capacitor is charged during the time between the leading edges of the binary start/stop signals. The capacitor can then be discharged at a slower rate and the time for this discharge measured with a slower clock.

Another method for the measurement of short intervals is the Vernier TDC, where two slightly detuned clocks are started by the leading edges of the start and stop signals. As the clocks are running, they will reach a point where the leading edges of the clocks coincide and this can be used to determine the interval between them with high precision. If the first clock has until then counted n_1 cycles with a frequency f_1, and the second n_2 cycles with a frequency f_2, then the time between the start and stop signals is given by

$$T = \frac{n_1 - 1}{f_1} - \frac{n_2 - 1}{f_2}.$$

A comparator outputs a voltage cor-responding to logical 1 or "on" if the voltage is larger than a threshold volt-age, or a voltage corresponding to logical 0 or "off" if it is below.

At this point the device is a time-to-amplitude converter (TAC). The first TACs have been developed by Bruno Rossi to measure the muon life-time [434].

Like in every other digitisation operation the error introduced by the digitisation is given by the smallest digitisation step divided by $\sqrt{12}$, i.e. for a simple clock-counting TDC with a cycle period of T the average

RMS error is given by $T/\sqrt{12}$ (with a square distribution). Other errors, for example from electronic noise, will be added on top of this.

One common issue with start and stop signals with finite rise time and varying amplitude is 'time walk'. A regular comparator fires at a fixed threshold, and this means that signals cross the threshold at different times, depending on their amplitude. A clever way to overcome this, in particular for very fast signals as typical in nuclear and particle physics, is the constant fraction discriminator (CFD). Here an inverted and delayed copy of the signal is added to the original signal. The delay is chosen so that the delayed signal starts approximately at the maximum of the original signal. The discriminator then fires at the point of zero-crossing, which will always be at the same time. For signals with a length of a few ten to hundreds of ns the delay can conveniently be achieved by a simple cable of appropriate length. For fast particle detectors the delays of $\mathcal{O}(1\ \text{ns})$ can be created using a chain of gates on the readout ASIC.

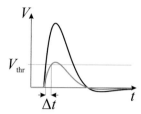

Figure 3.16: Time walk.

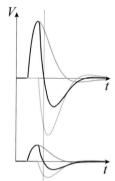

Figure 3.17: Operating principle of a Constant Fraction Discriminator. The dark curve is in both cases the sum of the original signal and the inverted and delayed copy (grey). Note that for both pulseheights the time of transition through zero is the same.

ANALOGUE-TO-DIGITAL CONVERSION

Conversion of the magnitude of a signal (typically a voltage) to digital data is a common task in the processing of sensor outputs, and a wide range of different technologies have been developed for this (see, for example, [402]). These technologies differ in speed, dynamic range, linearity, noise performance, power consumption and cost.

One of the conceptually most simple, and thus very fast, conversion technologies is the Flash-ADC (FADC). It is thus the technology used in ultra-fast sampling ADCs (up to GHz).

In the FADC the analogue signal is compared to an array of threshold voltages that are created by a voltage divider chain from a reference voltage. The comparison is done by comparators, the output of which is then encoded to digital information.

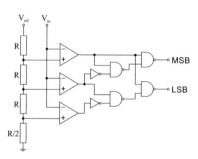

Figure 3.18: Schematics for a 2-bit FADC. MSB (LSB) stands for Most (Least) Significant Bit.

The strength of this technology is that the result of the comparator chain is instantly available and the digitisation time is only given by the bandwidth of the comparators and the time needed to encode the output of the comparators. The challenges are that the effort (number of components, cost, etc.) grows exponentially with the dynamic range of the ADC. The most serious performance limitation for flash ADCs is the linearity of the threshold voltages in the comparator chain due to variations of the resistance values in the resistor chain, and, in high speed applications, the propagation of the comparator outputs through the encoding logic into memory. If extremely high speed is not required, other ADC technologies have significant benefits (linearity, DNR, digisation noise etc.) [402].

3.9 RADIATION EFFECTS ON ELECTRONICS

Radiation effects can induce permanent damage in bipolar and CMOS electronics. In addition, radiation can cause temporary effects, which are called single event effects (SEEs).

RADIATION EFFECTS ON BIPOLAR ELECTRONICS

For bipolar electronics the main effect of irradiation entails bulk damage in the base of a transistor (for both *npn* or *pnp* transistors). This results in an increase in leakage current in the same way as for any semiconductor. The leakage current results in a decrease of the DC current gain β of the transistor as the base current is increased. Smaller devices operating with the same collector current will be more radiation-tolerant as the leakage current is proportional to volume. The radiation tolerance will be greater if the transistors can be operated at higher collector currents because then a given increase in base current will reflect in a smaller change in the current gain $\beta = i_c/i_b$, where i_c is the collector current and i_b is the base current. However, this results in increased power consumption. Newer technology has faster transistors with thinner base regions and therefore smaller volumes. This results in lower base leakage currents after radiation damage. This implies that there is a smaller degradation of the DC current gain β and therefore less sensitivity to radiation damage. However, bipolar devices will always show a decrease in β with radiation, but some decrease in β can be allowed for in the circuit designs.

We will discuss diode leakage current in the context of silicon detectors in more detail in section 9.5, and how it is affected by radiation in section 9.11.

RADIATION EFFECTS ON CMOS ELECTRONICS

As CMOS transistors rely on the transport of majority carriers, they are less sensitive to bulk damage of the silicon lattice. However, they are sensitive to radiation damage from ionising dose. Ionising radiation creates electrons and holes in the gate oxide. The holes diffuse slowly and can be trapped in the SiO_2 passivation layer present in CMOS technology. Ionising damage also creates traps at the interface of the silicon and SiO_2. If this charge is negative, it can eventually neutralise the positive charge from the trapped holes but only over very long timescales. If the charge trapped is Q_{Ox} and the capacitance is C_{Ox} then this causes a change in threshold voltage $\Delta V_{th} = Q_{Ox}/C_{Ox}$. If the change in threshold voltage is too large it can cause a transistor to be inoperable.

With modern deep sub-micron CMOS technology, the thickness of the gate oxides has become so thin that quantum tunnelling will result in the trapped charges being neutralised. This has enabled very radiation-tolerant ASICs to be designed using commercial CMOS processes. However, there are other oxide layers used in CMOS processes and these can be thicker and therefore not benefit from the release of trapped charge through quantum tunnelling. Radiation-induced trapped charge at oxide layers can cause increased leakage currents that limit the radiation tolerance. Improvements can be made using 'Hardness By Design' techniques

The ionising dose is described by the energy absorbed per unit mass and the SI unit is Gray (Gy, 1 Gy = 1 J/kg). The ionising dose needs to refer to a particular element or molecule. It is conventional to use silicon so the dose is specified in units of Gy(Si). The older unit of Rad is also still used (1 Gy = 100 Rad). In LHC silicon detectors the lifetime doses are in the range from 100 kGy(Si) to 1 MGy(Si).

The threshold voltage in CMOS devices is the minimum gate-to-source voltage V_{gs} that is needed to create a conducting path between the source and drain terminals.

'Deep sub-micron' refers to the feature size on the silicon wafer, which is typically smaller than 0.18 μm for these technologies.

to overcome these problems. One example is replacing conventional linear *n*-type MOS (NMOS) transistors with enclosed layout transistors. In the enclosed layout the source surrounds the gate which surrounds the drain. This approach has been very successful in designing radiation-tolerant electronics using commercial CMOS technologies but it does require more silicon area and it is therefore only used for the most sensitive areas [239].

SINGLE EVENT EFFECTS

If too much energy is deposited in a transistor, this can flip the state of a digital logic cell. The energy deposited by a MIP is generally too small for this to occur but more heavily ionising particles such as heavy ions can cause this effect.

Older CMOS processes were not sufficiently radiation-tolerant for LHC applications, but there were proprietary radiation-tolerant processes developed for military applications. These were used for the first prototypes of radiation-tolerant electronics for LHC experiments.

In accelerator applications, the main cause of SEE is particles undergoing nuclear reactions creating heavily ionising fragments. One typical type of SEE will result in a bit flip in a memory cell. This type of Single Effect Upset (SEU) might be rather benign if the cell contained transient hit data from a detector. However, if the cell was used as a register that affected the operation of an ASIC, it could result in a detector element being inoperable until it was reset. It is impossible to avoid all SEEs and it is therefore essential to have effective mitigation strategies. The simplest one is to have periodic resets for all registers of the on-detector ASICs. A more effective strategy is to detect the SEEs and issue resets when errors are detected.

This is a major issue for electronics in space missions.

A more fundamental approach is to use triple-event redundancy for critical registers. In this approach there are three instances of the register on the ASIC, and a 'voting' circuit to determine the correct value. So if one register is corrupted by an SEE, it will be out-voted by the other two copies of the register. This is a very powerful mitigation strategy but it obviously requires more silicon area on the ASIC, so it should only be used when needed.

It is essential that the three copies of the same register should be sufficiently separated so that one nuclear interaction can't deposit energy in more than one copy of the register.

Key lessons from this chapter

- We can calculate the induced currents on a conductor caused by moving charges using the Ramo-Shockley theorem. They are proportional to the product of velocity of the moving charge and the weighting field. Thus, a current is only induced while the charge is moving.

- The most important fundamental noise sources in a combined detector and amplifier system are current noise from sources parallel to the detector and series noise from the amplifier.

- In a well-designed system it is the noise of the first amplifier that determines the system S/N performance.

- The signal-to-noise ratio can be optimised using pulse shaping. The optimal shaping time occurs when series and parallel noise are equal. This optimisation can be affected by other considerations, like high-rate operation.

- The origins of the noise in BJT and FET amplifiers are distinct and this results in different optimisations for systems using the two technologies.

- Radiation damage in bipolar electronics results in a decrease in the DC current gain (β) and this limits the use at very high fluences.

- Radiation damage in modern deep sub-micron CMOS electronics is greatly reduced and the residual effects can be accommodated in the design.

EXERCISES

1. *Signal and noise.*

   Consider a Gaussian signal with mean μ and a standard deviation σ in the presence of a Gaussian noise source. The noise distribution has $\mu = 0$ and the same value of σ as the signal. We wish to set a threshold value t, such that signals $x > t$ are genuine (from the signal source) and not noise. We define the true efficiency to be the fraction of signal events which are above this threshold and the background efficiency to be the probability of a signal above threshold to arise from background alone. Let $\mu = 1$ and consider the cases $\sigma = 1$, 3 or 10. For these three cases sketch graphs of signal efficiency versus threshold and background efficiency versus threshold. Comment on the significance of these results.

2. *Laplace transform.*

   Show that the Laplace transform of the function $f(t) = (t/\tau) \exp(-t/\tau)$ is given by eq. (3.14). If you are familiar with complex analysis, you can perform the direct calculation of the inverse Laplace transform of eq. (3.14).

3. *Calculation of induced current using mirror charges.*

   Consider a point charge of magnitude Q at a vertical distance z above an infinite grounded conductive plane.

   a) Calculate the electric field just above the grounded plane as a function of the horizontal distance r and hence the surface charge density. Hint: use a mirror charge.

b) The total induced charge on the grounded plate is $-Q$ and clearly doesn't change as the point charge moves. The grounded plane is divided into strips of width w in the x-direction (infinitely long in the y-direction). One strip is centred below the point charge. Show that the charge induced on the central strip is given by

$$q(z) = -\frac{2Q}{\pi} \arctan\left(\frac{w}{2z}\right).$$

c) Using the result from part b) show that if the point charge moved at a velocity v in the vertical direction towards the grounded plate, the induced current in the central strip is

$$i(t) = \frac{4Qw}{\pi(4z^2 + w^2)} v.$$

4. *Common source amplifier and cascode.*

This question calculates the gain of a simple common source FET amplifier and shows how it can be increased using the cascode circuit.

a) Explain the small signal equivalent of a common source amplifier (compare Figure 3.6).

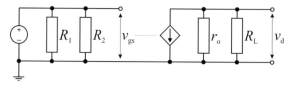

b) Derive eq. (3.9) for the gain of the common source amplifier.

c) Find an expression for the output impedance of the common source amplifier.

d) The cascode amplifier has a common source stage with a common gate stage as the load for the common gate (see Figure 3.7). The input voltage is applied to the gate of the common source amplifier. Calculate the gain and output resistance of this cascode circuit.

5. *Optimal shaping time.*

Using eq. (3.17) determine the optimal shaping time and resulting noise. Assuming the use of a MOSFET amplifier operating at a current $I_d = 2$ mA and $g_m = 6$ mS, determine the optimal shaping times and resulting ENC values for a detector element with a capacitance of 10 pF and with a leakage current of (a) 1 nA and (b) 1 μA. Comment on the significance of these results.

6. *$1/f$ noise from charge trapping.*

a) Consider a charge trapping/de-trapping process with a mean time between events of τ. If the events are distributed in time according to an exponential distribution, use Fourier theory to determine the form of the resulting power spectrum.

b) Now consider N such processes, with different time constants such that $\tau_{i+1} = 10\tau_i$, evaluate the resulting spectrum numerically for $N = 2$, 4 and 8 and plot a graph of the power spectrum and compare this with a $1/f$ spectrum.

7. *Noise from bias resistor.*

An often used component that contributes noise that we have ignored in our discussions in this chapter is a bias resistor, which is often required to connect a high voltage to the detector.

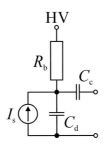

For small signals the DC high voltage appears as ground, and thus for the noise analysis this resistor appears in parallel with the detector capacitance.

a) Why can we ignore the coupling capacitor C_c as a source in the noise analysis?

b) Show that the noise spectral density due to R_b is given by

$$|e_{n,p}|^2 = \frac{4k_B T R_b}{1 + (\omega R_b C_d)^2}.$$

c) Hence, show that the equivalent noise charge due to this contribution, when integrated over all frequencies, is given by $Q_n^2 = kTC_d$. (This is the reason why this noise is often called 'kTC' noise, but beware, the noise does not originate in the capacitor. Capacitors are noiseless!)

d) Where in eq. (3.12) would we have to add this contribution? Hence, will it increase the parallel or the series noise component?

4 Movement of charges and internal amplification

In chapter 2 we have seen that the interaction of an incoming charged particle with the detector material results in a number of energy transfers along the track that can lead to the ionisation of the detector material, creating pairs of negative (usually electrons) and positive (positive ions or holes) charge carriers. In chapter 3 we have seen how the movement of charges creates a current signal on a system of readout electrodes of an electronic particle detector. To create this signal, we have first, to separate the charges, and then, second, to keep them moving, as the induced current depends on their drift velocity in the external fields. To collect these charges for electrical readout, they need to be mobile, with a low probability for capture and recombination. Viable detector materials are therefore non-polar pure gases or liquids, or pure or lightly doped semiconductors.

In semiconductors we can also use diffusion to create a signal (as used in some CMOS image sensors for cameras) but the resulting signal is usually too slow for use in particle detectors.

As the charges drift, they will collide with atoms in the material. Elastic collisions will affect their trajectory, affecting their position in a process that we describe macroscopically as diffusion. In addition, once the energy acquired by the electrons between collisions becomes sufficiently large, inelastic collisions will occur. One type of inelastic collision is ionisation of the atom with which the electron collides, resulting in an avalanche growth of the drifting charge cloud. This can be used to internally amplify the signal of a particle detector, thus reducing the demands on external electronic amplification and noise performance.

4.1 CHARGE DRIFT IN GASES

DRIFT VELOCITY

When a charged particle is moving through a gas, it will undergo collisions with the molecules in the medium. If we apply an electric field, the charge will be accelerated between collisions. As a consequence, the particle will have a varying microscopic drift velocity $\vec{v}(t)$. Macroscopically, the movement can be described by the Langevin equation

$$m\frac{d\vec{u}}{dt} = e\vec{E} + \vec{A}(t),\tag{4.1}$$

where $\vec{u}$ is the macroscopic drift velocity, and $\vec{A}(t)$ is a force term describing the interaction between the drifting particle and the gas molecules. On average, the latter can be described by $\langle\vec{A}(t)\rangle = -m\vec{u}/\tau$, where τ is the average time between collisions. In the steady state this leads to

$$\vec{u} = \frac{e\tau}{m}\vec{E} = \mu\vec{E},\tag{4.2}$$

Due to the $1/m$ dependency the drift velocity for electrons in gases is about 10^3 times larger than for ions.

where we have introduced the mobility μ.

DOI: 10.1201/9781003287674-4

Ions, being of similar mass than their collision partners, will lose a significant fraction of their energy in collisions. Consequently, they will acquire only limited amounts of energy from the field, and as long as the fields are not becoming very large, most of their energy will be thermal, $\varepsilon = 3k_B T/2$ (about 0.038 eV at room temperature). Hence we can estimate the microscopic relative speed v_{rel} between ions and the molecules from $m^* \langle v_{rel}^2 \rangle/2 = 3k_B T/2$, where m^* is the reduced mass of the ion/gas molecule system. Thus,

$$u_{ion} = \frac{e\tau}{m^*}E = \frac{e}{m^* n\sigma v_{rel}}E \simeq \frac{e}{n\sigma \left(3k_B T m^*\right)^{1/2}}E,$$

n is the volume number density of the molecules. The mean free path is $\lambda = 1/(n\sigma)$ and therefore the mean time between collisions is $\tau = 1/(n\sigma v_{rel})$.

with the momentum transfer cross-section σ. The macroscopic ion drift velocity for low fields is thus proportional to the field E. It is also proportional to n^{-1}, which is proportional to the inverse of the pressure, p^{-1}, and thus the drift velocity scales with E/p, implying that the mobility is constant at a given pressure. Typical ion mobilities are between 1 and 10 cm^{2}V^{-1}s^{-1}.

E/p is also referred to as the 'reduced field'.

In gas mixtures the charge of ions with a high ionisation potential gets transferred in collisions to ions with lower ionisation potential, so that it will be these that will define the ion mobility.

On slightly longer timescales ions and atoms will also form charged clusters. This will be discussed in section 7.3.

For electrons the situation is generally different. In steady state there is a balance between energy obtained from the field over a distance x, eEx, and the energy lost in N collisions over that distance, given by $N\varepsilon\Lambda(\varepsilon)$, where ε is the energy of the electron and $\Lambda(\varepsilon)$ is the average relative energy loss per collision. Hence,

Strictly speaking, the energy here is only the energy due to the electric field, without the thermal energy. However, for electron drift in gases the thermal energy is usually negligible.

$$eEx = N\varepsilon\Lambda(\varepsilon) = \frac{x}{u\tau}\varepsilon\Lambda(\varepsilon) = \frac{x}{(eE\tau/m)\tau}\varepsilon\Lambda(\varepsilon) \simeq$$

$$\simeq \frac{x(n\sigma(\varepsilon)\langle v\rangle)^2}{eE/m}\frac{m\langle v^2\rangle}{2}\Lambda(\varepsilon),$$

where we used $\tau = (n\sigma(\varepsilon)\langle v\rangle)^{-1}$ and $\varepsilon = m\langle v^2\rangle/2$. The solutions of these equations for the macroscopic drift and microscopic velocities are then

We approximate $\langle v\rangle^2 \simeq \langle v^2\rangle$.

$$\langle v\rangle = \sqrt{\frac{eE}{mn\sigma(\varepsilon)}}\sqrt{\frac{2}{\Lambda(\varepsilon)}} \quad \text{and} \quad u = \sqrt{\frac{eE}{mn\sigma(\varepsilon)}}\sqrt{\frac{\Lambda(\varepsilon)}{2}}.$$

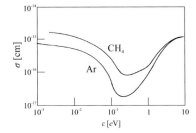
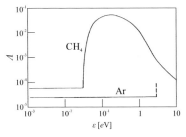

Figure 4.1: Collision properties far argon and methane as a function of the energy (modified from [148]). Left: Effective collision cross-section. Right: Fractional energy loss.

For electrons the momentum transfer cross-section and the fraction of energy lost per collision are both strongly dependent on the energy of the electron and the type of molecule. In gases with molecules with a high polarisability the cross-section shows a pronounced dip due to the

The polarisability α is defined as the ratio of the induced electric dipole moment of a molecule, and the field generating it, $\alpha = p_{ind}/E$.

Ramsauer effect. As the electron energy increases from zero, the electron becomes more localised, resulting at first in higher polarisation of the molecule, which increases the magnitude of the scattering potential, resulting in an increase of the partial wave phase shift δ_0. Only at higher energies this eventually decreases due to short-range repulsive interactions, leading to a condition where the scattering length and thus the cross-section goes through a minimum [104]. Gases with high molecular polarisability are the heavy noble gases (argon, krypton and xenon, but not helium or neon) because of their large electron cloud, and spherically symmetric organic molecules, like methane or neo-pentane.

We are looking for non-polar, but polarisable molecules.

The energy dependence of the fractional energy loss reflects the additional degrees of freedom available to three-dimensional molecules that turn on at low energies.

As a consequence of these variations, drift velocities vary significantly for different gases and different values of E/p. Typical electron drift velocities are between 0.1 and a very few 10 cm/μs.

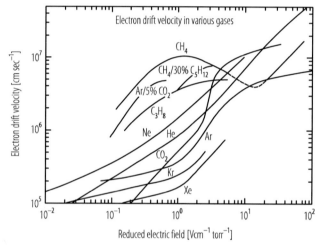

Figure 4.2: Electron drift velocities for different gases as a function of the reduced field [291].

Similar arguments will also apply for ions at higher fields, but because of the smaller energies retained in the collisions without the dramatic variations of the cross-section and the fractional energy loss. Nevertheless, at high fields the dependency of the ion drift velocity on the field will change from $\propto E$ to $\propto \sqrt{E}$, reflecting the transition from the thermal regime to the field-dominated regime.

DIFFUSION

The collisions with the gas molecules are stochastic. Macroscopically, this results in slight variations of the drift velocity, which manifest themselves as diffusion for an ensemble of drifting charge carriers. From the diffusion equation a Gaussian distributed cloud of charge carriers will expand with time in one dimension as

$$n(x) = \frac{n_0}{\sqrt{4\pi Dt}} \exp\left(-\frac{x^2}{4Dt}\right), \qquad (4.3)$$

where D is the diffusion constant. The width of the cloud can be described by its standard deviation, which increases as $\sigma_x = \sqrt{2Dt}$. This spreading

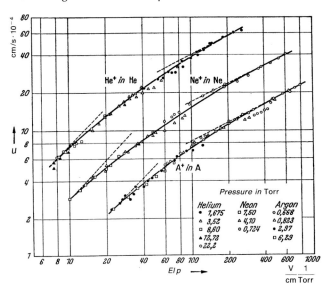

Figure 4.3: Ion drift velocities for different ions and gases as a function of the reduced field [148].

also applies if there is a superimposed average macroscopic drift velocity, so that $x \to x - ut$.

After $N = t/\tau$ collisions, each with a displacement ξ_i, the total displacement is given by $x = \sum_{i=1}^{N} \xi_i$. Hence,

$$\sigma_x^2 = \langle x^2 \rangle = N\langle \xi^2 \rangle = N\langle v_x^2 \rangle\langle t^2 \rangle = \frac{N}{3}\langle v^2 \rangle\langle t^2 \rangle = \frac{2N}{3}\langle v^2 \rangle\tau^2 = \frac{2}{3}\langle v^2 \rangle\tau t,$$

where τ is the mean time between collisions, and thus

We use $\langle v^2 \rangle = \langle v_x^2 \rangle + \langle v_y^2 \rangle + \langle v_z^2 \rangle = 3\langle v_x^2 \rangle$ because of isotropy, and $\langle t^2 \rangle = 2\tau^2$, for the collision time distribution $f(t) = \tau^{-1}\exp(-t/\tau)$.

$$D = \frac{\langle v^2 \rangle \tau}{3} = \frac{2\langle \varepsilon \rangle}{3m}\tau,$$

where $\langle \varepsilon \rangle = m\langle v^2 \rangle/2$ is the average energy of the electrons.

In the case that the energy of the electrons is given by the thermal limit $\langle \varepsilon \rangle = 3k_B T/2$,

$$D = \frac{k_B T}{m}\tau = \frac{k_B T}{e}\mu, \tag{4.4}$$

This equation is known as the Nernst-Townsend or Einstein equation.

and the standard deviation of the diffusion cloud is growing like

$$\sigma_x = \sqrt{\frac{2k_B T \mu t}{e}} = \sqrt{\frac{2k_B T l}{eE}}, \tag{4.5}$$

where l is the drift distance. In real gases at high reduced field E/p, $\langle \varepsilon \rangle \gg k_B T$ and

$$\sigma_x = \sqrt{\frac{4l\langle \varepsilon \rangle}{3eE}}. \tag{4.6}$$

Gases, which stay close to the thermal limit up to high values of the reduced field E/p (for example CO_2) are called 'cold' gases, the opposite being 'hot' gases (for example argon).

For a cloud of drifting electrons diffusion in the direction of the drift field is reduced as electrons in the leading edge of the cloud have a higher average speed and hence a higher collision rate. Similarly the mobility of

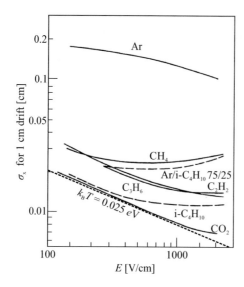

Figure 4.4: Standard deviation of diffusion after 1 cm drift in different gases for different electric fields at atmospheric pressure (modified from [396]). Full lines are from experiment, dashed lines are calculated. The dotted line is the thermal limit for room temperature

the trailing edge is enhanced, and consequently, while the diffusion equation still holds in principle, the half width of the pulse in the field direction is characterised by a reduced diffusion coefficient [398].

DRIFT IN MAGNETIC FIELDS

In a magnetic field the Langevin equation becomes

$$\frac{d\vec{u}}{dt} = \frac{e}{m}\vec{E} + \frac{e}{m}\left(\vec{u} \times \vec{B}\right) - \frac{1}{\tau}\vec{u}. \tag{4.7}$$

In steady state $d\vec{u}/dt = 0$, and the solution for the drift velocity of a drifting electron is

$$\vec{u} = \frac{e}{m_e}\frac{\tau}{1+\omega^2\tau^2}\left[\vec{E} + \frac{\omega\tau}{B}\vec{E} \times \vec{B} + \frac{\omega^2\tau^2}{B^2}\left(\vec{E}\cdot\vec{B}\right)\vec{B}\right], \tag{4.8}$$

with the cyclotron frequency $\omega = eB/m_e$.

In tracking detectors the E and B fields are often perpendicular, as the electron drift, which is providing the accurate position information, and the bending of the incoming particle track are both in the same direction. In this case the solution reduces to (using $\vec{E} = (E,0,0)$ and $\vec{B} = (0,0,B)$)

$$u_x = \frac{e}{m_e}\frac{\tau}{1+\omega^2\tau^2}E,$$

$$u_y = -\frac{e}{m_e}\frac{\tau}{1+\omega^2\tau^2}\omega\tau E,$$

$$u_z = 0.$$

For a detailed discussion of drift chambers see section 7.5.

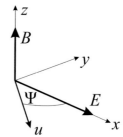

Figure 4.5: Lorentz angle in orthogonal electric and magnetic field.

Hence, the motion of the electron is rotated around the direction of the B field by an angle $\Psi = \operatorname{atan}(u_y/u_x) = \operatorname{atan}(-\omega\tau)$. Ψ is called the 'Lorentz angle'. The speed of the drifting electron is given by $u = (e/m_e)\tau E\cos\Psi$, hence it is determined by the electric field in the drift direction. Similarly to the drift velocity, the Lorentz angle depends on the drift gas and the reduced field E/p.

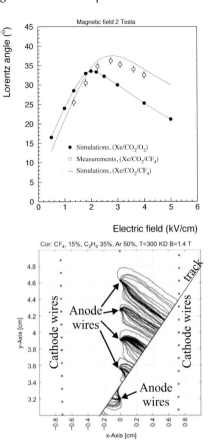

Figure 4.6: Lorentz angle in different gases [102].

Figure 4.7: Electron drift in the CDF outer tracker simulated with MagBoltz and GARFIELD (modified from [25]. Argon/ethane/CF$_4$ 50/35/15 in a drift field of 2.5 kV/cm and for a Lorentz angle of 35°, wire diameter 40 μm.

Another important field configuration is a parallel E and B field. In that case the macroscopic drift velocity is not modified by the magnetic field. However, microscopically the trajectories of the drifting electrons between collisions become helices around the field direction, with a radius $\rho = (v/\omega)\sin(\theta)$, where θ is the angle between the electron trajectory and direction of the magnetic field. The position of an electron emerging from a collision at the origin with an original direction Φ moving for a trajectory length l in the magnetic field is described by

$$x(l) = \rho \left[\sin\left(\frac{\omega l}{c} - \Phi \right) + \sin\Phi \right],$$

$$y(l) = \rho \left[\cos\left(\frac{\omega l}{c} - \Phi \right) - \cos\Phi \right],$$

$$z(l) = l\sin\theta.$$

The mean square displacement in one direction can be found from the integration over the collision length distribution,

$$\langle x_i^2 \rangle = \frac{1}{4\pi\lambda} \int_0^\infty x_i^2(l)\, e^{-l/\lambda} \, \sin\theta \, d\theta \, d\phi \, dl,$$

with the mean free path λ, which yields

$$\langle x^2 \rangle = \langle y^2 \rangle = \frac{2}{3} \frac{\lambda^2}{1 + \omega^2 \lambda^2 / c^2} = \frac{2}{3} \frac{\lambda^2}{1 + \omega^2 \tau^2},$$

$$\langle z^2 \rangle = \frac{2}{3} \lambda^2,$$

and thus the transverse diffusion of a cloud drifting in the field direction is reduced, $D_t(B)/D_t(0) = (1 + \omega^2 \tau^2)^{-1}$. The longitudinal diffusion is unaffected.

ELECTRON ATTACHMENT

The tendency of a molecule to capture an electron is given by the electron affinity, which is the energy released or required in the capture process. Attachment is relevant for gases with high electron affinity (e.g. halogens, but also oxygen or water), whereas it is absent for gases with a small or even negative electron affinity (e.g. noble gases, nitrogen, CO_2). To not lose electrons in the collection process, detector gases usually are chosen to have low electron affinity, but gases with high electron affinity can be present as contaminants.

A practical parameter to describe electron attachment is the attachment coefficient, which describes the probability for attachment for a collision, which depends on the electron energy. For electron gases with high electron affinity this will be large at low energies and decreases for higher energies. For negative electron affinity gases the attachment coefficient turns on at a threshold energy and then increases slightly with energy.

COMPLETE TRANSPORT THEORY

A complete solution for all drift properties can be found from the solution to the Boltzmann transport equation,

$$\frac{\partial f}{\partial t} + \vec{v} \cdot \vec{\nabla}_r f + \frac{q_e}{m_e}\left(\vec{E} + \vec{v} \times \vec{B}\right) \cdot \vec{\nabla}_v f = -Jf = \left(\frac{\partial f}{\partial t}\right)_{\text{coll}}, \qquad (4.9)$$

where $f = f(\vec{r}, \vec{v}, t)$ is the six-dimensional phase-space probability density function for a single particle, and J is the collision operator. J is a local operator in $\vec{r}$ and t following the assumption that both the range and the duration of collisions are negligible. The operator J depends functionally on the neutral distribution and interaction cross-sections, which comprise elastic and inelastic (excitation and ionisation) scattering cross-sections.

For a detailed discussion of the Boltzmann transport equation applied to the problem of electron drift see [407, 323].

The distribution function $f(\vec{r}, \vec{v}, t)$ can then be expanded in a power series of the spatial gradient of the number density $n(\vec{r}, t)$,

$n(\vec{r}, t) = \int f(\vec{r}, \vec{v}, t) d^3 v$.

$$f(\vec{r}, \vec{v}, t) = \sum_{k=0}^{\infty} f^{(k)}(\vec{v}) \otimes (-\vec{\nabla})^k n(\vec{r}, t), \qquad (4.10)$$

where $(-\vec{\nabla})^k$ represents a k-fold outer product of the gradient operator with itself and $\otimes$ indicates a k-fold inner-product operation. The coefficients in the expansion, $f^{(k)}(\vec{v})$, are tensors of rank k.

Under 'hydrodynamic conditions', a related equation

When the particle swarm evolves without memory of initial conditions and unaffected by boundary constraints.

$$\frac{dn(\vec{r},t)}{dt} = \sum_{k=0}^{\infty} \omega^{(k)} \otimes (-\vec{\nabla})^k n(\vec{r},t) \qquad (4.11)$$

can be used. Here, the tensor parameters $\omega^{(k)}$ can be associated with the relevant observables for the transport process: for a uniform density $-\omega^{(0)}$ is the reaction rate, the vector $\omega^{(1)}$ with the drift velocity, and the rank-2 tensor $\omega^{(2)}$ with the diffusion coefficient.

Inserting eq. (4.10) into the Boltzmann transport equation, eq. (4.9), rearranging and using eq. (4.11) yields

$$\left[\frac{q_e}{m_e}\left(\vec{E} + \vec{v} \times \vec{B}\right) \cdot \vec{\nabla}_v - J\right] f^{(k)}(\vec{v}) = \vec{v} f^{(k-1)}(\vec{v}) - \sum_{j=0}^{k} \omega^{(j)} f^{(k-j)}(\vec{v}),$$

and we can, in principle, find $\omega^{(k)}$ and $f^{(k)}$ from all the previous $\omega^{(k-1)}$ and $f^{(k-1)}$. For this the collision operator J has to be found from the elastic and inelastic cross-sections. The calculations are quite elaborate and typically involve expansion of the $f^{(k)}$ using Legendre polynomials, which are truncated at some suitable low order. For the details see [323].

These cross-sections generally can depend on the scattering angle.

Today a more practical approach is based on Monte Carlo calculation, which tracks the microscopic path of a particle through a large number of collisions [251]. The path of the particle is split into path segments and the probability for collisions within each path segment and their outcome (elastic collision, excitation, ionisation, etc.) are calculated from cross-section data. The main challenge of the calculation is to find appropriate lengths for the path segments, to obtain convergence of the results. This is the approach used by the MagBoltz software to calculate transport properties of electrons in a wide range of gases [137].

MagBoltz can be accessed through the GARFIELD software package (see section 7.8).

A lot more experimental data are available for transport parameters in a wide range of gases than direct measurements of the collision cross-sections. Thus, the cross-section data required for these calculations are reverse-calculated from the measured transport properties, either from the solution of the Boltzmann transport equation, or from the application of the Monte Carlo method.

An open-access website with electron and ion scattering cross-sections is [352].

4.2 CHARGE DRIFT IN LIQUIDS

For a more detailed summary of charge drift in liquids see [231].

Charge transport in liquids is only possible in liquids with non-polar molecules. The main challenge is therefore to avoid contamination with substances with high electron affinity, as the high density of attachment centres in the liquid will lead to a quick depletion of the ionisation charge.

Polar molecules have a positively charged end that will capture electrons.

In a pure dielectric liquid the conductivity is low ($<10^{-18}$ Ω^{-1}cm^{-1}), because the energies of electrons bound in the liquid and the energy of excess electrons is separated by about 6 to 9 eV. This gap is slightly smaller than in gases, because polarisation effects tend to increase the energy of bound electrons by 1 to 1.5 eV, whereas excess electrons move in a conduction band, which is below the potential energy of a free electron by about 0.2 to 1 eV.

A conduction band will emerge when the mean free path becomes comparable to the wavelength of the electron.

In liquids the detectable ionisation charge is smaller than the generated charge, because the time needed for thermalisation is much shorter than in gases. Hence there is a significant probability for recombination of a newly created electron-ion pair, and the charge yield (defined as the number of electron/ion pairs created by an energy loss of the incoming particle of 100 *e*V), increases with the electric field. Typical values for the charge yield are from 2 to 4 for argon from zero field to 10 kV/cm, and about half of that for dielectric organic liquids.

This is referred to as 'geminate recombination', in contrast to 'volume recombination', if charges recombine after some drift.

Still, because of the higher density the passage of charged particles in liquids generates more ionisation charge per unit track length than in gases. Nevertheless, we would not expect this to result in larger induced current signals, as we would presume based on eq. (4.2) that the mobility of the charge carries would decrease proportionally to p^{-1}, because of the decrease of the mean free path. However, this is not true in liquids with molecules with high molecular polarisability, where the density increase weakens the long-distance polarisation attraction between the electron and a molecule in the liquid, and thus the electron free path becomes larger [104].

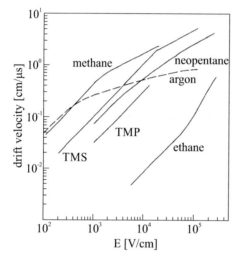

Figure 4.8: Electron drift velocity as a function of electric field in different liquids (methane at 111 K, ethane at 200 K, others at 295 K, modified from [288]).

The calculation of transport properties in liquids is more complicated than for gases, because of the small distance between molecules and their highly correlated separations. Nevertheless, a multi-term approach to the solution of the Boltzmann equation with liquid phase cross-sections can be used for this calculation [160].

The classical liquids used in particle physics are liquid noble gases, and in particular argon, due to its high charge yield and electron mobility, and moderate cost. Purification of cryogenic liquids is a by-product of the liquefaction process, but the obvious drawback of these liquids are the low temperatures required for liquefaction, with the associated need for thermal insulation and substantial cryogenic plants.

Liquefaction temperatures are −186 °C for argon, −153 °C for krypton, −108 °C for xenon (at atmospheric pressure).

Examples for organic liquids with high electron mobilities used in particle detectors at room temperature are tetramethylsilane (TMS) or 2,2,4,4-Tetramethylpentane (TMP). The main challenge for these liquids is the purification, but the viability of the technology for large scale experiments has been demonstrated for the KASCADE experiment [409].

4.3   CHARGE DRIFT IN SEMICONDUCTORS

In solids electron states occur in energy bands. Typically most outer electrons reside in a filled valence band. These electrons cannot drift under the influence of an external electric field because of the Fermi exclusion principle. To move, they have to be in the higher-energy conduction band. We distinguish three cases:

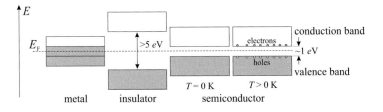

Figure 4.9: Band structure in solids.

- In metals, the conduction and the valence band overlap. The number of mobile electrons far exceeds the number of electrons generated by ionisation, and the ionisation signal is unobservable.

- In semiconductors, there is a finite, but small band gap (a few *e*V). Some electrons are lifted to the conduction band thermally, but this population can be kept small by lowering the temperature or reducing charge carrier concentrations by doping and biasing. Electrons can then be raised from the valence to the conduction band by the interaction with the incoming particle. This then leads to an 'electron' in the conduction band and a 'hole' in the valence band. The hole behaves effectively like a mobile positive charge. The resulting electrons and holes can be seen as moving to the readout electrodes under the influence of an electric field.

See section 9.1.

Both 'electrons' and 'holes' in the semiconductor are intuitively simple concepts that are widely used, but should not be confused with free negative or positive charges. They describe a quantum mechanical state of the lattice with its electrons as a whole, similar to electrons in an atom that cannot be described individually, but are part of the wavefunction of the whole atom. Microscopic classical pictures of the transport process are of limited validity.

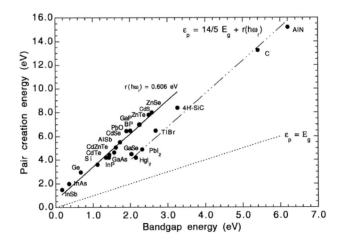

Figure 4.10: Energy used per electron-hole pair vs band gap in different materials [395]. The difference between the energy used per electron-hole pair and the band gap is due to lattice excitations (phonons) to conserve 4-momentum, and damage to the lattice [309]. The dotted line is the line where the energy to create an electron-hole pair equals the band gap. Other lines are the result of a linear fit to two groups of data following the parameterisation given in the figure and explained in the references. In silicon, the band gap is 1.12 eV, but the average energy used per electron-hole pair is 3.6 eV.

- In insulators the band gap is larger (>5 *e*V). Insulators can be used as particle detectors (for example diamond with a band gap of 5.5 *e*V), but only if the material is very pure. Normal insulators have too many interstitial states, which will trap the electrons and result in local de-excitation, strongly limiting the range of the electrons, so that a useful signal cannot be generated.

The drift of charge carriers in semiconductors is, like in gases or liquids, governed by collisions in the material. In semiconductors two sources of collisions are relevant: Scattering on the lattice, i.e. interactions with phonons, which are acoustic at low energies and optical at high energies, and scattering on ionised impurities.

It can be shown that the mobility for scattering on acoustic phonons μ_l is proportional to $T^{-3/2}(m_c^*)^{-5/2}$, where m_c^* is the conductivity effective mass [113]. However, at larger field strengths the drift velocity saturates. In materials without accessible higher bands (for example silicon), the cause for the saturation is inelastic scattering of the charge carriers with the emission of optical phonons. If the energy of the phonons is given by E_{phonon}, then the speed, at which this occurs will be given by $v \simeq (2E_{\mathrm{phonon}}/m_c^*)^{1/2}$. In silicon, the conductivity effective masses of electrons and holes are similar ($0.26\,m_e$ and $0.39\,m_e$, respectively), and the ratio of saturation velocities is close to 1 ($v_n^{\mathrm{sat}}/v_p^{\mathrm{sat}} \simeq 5/4$). In GaAs the electron conductivity effective mass ($0.067\,m_e$) is much less than the hole effective mass ($0.34\,m_e$), resulting in a much higher saturation velocity for electrons at moderate electric fields. At higher fields the mobility in GaAs actually goes down, due to the existence of two local valleys in the band structure, one being a high-mobility region, and the other a low mobility region at higher energy. At high fields the charge carriers can be lifted to the low-mobility valley, resulting in a decrease of the drift velocity.

Acoustic phonons are coherent movements of atoms of the lattice out of their equilibrium positions, resulting in denser and less dense areas if the displacement is in the direction of propagation, like in a sound wave. Optical phonons are opposite displacements of adjacent atoms in the lattice. They are called optical because in ionic crystals such fluctuations in displacement create an electrical polarisation that couples to the electromagnetic field.

This is called the 'transferred electron' effect.

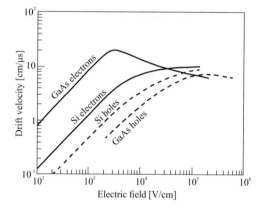

Figure 4.11: Drift velocity dependence on the electric field in high-purity silicon and GaAs at 300 K (modified from [468]).

The contribution to the mobility from impurity scattering is $\mu_i \propto T^{3/2}m_c^{*-5/2}n_i^{-1}$, where n_i is the impurity density [211]. For typical doping concentrations in the active areas of silicon detectors ($n_i < \mathcal{O}(10^{17})$ cm^{-3}), the electron and hole mobilities do not vary significantly with impurity concentration [70].

The combined mobility due to impurity and lattice scattering can be found from $1/\mu = 1/\mu_i + 1/\mu_l$.

Thermal movement of the lattice manifests itself in oscillations of the scattering centres around their equilibrium positions, and thus the mobility for scattering on the lattice for both charge carrier types depends on the temperature.

A practical parameterisation of the temperature dependence is

$$\mu_t^{\mathrm{L}}(T) = \left(\frac{T}{T_0}\right)^{-\theta} \mu_t^{\mathrm{L}}(T_0), \quad t = \mathrm{n, p,}$$

where T_0 is a reference temperature, usually 300 K, and θ is 2.4–2.5 for electrons and $\theta \simeq 2.2$ for holes in silicon.

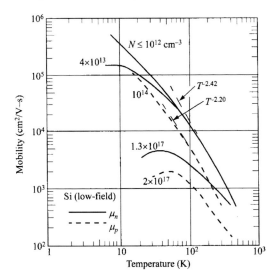

Figure 4.12: Mobility of electrons and holes in silicon as a function of temperature [468].

A widely used parameterisation for the mobility as a function of the electric field is [181]

$$\mu(E) = \mu_0 \left[1 + \left(\frac{E}{E_e} \right)^{\beta(T)} \right]^{-1/\beta(T)} ,$$

where μ_0 is the low-field mobility, E_e an empirical fit parameter, and $\beta(T)$ is a temperature dependent parameter, which has a value of order unity. Diffusion again increases like the square root of the drift distance, and the diffusion coefficient can be found from the Nernst-Townsend equation, eq. (4.4).

For electrons in Si at $T = 300$ K, $\beta = 1.3$, $\mu_0 = 1450$ V/(cm²s), and $E_e = 7240$ V/cm.

In the presence of a magnetic field, the drift is described similarly to eq. (4.8) in gases,

$$\vec{u} = \frac{\mu}{1 + \mu_H^2 B^2} \left[\vec{E} + \mu_H \vec{E} \times \vec{B} + \mu_H^2 \left(\vec{E} \cdot \vec{B} \right) \vec{B} \right] ,$$

but with the Hall mobility $\mu_H = r_H \mu$, with the Hall scattering factor r_H, which is about 1.15 for electrons and 0.7 for holes in silicon [468]. Because of the similar expression for the drift velocity the direction of the charge carriers in orthogonal electric and magnetic fields is again rotated by the Lorentz angle from the direction of the electric field.

4.4 INTERNAL AMPLIFICATION

Often the ionisation charge produced in the detector material is too small to be sensed amongst the irreducible electronic noise that is present in an electronic readout system. This is particularly true for gaseous detectors, where the primary ionisation can be as low as a few ten to a hundred electron/ion pairs, but even in semiconductor detectors, where the ionisation from a minimum-ionising particle can be a few thousand electron/hole pairs, a primary signal increase can be beneficial for small signals (for example in response to low intensity visible light).

In these cases internal charge multiplication as part of the creation of ionisation signal in the detector is required, which amplifies the charge generated in the detector before the signal is collected on the electrodes. This is achieved by impact ionisation in high electric fields in the detector. The primary electrons get accelerated in the high electric fields in between collisions with atoms inside the detector, so that they get sufficient energy to ionise another atom in the next collision. The same will happen for holes, but at higher electric fields, due to their lower mobility. The difference between the ionisation rates for electrons and holes is very important for Avalanche Photo Diodes (APDs) as discussed in section 9.9.

We frequently use the term charge collection, but as we have seen in chapter 3 the current signal is induced during the motion of the charged particles until they arrive on the electrodes.

For a detailed mathematical description of avalanches in thin silicon layers see [500].

PHOTOMULTIPLIER TUBES

A particle detector that exploits impact ionisation on a macroscopic scale is the photomultiplier tube (PMT).

For a comprehensive discussion of PMTs see [405].

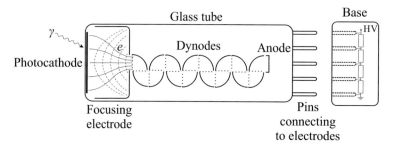

Figure 4.13: Schematics of a box and grid type photomultiplier tube.

A PMT is a device to detect photons in the visible and near-UV range. The sensitive element in the PMT is a photocathode, in which the incoming photons create photoelectrons. A critical figure of merit for the photocathode is the quantum efficiency (QE), which is defined as the probability of an incident photon resulting in an emitted photoelectron. There are three stages which affect the QE [316]:

1. A photon is absorbed in the cathode by an atom with the emission of an electron ('photoelectron');

2. The photoelectron moves to the surface of the cathode;

3. The photoelectron escapes from the cathode into the vacuum.

The efficiency for the first stage depends on the reflection at the cathode surface and the probability of photon absorption within the finite thickness of the cathode. In the second stage the photoelectron will lose energy E_{loss}. In order for an electron reaching the edge of the cathode to escape, it must have a kinetic energy (T) greater than the barrier height ($e\Phi_A$) at the photocathode surface.

This happens by electron-electron and electron-phonon interactions.

For a metal, there are electrons in the conduction band. The barrier height is given by $e\Phi_A = E_{\text{vac}} - E_{\text{F}}$, where E_{vac} is the energy of an electron in the vacuum and E_{F} is the Fermi energy. In a semiconductor, the photon absorption comes from interactions with electrons in the valence band. Therefore the barrier height is $e\Phi_A = E_G + e\chi_s$, where E_G is the bandgap energy and $e\chi_s$ is the energy difference between the bottom of the conduction band and the vacuum [316]. The photoelectron initially has an

Note this is different to the case of single isolated atoms. Here we are considering electrons in a solid. Φ_A is called the electron affinity. Semiconductors can be engineered to have negative electron affinity and therefore lower barrier heights and therefore better QE at longer wavelengths [316].

energy $T = h\nu$, where ν is the frequency of the photon. Therefore for the photoelectron to escape $h\nu - E_{loss} > e\Phi_A$.

E_{loss} is the energy the photoelectron loses before it reaches the surface.

The QE for metals is too low to make useful photocathodes. High quantum efficiency photocathodes are therefore usually made using compound semiconductors. These will have much lower reflection than metals, and because the electron density in the conduction band is very low, there will be very little electron-electron scattering. Semiconductors with high QE are usually made from a combination of antimony (Sb) and one or more alkali metals, e.g. $(NaK)_3Sb$. These semiconductors have small band gaps and very high absorption coefficients for UV photons. They also have small values of the electron affinity. The QE can be improved by appropriate doping to change the Fermi energy level. In addition a very thin layer of a few atoms of Cs can lower the energy of the top of the conduction band to be below the vacuum energy level ('negative electron affinity'). Photoelectrons are emitted by quantum tunnelling through a very thin surface barrier [316].

Metals have a very high reflectivity for the incoming photons. Also the large density of free electrons results in very rapid energy losses of the photoelectron by electron-electron scattering. Therefore the probability of a photoelectron reaching the surface with sufficient energy to escape is very low.

Photocathodes using two alkali elements are called 'bi-alkali' photocathodes. These can be selected to achieve higher QE at longer wavelengths than 'mono-alkali' photocathodes. For good QE for visible photons multi-alkali photocathodes are used.

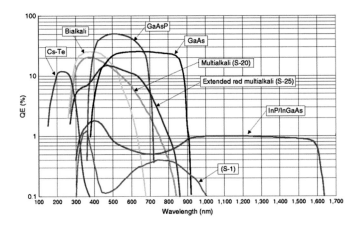

Figure 4.14: Quantum efficiency for different photocathode materials as a function of wavelength [317]. Bialkali photocathodes have a good sensitivity to the wavelengths of most commonly used scintillators. Multialkali photocathodes can be used if sensitivity at longer wavelengths is required. GaAsP and GaAs are more expensive and are only used for very small area photocathodes.

The photocathode only needs to be a very thin layer, and is usually created by vacuum deposition on the rear of the photon entrance window.

The typical mean free path in the photocathode for a photon with 1 to 3 eV is less than 1 μm (see section 2.1).

In the PMT the photoelectrons emerging from the photocathode are guided by electric fields onto a dynode. The field is designed to maximise the fraction of collected electrons, and to accelerate the electrons to sufficient energy, so that the impact of the electron on the dynode liberates additional electrons from the surface. This process continues then on a set of typically 10 to 14 additional dynodes that are held at progressively lower potential by a voltage divider network. The electric field between the dynodes accelerates the electrons, so that more electrons are ejected in the impact on the following dynode. A modest 'gain' in the number of electrons at each dynode can result in a very large overall gain by using multiple dynodes. To allow for the travel of the electrons and to protect the photocathode and dynode surfaces the photomultiplier is contained in an evacuated glass tube.

A dynode is an electrode that is optimised for secondary electron emission. Unlike the photocathode the dynodes do see significant currents, and thus typically consist of an alkali on a good conductor (metal).

Typically 3-5 secondary electrons are ejected by an impinging electron with an energy of 100 to 200 eV.

Typical total gains can be as large as $\mathcal{O}(10^6)$.

Photomultipliers are cheap, yet efficient and fast photon detectors with a large acceptance. They are therefore ideal for the detection of Cherenkov photons in large neutrino experiments, or for the optical readout of large calorimeters. The high speed makes PMTs very capable timing detectors with good time resolution (small 'time jitter'), typically at the level of ns. The time resolution is dominated by the variations of the transit time for photo- and secondary electrons emitted in different directions. The main drawbacks of PMTs are their large size and hence limited position resolution, and that they are strongly affected by magnetic fields. They are therefore often enclosed in a layer of 'μ metal', a soft, ferromagnetic nickel-iron alloy.

Another contribution to the resolution of the time measured when the signal crosses a threshold is from pulseheight variations due to the stochastic amplification process. This can be overcome by the use of constant fraction discriminators (see section 3.8).

The dominant source of noise in a PMT is thermionic emission. If this happens in the photocathode or the first dynodes, the emission will result in an output signal that cannot easily be distinguished from a photon signal (this is referred to as 'dark current'). If purity of the signal is required noise signals need to be rejected by coincidence techniques (see section 13.2).

μ metal shielding works very effectively for low magnetic fields but there is a saturation effect so they do not provide sufficient shielding for the large magnetic fields often used in high energy physics experiments. Therefore alternative detectors such as avalanche photodiodes (see section 4.6) are used for readout of scintillators in regions of high magnetic field.

A large number of PMT models with different electrode configurations are available, providing a wide range of sizes, acceptances, timing performances and spectral sensitivities.

Figure 4.15: Different PMT geometries [230]. (Courtesy of Hamamatsu Photonics K.K.) The diameter of the largest tube in the picture is about 50 cm.

MICROCHANNEL PLATES

A microchannel plate (MCP) detector consists of one or more plates made of resistive material (typically glass), perforated with small holes (typically 5 to 20 μm diameter).

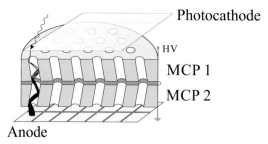

Figure 4.16: Schematics of a chevron MCP.

The inside of the holes is coated with a semiconductor material, and the faces of the plate are metalised. A voltage is applied between the surfaces of the plate, which accelerates electrons in the holes and impact ionisation takes place on the inner hole walls. To increase gain, the holes are not exactly perpendicular to the faces of the plate (typically they are inclined by 10°). Consecutive plates have opposite inclinations.

An MCP is a continuous-dynode electron multiplier.

For two plates this is called a 'chevron' MCP, for three a 'Z stack' MCP.

Because of the localisation of the amplifying holes MCPs have a high spatial resolution and are often used in image intensifiers in night-vision devices. The small size of the amplification channels results in very good time resolution. In experimental physics they are therefore the ideal detectors for time-of-flight mass spectrometry, and they are used in particle detectors using precision time-of-flight measurements for particle identification, e.g. TORCH [399].

See section 12.4.

4.5 AVALANCHE MULTIPLICATION IN GASES

Due to the low density, the amount of primary ionisation in gaseous detectors is small (typically a few 10 electron-ion pairs/cm). Internal amplification is therefore a must. This is achieved by very strong electrical fields (>100 kV/cm), so that an electron drifting through the gas acquires enough energy between collisions with the gas molecules that it will ionise the next molecule it collides with.

Once this condition is met, the number of charge carriers will grow exponentially into an avalanche of electrons. At this point we assume that the increase of electrons in the avalanche is proportional to the number of electrons already present, $d\langle N_e \rangle = \alpha \langle N_e \rangle dx$, where the proportionality constant α is called the '(first) Townsend coefficient'. Predominantly, every collision will create one additional electron-ion pair, and thus $1/\alpha$ corresponds to the mean free path between collisions. As the Townsend coefficient depends on the field and the mean free path between collisions, it scales with the reduced field E/p, where E is the field strength, and p the pressure of the gas, once the field strength exceeds the threshold for avalanche creation.

The second Townsend coefficient gives the number of electrons, which can be released on average from a surface by an incoming ion.

The mean free path is proportional to the molecular density n, which in a gas is given by $n \simeq p/(k_B T)$.

In general, we also have to allow for a competing mechanism, electron attachment. The effective increase in the number of electrons is

$$d\langle N_e \rangle = \langle N_e \rangle (\alpha - \eta) dx,$$

with an attachment coefficient η. This equation can be integrated to get

In principle, this should also include recombinations, although in practice these are less relevant, because ion densities are usually still low, and the electric fields tend to separate electrons and ions in the avalanche.

$$\langle N_e(x) \rangle = N_0 e^{(\alpha - \eta)x}. \tag{4.12}$$

$\alpha^* = \alpha - \eta$ is called the effective ionisation coefficient.

$A = \langle N_{e,\text{final}} \rangle / N_0$ is called the 'gas gain'. In general, both the Townsend and the attachment coefficient do depend on the local field strength. This will vary with location for general field geometries, and in addition the local field strength will be modified by the space charge created in the avalanche itself, if the number of electrons in the avalanche is large. The exact avalanche properties can thus in general not be calculated analytically, but they can be obtained using Monte Carlo techniques (see for example [338]).

Similarly, for the number of positive ions $d\langle N_{\text{ion}}^+ \rangle = \langle N_e \rangle \alpha dx$, and

$$\langle N_{\text{ion}}^+(x) \rangle = \frac{\alpha N_0}{\alpha - \eta} \left(e^{(\alpha - \eta)x} - 1 \right).$$

The number of negative ions can be found from

$$\langle N_{\text{ion}}^-(x) \rangle = \langle N_e(x) \rangle - \langle N_{\text{ion}}^+(x) \rangle.$$

Due to the absence of rotational and vibrational modes, ionising collisions in noble gases have no competition, and thus avalanche multiplication in noble gases starts at lower field strengths than for more complex molecules. This and the negative electron affinity makes noble gases attractive detector gases. Argon provides a larger number of primary electrons than helium and neon, and is more affordable than krypton or xenon, and thus is chosen as the main gas component for many standard gaseous detectors.

Helium is used in gaseous detectors if low material (multiple scattering) is a design driver. Xenon is used when detection of X-ray photons is required, for example in transition radiation detectors (see section 12.5).

In mixtures of noble gases and gases with low ionisation potentials (in gaseous detector typically the quencher, see below) the gain is increased by Penning transfers, where excited states of the noble gas de-excite and the energy is used to ionise the other gas,

For example, in Ar/CO_2 mixtures excited states of Ar from $3p^5 3d$ and higher can ionise CO_2.

$$A^* + B \rightarrow A + B^+ + e^-.$$

The increase in gain can be described by a modification of the Townsend coefficient

$$\alpha_{\text{Penning}} = \alpha \left(1 + r_{\text{Penning}} \frac{f_{\text{exc}}}{f_{\text{ion}}} \right),$$

where α is the Townsend coefficient without transfer, i.e. only due to the collision cross-sections, r_{Penning} the Penning transfer rate, which is the probability that an excited noble gas atom A ionises a molecule B, f_{exc} is the sum of the production rates for the excited states of the noble gas A that have a larger excitation energy than the ionisation threshold of the molecule B, and f_{ion} the sum of the production rates of the direct ionisations [438]. Similar arguments apply for ternary mixtures [437].

Our use of averages reflects the fact that the avalanches are created by stochastic processes, and thus, for the same initial charge, their size will vary. In the absence of attachment and space charge effects, and for moderate fields, the ionisation coefficient can be assumed to be constant. In that case it can be shown that the probability density for an avalanche of size N_e after a distance d, starting from a single electron, will fall exponentially,

See problem 3.

$$P(N_e|d) = \frac{1}{\langle N_e(d) \rangle} e^{-N_e(d)/\langle N_e(d) \rangle}. \tag{4.13}$$

The standard deviation of the avalanche size in this case is equal to the mean, $\langle N_e(d) \rangle$.

Including attachment does not change the shape of the distribution (for sufficiently large $\langle N_e \rangle$), but there is a finite probability that the avalanche fizzles out ($N_e = 0$), if the few electrons at the start of the avalanche are all getting attached. The exponential distribution of the remaining avalanches adjusts to still satisfy eq. (4.12) [328].

If there is a number of m initial electrons the total amount of charge after amplification is the result of the superposition of individual avalanches, if the gas gain is sufficiently low that the space charge created in the amplification does not distort the local field. For large m the avalanche size distribution will approach a Gaussian, in accordance with the central limit theorem.

However, the assumption of a constant ionisation coefficient is not correct, because the ionisation cross-section depends on the energy of the electron. As the electron typically emerges from the previous collision at a low energy, and only gradually increases its energy in the accelerating field, the probability for an ionising collision will vary along its path. In addition, there is competition from other inelastic processes (excitation), which will dominate at lower electron energies. The energy spectrum of electrons in the avalanche shifts to higher energies for higher values of the reduced field E/p, and thus the probability for ionisation increases.

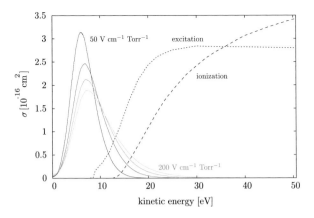

Figure 4.17: Energy distribution (arbitrary scale) of avalanche electrons in inelastic collisions, and inelastic cross-sections. Energy distributions are computed using Monte Carlo simulation for different homogeneous fields in methane (modified from [447]).

Several attempts have been made to model this behaviour, most of them parameterising the ionisation coefficient as a function of the distance since the last collision. The results of these models are typically described as Pólya distributions,

For an overview see [43].

$$P(N_e|d,\theta) = \frac{(\theta+1)^{\theta+1}}{\langle N_e(d)\rangle \Gamma(\theta+1)} \left(\frac{N_e(d)}{\langle N_e(d)\rangle} \right)^\theta e^{-(\theta+1)\frac{N_e(d)}{\langle N_e(d)\rangle}}, \qquad (4.14)$$

for large N_e. The parameter θ is usually determined from a fit to the data. The agreement with data is satisfactory, but the exact physical interpretation of this parameter is under debate [212, 43]. For $\theta = 0$ eq. (4.14) reverts to eq. (4.13), for $\theta \to \infty$ it becomes a Poisson distribution.

Again, Monte Carlo simulation is the most powerful tool to deal with the complexities of the physics [447].

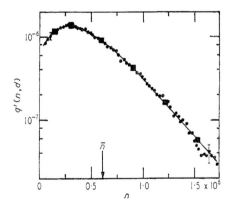

Figure 4.18: Avalanche size distribution for methane [212]. Circles: measured for $E/p = 156\ Vcm^{-1}Torr^{-1}$, $E/(\alpha W_{ion}) = 5.3$. Squares: Pólya distribution for $\theta = 1$ (arbitrary ordinate units).

In the avalanche ionisation is again not the only possible outcome of the collisions. Sometimes, a hit atom will end up in an excited state, and subsequently de-excite with the emission of a photon. Gaseous detectors are typically optically transparent, and this photon can travel macroscopic distances. At some point, it will interact with the gas or the detector walls, releasing a photoelectron, which in turn will become the start of a new avalanche. When this process repeats, it leads to a continuous discharge in the detector, rendering the detector useless.

This is called 'photon feedback'.

There are two ways around this issue:

- The high voltage for the amplification field is supplied via a resistor. When the detector goes into discharge, the increase in current will increase the voltage drop across this resistor, so that the field in the detector drops to levels below what is required for avalanche multiplication. This is the principle of the Geiger-Müller counter.

  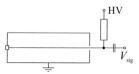

  The drawback of this approach is that the removal of the charge through the large resistor takes a long time ($>\mu$s), during which the detector is blind to other particles ('dead time').

- A polyatomic gas like CH_4 or i-C_4H_{10} is added that has a large absorption coefficient for photons in the few-eV range, by excitation of rotational and vibrational levels. Ultimately, the energy is then safely absorbed in relaxation, dissociation or elastic collisions of these molecules. Such an additive is called a 'quencher gas'. A secondary function of a good quencher is that its atoms exchange charge with ions of the main gas so that these ions cannot strike the cathode surface, where they can release secondary electrons. For this its ionisation potential needs to be lower than for the main gas. The ionised quencher molecule does not emit a secondary electron at the cathode, but instead neutralises and de-excites or dissociates.

i-C_4H_{10} is the empirical formula for iso-butane, $(CH_3)_2CHCH_3$.

A historic approach to limiting the discharge in a gaseous detector by electronic means were spark or streamer chambers. Here, the high voltage was turned on inside the detector for a short time in response to an external trigger signal that indicated the passage of a particle through the chamber. The detector created visible (and audible) sparks, which were photographed. This type of detector was used from the 1930s to the 1960s (for more information see [428]). Today it is sometimes used as a demonstration device.

Equation (4.12) implies the proportionality of the signal charge with the primary charge generated in the gas volume by the incoming particle. Typical gas gains for proportional avalanche multiplication are 10^4 to 10^6. However, at higher fields so many secondary electron-ion pairs can be created in the avalanche that their collective space charge modifies the local electric field. The field is increased in the regions of large charge densities at the front and the tail of the avalanche, and reduced in between. At this point the proportionality starts to break down. At even higher fields and thus gas gains space charge effects together with photons from recombinations in the region of reduced field can create macroscopic discharge channels or 'streamers', for which the signal becomes completely independent of the amount of primary charge. This is called a 'limited streamer mode'. In this mode the typical gas gains are between 10^7 and 10^9.

This is called the 'limited proportionality' regime.

Typically this happens once the space charge-induced electrical field becomes of similar strength as the external field.

In geometries with strongly varying external field (for example around a wire) the growth of the streamer will be stopped once the external field becomes too small to sustain the streamer.

Finally, at extremely high fields the detector will go into plasma discharge, at which operational stability can only be restored by external reduction of the voltage (Geiger-Müller mode). Empirically, the discharge happens once the avalanche from a single electron has reached $\exp(\alpha x)$ particles with $\alpha x \simeq 20$, or about 5×10^8 particles in the avalanche.

This is known as the Raether limit [416].

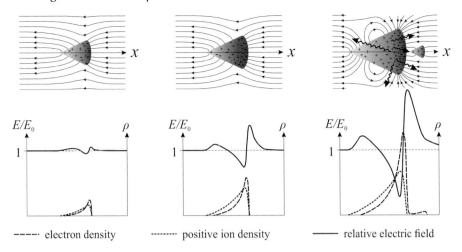

----- electron density   ······· positive ion density   —— relative electric field

Figure 4.20: Schematics of avalanche development in gases for small avalanches and proportional mode (left), medium avalanches and limited proportionality (middle), and large avalanches and streamer formation (right). x is in the direction of the field. E_0 is the field in the absence of the avalanche space charge.

4.6   AVALANCHE MULTIPLICATION IN SEMICONDUCTORS

Unlike in gases, impact ionisation in semiconductors can occur for both types of charge carriers, electrons and holes. If the mean free path for a charge carrier is λ, then a charge carrier drifting in an electric field of strength E will on average acquire an energy λE. If this energy is sufficiently large, the carrier can cause further ionisation, hence we have the possibility of charge multiplication in a similar way as in gases. Generally, multiplication factors for electrons and holes are not the same and it does depend on the type of semiconductor which one is larger at a given field strength. For silicon, the multiplication factor for electrons is much larger than for holes.

The difference between impact ionisation rates for electrons and holes is important for low noise semiconductor avalanche detectors.

We will look at practical application of avalanche multiplication in semiconductors in chapter 9, once we have discussed the principles of doping of semiconductors.

Figure 4.21: Ionisation rates at 300 K versus reciprocal electric field for Si, GaAs, and some IV-IV and III-V compound semiconductors [468].

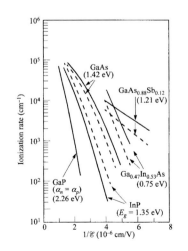

Key lessons from this chapter

- The movement of charge carriers is caused by electric fields and governed by collisions with the detector material.

- In principle the drift velocity is proportional to the electric field, with the mobility as the proportionality factor. In practice the mobility is not a constant, but depends on the physics of the collisions.

- In gases electrons can acquire significant energy between collisions. The large range in energy, together with strong variations of the collision cross-section (Ramsauer effect) and the energy loss fraction leads to non-trivial behaviour of the drift velocity. For ions, the drift velocity is simpler, but much slower (by about three orders of magnitude).

- The size of a cloud of electrons is increased by diffusion. It generally grows with the square root of the drift distance. In gases, the diffusion coefficient is at the thermal limit in 'cold' gases like CO_2 and larger in 'hot' gases like argon.

- In liquids high electron drift velocities can be obtained for materials with molecules with a high polarisability (typically these have a high degree of spherical symmetry). Typical examples are larger noble gases and some organic liquids. The challenge is the required purity.

- Typical solid particle detectors are made of semiconductors, because of the availability of pure material, and the band gap resulting in only moderate amounts of charge carriers in the conduction band, which can be removed by appropriate means. In solids, electrons and holes can contribute to the net movement of charge.

- Magnetic fields affect the drift of charge carriers. If the magnetic field is orthogonal to the electric field, the drift direction gets rotated by the Lorentz angle. In parallel fields, the transverse diffusion is reduced.

- If the energy acquired between collisions is large enough, secondary ionisation occurs, leading to an avalanche growth of the charge in the detector.

EXERCISES

1. *Transport properties in gases.*

   Estimate the mean free path for electrons in a gas at STP. (You might find useful that the molar volume for an ideal gas is 22.4×10^{-3} m^3/mol and the diameter of atoms is of the order 10^{-10} m).

   Assuming that the thermal component of the electron velocity dominates, what is the mean velocity of the electrons and the time between two collisions?

2. *Drift in magnetic fields.*

   Show that eq. (4.8) is a solution to the Langevin equation, eq. (4.7).

3. *Avalanche size distribution.*

   In this problem you will derive the avalanche size distribution $P(n|d)$, that the avalanche comprises n electrons after a distance d (eq. (4.13)).

Assume that the Townsend coefficient varies with the location, $\alpha = \alpha(x)$. (It is sufficient to think about this in 1D.) The starting point for the avalanche is one electron at $x = 0$, i.e. $P(1|0) = 1$.

a) First, find the probability $P(1|x)$ that a charge has not created a secondary ionisation at all between $x = 0$ and $x = x$.

b) If the probability that the avalanche contains $n - 1$ electrons at $x = x'$ is $P(n-1, x')$, show that for small $\Delta x'$ the probability that one and only one of these electrons will ionise between x' and $x' + \Delta x'$, so that there are n electrons, can be written as $(n-1)\alpha(x')\Delta x'$.

c) Show that the probability that none of these n electrons will ionise in the region between $x' + \Delta x'$ and x is $\exp[-n(\int_0^x \alpha(x'')\, dx'' - \int_0^{x'} \alpha(x'')\, dx'')]$.

d) Find an equation for $P(n|x)$ by multiplying these expressions and integrating over x'. Using

$$\overline{N}_e = \sum_{n=1}^{\infty} nP(x|d) = e^{\int_0^d \alpha(x)\, dx},$$

reproduce eq. (4.13).

4. *Efficiency and dark current of a PMT.*

The typical quantum efficiency for a PMT with semitransparent bi-alkali photocathode is close to 25% in the spectral range between 350 and 400 nm.

a) If 10 photons impinge on the photocathode, what is the probability that no photoelectron is produced?

b) If the PMT is coupled to a slab of plastic scintillator ($\rho = 1.05$ g/cm^3 and a light yield of 10 photons/keV), and the probability of a scintillation photon from the scintillator to reach the photocathode is 10% (due to collection and coupling efficiencies), how thick does the slab have to be to achieve 95% detection efficiency for minimum ionising particles?

c) Estimate the transit time (the time from the creation of the photoelectron to the arrival of the signal at the anode), if the PMT has 10 stages with a separation of 10 mm, and the inter-dynode voltage is 200 V. Assume that at each dynode the electrons are created at rest.

d) The PMT has a gain of 10^6 and displays a dark current of 1 nA. What is the rate of thermal electron emission from the photocathode? Why is this an issue, and what can you do to get this under control?

5. *Gain stability of a PMT.*

The gain of a dynode in a 10-stage PMT as a function of the inter-dynode voltage V_{dyn} can be parameterised as $G_{dyn} \propto V_{dyn}^{0.6}$. If the overall voltage with which the PMT is operated is 1 kV, what is the acceptable operating voltage fluctuation, if the overall gain of the PMT is to stay within 1%?

5 Response to excitation

So far, we have discussed signals from the collection of ionisation charge produced by an incoming charged particle. As long as the volume is large enough to provide a sufficient amount of charge, the detection elements in such detectors can be made small, leading to detectors that are capable of precisely locating the passage of charged particles. However, sometimes position resolution is not the primary target, but a large dense detector volume or cheap instrumentation of a large volume is required. In that case it is often easier to collect optical photons, which have a long range in transparent materials. Such photons can be generated in detector materials called 'scintillators' from the relaxation of excited states created by the electromagnetic interaction of the incoming charged particle with the detector.

For an early, but comprehensive introduction to scintillation see [145].

To achieve this long range, the energy of the excitations needs to be degraded, so that the energy of the photon is insufficient to cause further excitations in the detector material, resulting in the loss of the photon.

The other source of optical photons in particle detectors is Cherenkov radiation, as discussed in section 2.8. Section 5.7 will compare Cherenkov and scintillation radiation.

Depending on the nature of the material, we distinguish between organic and inorganic scintillators.

5.1 ORGANIC SCINTILLATORS

In organic scintillators it is the de-excitation of molecular electrons that generates these photons. Organic substances that contain aromatic rings, such as polystyrene (PS) and polyvinyltoluene (PVT), do display this effect. Because scintillation in organic scintillators is happening within molecules, the actual state of the material is not that relevant, and there are plastic, crystalline, liquid and gaseous organic scintillators.

The molecules typically have a spectrum of electronic singlet and triplet states, with a typical separation at the level of eV, plus a fine structure of vibrational levels with a ten times smaller spacing.

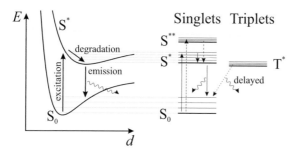

Figure 5.1: Schematic operating principle of an organic scintillator. Left: Potential energy as a function of atomic separation. Right: Energy levels.

At room temperature most electrons are in the S_0 singlet ground state. The electromagnetic interaction with an incoming particle lifts the electrons into an excited state somewhere in this spectrum. The singlet excitations decay very quickly (< 10 ps) to the S* first excited state (internal degradation), which subsequently does not decay to the electronic ground state, but typically to one of the vibrationally excited S_0 levels, with the emission of a photon on the scale of ns ('prompt fluorescence').

The shift in wavelength is called 'Stokes shift'.

DOI: 10.1201/9781003287674-5

Re-absorption of the emitted photon is suppressed, as its energy is less than the transition from the ground state $S_0 \rightarrow S^*$.

While singlet-triplet transitions are forbidden by electric dipole selection rules, a small number of molecules will become excited in a triplet state. Such states are metastable and the de-excitation to the ground state S_0 takes in the order of ms. The metastable state can also be lifted to an excited S^* state (by thermal excitation or another charge particle collision), which then decays promptly as discussed above. This process is called 'delayed fluorescence'.

This is called 'phosphorescence'.

To achieve larger macroscopic transmission distances, additional organic fluorescing compounds are added, which degrade the photon energy further. The first step in this degradation cascade usually happens on such small distance scales that the energy is not transferred by photons, but by a resonant dipole-dipole interaction (Förster energy transfer). The primary scintillator and the wavelength shifting additives are chosen so that the emission and absorption energies of the different stages match.

These substances are also called 'wavelength shifters'. For a detailed review of modern scintillator materials, see [129].

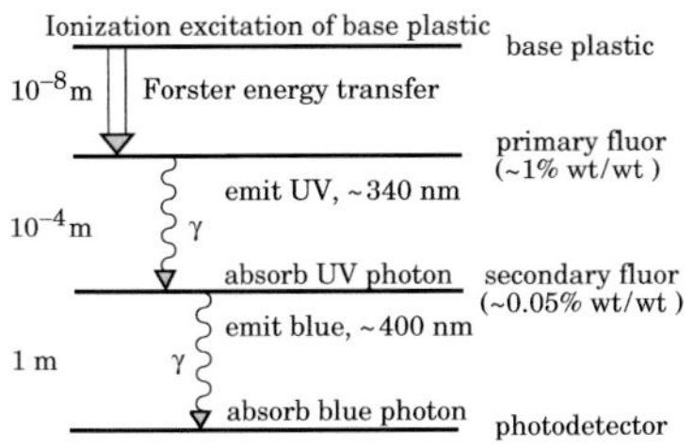

Ionization excitation of base plastic — base plastic
10^{-8} m — Forster energy transfer
primary fluor (~1% wt/wt)
10^{-4} m — emit UV, ~340 nm — γ
absorb UV photon — secondary fluor (~0.05% wt/wt)
1 m — emit blue, ~400 nm — γ
absorb blue photon — photodetector

Figure 5.2: Photon energy degradation in an organic scintillator [505]. In detectors in which the photon sensors are more than 1 m away from the scintillator, it is common to employ an additional wavelength shifter to convert the light from blue to green (this has a significantly longer attenuation length).

The prompt decay of the excited state has an exponential decay characteristics, $L \propto \exp(-t/\tau)$. Because organic scintillators involve relaxation of electronic states within one molecule the prompt fluorescence emission process can be fast (a few ns), and organic scintillators are often used in timing applications, like detectors used for triggers and detectors for time-of-flight measurements (see section 12.2). If delayed fluorescence occurs, another exponential with a longer time constant needs to be added.

L is the light output.

Only in very fast scintillator applications the finite risetime due to excitation and degradation can become relevant.

The light output of scintillators does saturate for high excitation densities. This effect is described by the empirical Birks' formula [144] for the light yield per unit length of the incoming particle's path,

Birks' law applies also for inorganic scintillators.

$$\frac{\mathrm{d}L}{\mathrm{d}x} = \frac{A\left(\frac{\mathrm{d}E}{\mathrm{d}x}\right)}{1 + kB\left(\frac{\mathrm{d}E}{\mathrm{d}x}\right)}, \quad (5.1)$$

where A is a proportionality factor that describes the proportionality of energy loss and light yield at low excitation density, k is the probability of quenching, and B is a proportionality factor between the energy loss and the density of already excited molecules. Together, kB is called the 'Birks' constant', which is a property of the scintillator and is found experimentally.

For example, for polystyrene $kB = 0.126$ mm/MeV.

For scintillators with more than one decay time constant the different components typically have different values of the Birks' constant, hence the signal shape will differ for particles with different energy loss.

The light yield of organic scintillators is usually fairly stable under temperature variations (between 0 and 3×10^{-3} per °C [403]). Generally, the light yield from scintillators decreases with temperature, due

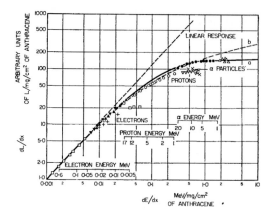

to thermal quenching, caused by non-radiative de-excitation or thermal ionisation of the luminescent centres. In the first case the state is excited to a level where the energies of the electronic states of the excited and the ground state cross, so that no energy is released in the transition between the two, with a subsequent de-excitation of the vibrational excitation within the ground electronic state. In the second case the excited state receives sufficient thermal energy that it becomes ionised before it can decay radiatively [306, 335].

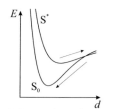

Figure 5.4: Non-radiative de-excitation.

A widely used type of organic scintillator are plastic scintillators. There the scintillator molecules get dissolved in a styrene or vinyltoluene monomer which is then polymerised to create a solid plastic. Plastic scintillators are produced by casting or extrusion and can be conveniently made in many different shapes and sizes. For use, they are typically wrapped in a reflective and light-tight coating. In principle plastic scintillators are robust, but mechanical stress can cause the development of micro-cracks, which cause photon loss.

Liquid organic scintillators can have good scintillation yields and long attenuation lengths. They are thus easy to use and affordable for filling out a large active volume, and thus they are an attractive detector medium for neutrino experiments (see section 14.2). High purity is required for good light yield and contamination with water, oxygen and solvents must be avoided.

5.2   SCINTILLATING FIBRES

If a good spatial resolution of the scintillation is required, the obvious approach is to reduce the size of the scintillator elements. A practical geometry for this are scintillating fibres. In principle, diameters down to a few $100\,\mu$m are possible, but because of the trade-off between granularity of the position information and light output, often larger diameters (up to 1 mm) are chosen. Scintillating fibres can be made of appropriately doped polystyrene or glass, or they can be liquid scintillator contained in capillaries.

For a comprehensive review of scintillating fibres see [331].

A benefit of the fibre geometry is that the light can be transferred along the fibre by total internal reflection to a junction with a photon detector. Mechanically, individual fibres are flexible, although this flexibility is reduced once they are collected in bundles. Active (scintillating) fibres can be spliced onto passive (light-guiding) fibres.

This is beneficial because clear fibres have longer attenuation lengths than scintillating fibres.

To provide a controlled surface for the internal reflection the fibres are usually clad with a thin ($\lesssim 5\ \mu$m) layer of transparent material with a lower index of refraction. Photons with an angle larger than $\Theta_{\text{crit}} = \arcsin(n_{\text{clad}}/n_{\text{core}})$ to the surface of the interface, where n_{clad} and n_{core} are the index of refraction of the cladding and the core, respectively, will be transported along the fibre. Theoretically, the fraction of collected light emitted on the axis of the fibre in one direction of the fibre is given by

$$f = \frac{1}{4\pi} \int_0^{\pi/2-\Theta_{\text{crit}}} 2\pi \sin\theta \, d\theta = \frac{1}{2}\left(1 - \arcsin\left(\frac{n_{\text{clad}}}{n_{\text{core}}}\right)\right). \qquad (5.2)$$

A typical combination is a polystyrene core ($n_{\text{core}} = 1.59$) with a cladding of poly(methylmethacrylate) (PMMA, $n_{\text{clad}} \simeq 1.49$), and the fraction of collected light in this case can be 3%. Double cladding with an additional outer low-n cladding layer (for example with a fluoroacrylic with $n = 1.42$) can improve the trapping of the light in the fibre to about 5%.

A concern for closely packed fibres is cross-talk, where light from one fibre is collected and propagated in an adjacent fibre. Usually, photons outside of the collection cone of a fibre cannot be collected by a parallel fibre. However, if the photon activates a wavelength shifter in the other fibre, the isotropically re-emitted shifted photon can be in the acceptance cone for propagation. Small-diameter scintillating fibres therefore usually contain only one wavelength shifter with a large Stokes shift (for example 1-phenyl-3-mesityl-2-pyrazoline, PMP) and thus little overlap between emission and absorption bands. The drawback is a smaller light yield. Another approach is to deposit a thin layer of aluminium over the surface of the fibres. This eliminates cross talk and also increases the light yield.

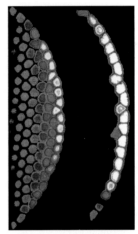

Figure 5.5: Cross-talk in bundles of scintillating fibres (modified from [215]). Left: scintillator with two wavelength shifters (p-Ter+POPOP). Right: scintillator with one wavelength shifter (PMP). Both bundles are illuminated with a UV laser from the right. The fibre diameter is about 30 μm.

5.3   INORGANIC SCINTILLATORS

Inorganic scintillators are transparent crystals, usually mixed with some dopant (for example NaI, doped with Tl). The interaction with the incoming charged particle lifts electrons well above the band gap and leaves deep holes in the valence band and below. The high excitations lose energy in inelastic collisions, while the holes gain energies from Auger transitions, until all charge carriers are contained in the valence and conduction bands.

Inorganic scintillators have been used from the early days of studying radioactivity. It were barium platinocyanide plates that allowed Röntgen to make his first observations of X-rays in 1895 [431].

These states then thermalise to the lower edge of the conduction band (electrons) and the top of the valence band (holes). They then form excitons (electron/hole states just inside the band gap),which travel together until they get captured as a whole by impurity centres ('activators'). These usually have been deliberately introduced, creating locations within the crystalline structure that are referred to as 'luminescence centres' or 'emission centres'. Finally, the excited activator states decay, emitting a visible or UV photon, which can travel long distances as their energy is smaller than the band gap. In addition, the wavelength of the light emitted by the activator is often better matched to the spectral sensitivity of the photon detector.

Because the process requires movement over significant distances, crystal scintillators are usually not very fast detectors.

For example, in pure NaI photons with a wavelength of 303 nm are created. The activator Tl changes the photon wavelength to 450 nm.

In some crystals the band structure does inhibit the thermalisation of deep hole states, so that there is a significant probability of electrons in

An example for such a material is BaF$_2$.

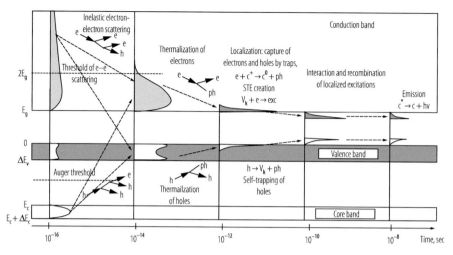

Figure 5.6: *Relaxation scheme for electronic excitations in an insulator: e, electrons; h, holes; ph, phonons; hν, photons; V_k, self-trapped holes; STE, self-trapped excitons; c^n, ionic centres with charge n. Occupation density is represented by grey and white areas for electrons and holes, respectively [326].*

the conduction band falling into these deep holes, resulting in a very fast emission of a UV photon. The challenge in exploiting this fast signal is the high energy of the photon, which is usually outside the sensitivity of most photocathodes.

This process is called 'cross-luminescence'.

Not all crystals require doping to achieve scintillation: Bismuth Germanium Oxide (BGO), for example, is a pure inorganic scintillator without any activator impurity. There, the scintillation process is due to an optical transition of the Bi^{3+} ion, a major constituent of the crystal.

The energy of the emitted photons from inorganic scintillators has a broad spectrum because of distortions of the emission centres due to lattice deformations, and because of temperature broadening of the optical transitions. Typically, between 10 and 100 eV are required per photon produced, depending on the band gap. Practically, the energy required per detected photon is larger than that, because of photon loss due to absorption and on surfaces, and inefficiencies in the detection.

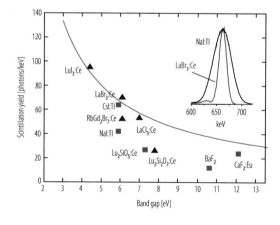

Figure 5.7: *Photon yield/keV of several inorganic scintillators as a function of the band gap [326].*

In the time domain the signal is again described by a decaying exponential, although the time constants are longer than for organic scintillators due to the time required for thermalisation of the charge carriers. Often crystal scintillators display two time constants: one for fast recombination (10^{-6}–10^{-9} s) from activation centres, and the other from delayed recombination due to trapping (10^{-3}–10^{-6} s).

Table 5.1: Some common scintillators (data from [505, 316, 512]).

Scintillator	Light yield [γ/keV]	Peak wavelength [nm]	Decay time [ns]	Density [g/cm^3]	dL/dT [%/°C]	Hygroscopic
NaI:Tl	43	410	245	3.67	−0.2	yes
CsI	1.5/0.5	310	30/6	4.51	−1.4	slightly
CsI:Tl	52	550	1220	4.51	0.4	slightly
LSO:Ce (Lu$_2$SiO$_5$:Ce)	27	402	41	7.40	−0.2	no
LYSO:Ce (Lu$_{1.8}$Y$_{0.2}$SiO$_5$:Ce)	33	420	40	7.10		no
BGO (Bi$_4$Ge$_3$O$_{12}$)	8.2	480	300	7.13	−0.9	no
PWO (PbWO$_4$)	0.13/0.03	425/420	30/10	8.30	−2.5	no
BaF$_2$	1.8/10	220/310	0.6-0.8/630	4.88	0.1/−1.9	no
LaBr$_3$:Ce	61	356	17-35	5.29		yes
Plastic scintillator (typ.)	10	425	2	1.03	0 to −0.3	no

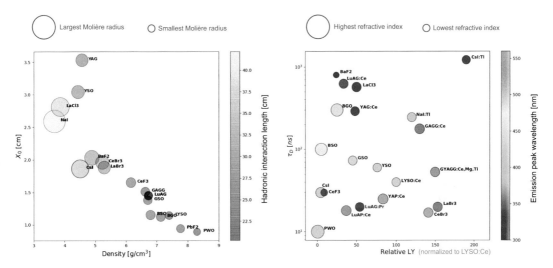

Figure 5.8: Material properties of inorganic scintillators [345].

The temperature dependence of the light yield for inorganic scintillators is typically larger than for organic scintillators, and is quite significant for BGO and PWO (1–2% per °C). In addition to the light yield also the decay time of the scintillation can change with temperature.

Temperature stabilisation is therefore needed for these crystals, if proportionality of the signal is required.

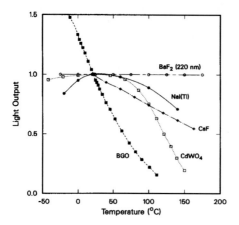

Figure 5.9: Light yield as a function of temperature for some inorganic scintillators (for BaF$_2$ only the 220 nm emission is shown) [367]. The curves are normalised to the light yield at 20 °C for each material.

Another complication is that some crystals are hygroscopic and the ingress of water can damage the crystal structure close to the surface, resulting in dead zones and poorer reflectivity of the surface.

Inorganic crystal scintillators can be very dense (for example, PbWO$_4$ has a density of more than 8 g/cm^3), which makes them well-suited for detectors where high stopping power is required, and in particular for the use in homogeneous calorimeters (see section 11.3).

5.4   SCINTILLATION IN LIQUID NOBLE GASES

Liquid noble gases can also be scintillators. An incoming particle can create excitons or electron-hole pairs in the liquid. Collisions with neutral atoms can then create diatomic excited or ionised molecules. In argon, for example, these can be an excited state Ar$_2^\star$ or an ionised state Ar$_2^+$. The ionised state can capture an electron, leading also to the neutral excited state Ar$_2^\star$. This state decays non-radiatively to an intermediate level, which then decays radiatively to the ground state, emitting a scintillation photon in the process. For liquid argon, the wavelength of this photon is centred around 127 nm. This is far in the UV region, beyond the sensitivity of standard photon detectors, and thus shifting of the wavelength is usually required. A suitable wavelength shifter is tetraphenyl-butadiene (TPB) [172].

These excited states are sometimes called 'excited dimers', or 'excimers'.

For liquid krypton the wavelength is about 150 nm, and for liquid xenon about 178 nm.

The liquid noble gas is transparent for its own scintillation light, because the typical distance between atoms in the liquid is about a factor 2 larger than the separation of the two atoms in the excimer, and thus the reverse of the emission process is highly unlikely.

This argument is only valid for pure argon. Contaminants can reduce light transmission.

The excimers can occur in singlet and triplet states, which have similar energies but very different decay lifetimes, resulting in two time constants, The lifetime for the singlet states is short (a few ns), but the lifetime for the triplet states can be considerably longer (up to 1 μs). The excitation produces approximately 40 × 10^3 γ/MeV in liquid argon.

46 × 10^3 γ/MeV in liquid xenon.

5.5   PHOTON DETECTION AND COUPLING

The actual number of detected photons from a scintillation detector is given by the light yield of the scintillator times the light collection efficiency and the quantum efficiency of the photon detector. The light collection efficiency is affected by the attenuation of the light within the scintillator and light guides due to finite transparency, as well as losses incurred during the total internal reflection on the inner walls of light-guiding elements.

In practical applications the scintillator is connected optically to the photon detector via a light guide, in which the light is conducted by internal reflection [506]. To couple efficiently, Liouville's theorem (Liouville's theorem demands that the volume in phase space stays the same) needs to be satisfied, which means that the cross-section of the light conductor should stay constant up to the photon detector. This target can lead to creative solutions for light guide geometries. If it cannot be achieved, the collection efficiency will be reduced by a factor up to A_{PD}/A_S, where A_{PD} and A_S are the photon detector and scintillator facing cross-sections, respectively. Large and complex scintillator and light guide geometries can lead to significant path length differences for the photons, degrading the timing resolution of the system ($c \simeq 20$ cm/ns in the material).

One approach to achieve reduction of aperture in a light guiding system is to couple the (large-area) scintillator via a small air gap to a smaller light guide, which is doped with a wavelength shifter (typically from blue to green). The re-emission from the wavelength shifter is isotropic, but with the smaller phase space due to the smaller dimensions of the light guide.

Good coupling of the light guide to the photon detector is achieved with optical grease or optical pads. In modern scintillator systems, the light transfer is often via optical fibres that are doped with wavelength shifters. The benefit of such a light transfer is a high mechanical flexibility of the fibres, that the optical connections can be distributed more uniformly within the scintillator, and that the optical path length can be more uniform.

Photon detectors commonly used are Photomultiplier tubes (PMTs) for large segmentation, and Avalanche Photodiodes (APDs) for small segmentation. Typical efficiencies for PMTs are about 30%, and for APDs up to 80%. Other, more recently developed photon detectors are Silicon Photomultipliers (SiPMs) or Multi-Pixel Photon Counters (MPPCs).

Ideally, the peak emission energy of the scintillator matches the peak quantum efficiency of the photodetectors used, typically 250–500 nm for photomultipliers (see also Figure 4.14) and 450–900 nm for solid state photodetectors (*p-i-n* diodes and avalanche photodiodes).

Surface imperfections can result in losses even for angles that would correspond to total internal reflection for a perfect surface.

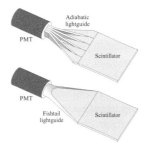

Figure 5.10: Light guide geometries. Top: Adiabatic light guide. Bottom: Fishtail light guide.

Optical grease is usually a transparent silicone grease with an index of refraction close to plastic scintillators ($n \simeq 1.465$). Optical pads are soft silicone pads with a few mm thickness with a similar index of refraction.

See section 4.4.

See section 9.9.

For a discussion of these devices, together with their strengths and weaknesses, see section 9.9.

Other photodiodes using InGaAs or InGaAsP as used in telecommunications have sensitivity up to 1550 nm.

5.6 RADIATION DAMAGE IN SCINTILLATORS

Radiation can affect scintillators in several ways: First and foremost, it can cause radiation-induced absorption by creating colour centres that trap photons in the scintillator and thus reduce the attenuation length. Second, it can introduce radiation-induced phosphorescence (afterglow), and finally, in organic scintillators it can damage the scintillation mechanism, degrading the light yield and changing the emission (and the absorption) spectrum.

In organic scintillators radiation can create free radicals, which strongly absorb UV photons. The radicals usually will anneal with time, but can polymerise with other radicals, in which case the radiation damage becomes permanent. Diffusion of oxygen into plastic scintillator during the annealing leads to the creation of carbonyl and hydroxyl groups, which absorb at longer wavelengths, resulting in reduced transmission. In plastic scintillators it is the base material that gets damaged, the dopants are usually stable [169].

In halide crystals the dominant damage mode is again oxygen/hydroxyl contamination. In oxide crystals it is structure defects, such as oxygen vacancies [512].

The radiation hardness can also sometimes be improved with suitable doping. For example, in $PbWO_4$, doping with niobium prevents the trapping of holes on oxygen near a lead vacancy [273].

Recovery of the light yield due to annealing occurs already at room or application temperature, but the radiation damage may also be reduced by heating the crystals (thermal annealing) or by exposing the crystal to light (optical bleaching). As a consequence of the annealing the radiation damage effects in most crystals are rate dependent. Under normal operation there is significant annealing such that for a given dose rate an asymptotic equilibrium between damage and annealing is obtained.

Radiation damage is less of a concern in liquid scintillators, as the medium usually can be exchanged, and contaminants need to be removed in any case, whether they are generated by radiation or by other sources.

5.7 COMPARISON OF SCINTILLATION AND CHERENKOV RADIATION

Both Cherenkov radiation and scintillation light comprise photons in the visible range, and they can be recorded by similar photon detectors. However, they are fundamentally different. Cherenkov radiation is directly generated by the incoming charged particle, and thus the emitted photons correlate with the direction and the velocity of the incoming particle. The relevant property of the detection medium is the index of refraction. The photons from Cherenkov radiation are created instantly when the incoming particle traverses the detector material.

A material often used similarly to inorganic crystal scintillators is lead glass (SiO_2/PbO), for example in homogeneous calorimeters. However, the photons created there are not from scintillation, but from Cherenkov radiation. Its light output is two to three times lower than crystal scintillators.

As we have seen in chapter 2, excitations are a prominent manifestation of charged particle energy loss, whereas Cherenkov radiation, when the conditions for its creation are met, constitutes only a small contribution ($\sim 10^{-3}$).

Scintillation light is coming from a secondary emission from the relaxation of excited states in the medium. It is thus isotropic, and the emission of scintillation light, in particular from inorganic scintillators, is delayed. The energy of the scintillation photons is governed by the energy levels in the molecules or lattice of the scintillator. Optimisation is therefore a matter of chemistry.

However, the two signals can occur concurrently in response to the passage of a charged particle in a scintillator, and if they do they are somewhat complementary, which can be exploited in detectors with dual readout (see section 11.11).

Key lessons from this chapter

- The key feature of a scintillation detector material is the shifting of the energy of the emitted photon from the energy associated with the energy gap involved in the photon creation (usually in the UV) to longer wavelengths. This is necessary to obtain photons that can traverse macroscopic distances in the scintillator, and to match the spectral sensitivity of the photon detector.

- This is usually achieved by doping.

  In organic scintillators this is achieved by doping with wavelength shifters.

  In inorganic scintillators the doping creates luminescence centres.

- In general, organic scintillators are faster, cheaper and are less dependent on temperature, and inorganic scintillators are denser and thus have a higher stopping power.

- The properties of different inorganic scintillators vary significantly. A fast, radiation-tolerant inorganic scintillator, $PbWO_4$, was developed for use at LHC.

- As usual, there is no universal solution. The choice of technology is driven by the specific requirements of the experiment.

EXERCISES

1. *Collection time for scintillation light.*

   How much time is required to collect 90% of the total light yield for a) NaI:Tl, b) $PbWO_4$, c) a typical plastic scintillator?

2. *Total internal reflection.*

   A scintillator is made of a slab of polystyrene ($n = 1.59$) of infinite area. Calculate the fraction of the light that escapes from either surface of the slab, and hence the fraction of the light that will be trapped in the scintillator.

3. *Scintillating fibres.*

   A minimum-ionising particle is passing through a scintillating fibre with a doped diameter of 300 μm. (The stopping power for a minimum ionising particle in polystyrene is 1.936 $MeVg^{-1}cm^2$.)

   a) Estimate the deposited energy and the number of scintillation photons created in the fibre. ($\rho_{core} = 1.05$ g/cm^3).

   b) Estimate the number of photons that will arrive at the end of a 1 m long scintillating fibre, if the index of refraction of the core and the cladding is 1.59 and 1.48, respectively, and the attenuation length is 2 m.

6 Detection of ionisation without charge movement

So far we have discussed particle detectors that ultimately produce an electrical signal from the collection of charges. However, there are detectors in which the ionisation charges produced by the passage of a charged particle cause a local change to the molecular structure of a photographic emulsion, or a local change to the physical state of the detector medium, resulting ultimately in an image that reflects the path of the incoming particle in the position of these localised changes. Interpretation of these images requires optical survey that produces the data that is then used in the analysis.

Detectors of this type have been historically very important, but today they have mostly been superseded with detectors with electronic readout, which are faster, can cope with high particle rates, and can easily be incorporated in electronic triggers. Nevertheless, these technologies still do provide some of the most accurate position measurements for charged particles available, and thus there are even today examples of these technologies used in experiments that do not involve high detection rates.

6.1 EMULSIONS

Nuclear emulsions, like conventional photographic film, consist of silver halide crystals (mostly bromide with a small admixture of iodide) immersed in a gelatinous or polymer carrier. The size of the crystals is typically between 0.1 μm and 1 μm. When a halogen atom in the crystal gets ionised, the electron can combine with a silver ion resulting in a neutral silver atom. Several silver atoms then coalesce and form a small cluster. Exposure to a development solution will make the silver clusters grow, until they cover a macroscopic area that can be resolved by eye. In a final development step another chemical, the fixer, removes the remaining silver halide crystals, but leaves the silver clusters in place. After a final wash to remove all chemicals the pattern of silver clusters representing the image is permanent. The gelatinous carrier maintains the position of the silver clusters throughout the process.

For a more thorough discussion of the chemical processes in emulsions see [67].

At this point we speak of a 'latent image'.

The average energy loss of a charged particle is insufficient to develop a silver halide grain, but ionisation events with a large energy transfer (δ-electrons) are usually contained within a grain and deposit all their energy there, which is sufficient to start the chemical process. Because of the statistics of these high-energy transfers, only a fraction of the grains along the path will get developed. To address this, nuclear emulsions differ from regular photographic film, in that they contain a higher, more uniform density of smaller silver halide grains, and are thicker. The grain density is about 10^{14} grains/cm^3, and each grain is essentially an independent detecting element.

The spatial resolution of a single grain can be described by the RMS of the response distribution, which can be idealised as a binary

DOI: 10.1201/9781003287674-6

distribution that is 1 for a spherical volume of the grain with diameter D, and 0 outside. It is found similarly as eq. (3.7), and is given by RMS $= \sqrt{\pi}D/8$. With a grain diameter of 0.2 μm, this results in a theoretical spatial resolution of 44 nm. The availability of a large density of high-precision space points is the strength of emulsion detectors.

Nuclear emulsions do display random noise, which is due to thermal noise, gelatine impurity and over-sensitisation. In addition, as there is usually a delay between the manufacture of the emulsion and its use in an experiment, there will usually also be a number of background tracks from that period. However, latent images disappear with time due to oxidation (a process called 'fading'). This process accelerates at high temperature and humidity, and by the addition of 5-methylbenzotriazole the disappearance of background tracks can be accelerated without affecting the sensitivity of the material [385].

This operation is called 'refreshing'.

Photographic emulsions have been used historically in particle physics (the first observation of subatomic particles has been made by Becquerel by accidentally exposing photographic plates to beta radiation from Uranium salts [124]). Today they are still in use where high position resolution for low charged particle densities and rates are needed (i.e. neutrino experiments).

Figure 6.1: Image of one of Becquerel's photographic plate after exposure to β radiation from uranium salts [125].

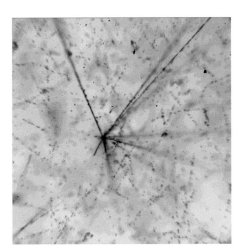

Figure 6.2: Image of a neutrino vertex in the emulsion target of the CHORUS experiment [232]. The image size corresponds to about 100×100 μm^2 (this is the typical scale of the size of one pixel in a silicon pixel detector).

Emulsion detectors have been developed into very high resolution tracking detectors with three-dimensional reconstruction capabilities by arranging emulsion layers into stacks, which are usually interleaved with passive material layers, like plastic or metal plates. Multi-layer stacks of emulsions are often employed as high-resolution active targets together with electronic detectors (scintillation counters, trackers or calorimeters), which give information to localise interesting vertices in the emulsions and complement the kinematic information of the events. An intermediate (double-)emulsion layer (referred to as a 'changeable sheet', CS) can be used to make a decision on whether a section of the ECC should be unpacked and developed.

Such an arrangement is called an emulsion cloud chamber (ECC).

The ECC target for the OPERA experiment was made of films with a total surface of 110,000 m^2 and 105,000 m^2 lead plates (target mass of 1.25 kt), relying on industrially produced, machine-coated emulsion films with very uniform layer thickness [385].

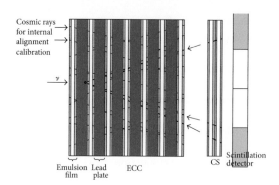

Figure 6.3: Schematic drawing of an ECC detector (OPERA) [232].

The other important technological development allowing for the exploitation of such large emulsion detectors is the development of automated scanning microscopes. Typically, the focal length of these microscopes is about 5–10% of the thickness of the emulsion, allowing for an automated recording of the three-dimensional track within the emulsion. The scanning of the events in the OPERA experiment was performed with 40 fully automated microscopes which allowed to scan about 500 m² of emulsion in 5 years.

Among all tracking devices used in particle physics, nuclear emulsion particle detectors feature the highest position and angular resolution in the measurement of tracks of ionising particles, in particular for the detection of short-lived particles. They are particularly useful when very high resolution is needed, but at low speed and cost (for example in τ neutrino experiments).

6.2   CLOUD AND BUBBLE CHAMBERS

Another class of detectors that historically has been very important for particle physics relies on local phase changes caused by the passage of a charged particle in a homogeneous detector medium that has been prepared in a metastable state (supercooled vapour or superheated liquid). The pattern of generated droplets or bubbles are then recorded using photography and analysed after scanning of the image.

For example, the discovery of strange particles in a cloud chamber [429], the confirmation of the quark model by the observation of the Ω^- particle in a bubble chamber [114], or the discovery of the neutral weak current in the Gargamelle bubble chamber [282].

BUBBLE CHAMBERS

A superheated liquid is unstable and will tend to minimise its Gibbs potential by evaporation. However, this process is resisted by the surface tension of the emerging vapour bubbles. For a bubble to grow it must have a minimum size given by the critical radius $r_c = 2\gamma(T)/\Delta P$, where $\gamma(T)$ is the surface tension at temperature T and ΔP is the pressure difference between the pressure of the vapour inside the drop and the pressure in the liquid. Therefore, to create a bubble locally a spike of energy deposited in the liquid must occur, which is sufficient to instantly create a bubble with a radius larger than the critical radius. Theoretically the energy required for this is between a few eV in helium to more than 100 eV in propane. The energy actually required is significantly higher and found from counting of bubbles, to be 0.5–1 keV, and thus requires a high-energy ionisation

See problem 1.

The energy must be deposited within a volume given by the critical radius.

event (δ-electron) from the passage of a charged particle through the liquid [452].

One side effect of the bubble formation is the creation of a sound wave in the liquid. This sound can be picked up with microphones, and from early on has been used to trigger the camera systems that take the photographic images of the bubbles in the chamber medium.

The main challenge with bubble chambers is that the superheated state is very fragile. Nucleate boiling commences instantly at any contamination and in particular at any surface feature of the containing vessel, conditions which locally reduce the threshold radius for bubble growth. This was an acceptable situation in accelerator-based bubble chamber experiments, as the active time did not have to exceed the duration of a bunch train from the accelerator (fractions of a second). In this case the cycling frequency of the bubble chamber was synchronised with the beam structure, typically up to a few Hz.

An example for the large last-generation bubble chamber experiments is the Big European Bubble Chamber (BEBC) at CERN [279]. This chamber was designed to operate with 35 m^3 of hydrogen, deuterium or oxygen as active liquid at temperatures between 25 and 36 K. Events in the active volume of 3.7 m diameter and 2 m height were recorded by four wide angle cameras on top of the experiment. The double-walled pressure vessel was designed for a pressure of 10 bar. Superheating was achieved by rapidly withdrawing a piston with 180 cm diameter and a mass of 1.8 t by 10 cm (an expansion of 0.7%$_v$). The maximum duration of an expansion/compression cycle was 45 ms, and the maximum acceleration 200 g. To allow for a momentum measurement of charged tracks in the chamber, the experiment was placed between two superconducting Helmholtz coils, which produced a vertical field of 3.5 T in the tracking volume.

At the time this was the largest superconducting magnet system in the world.

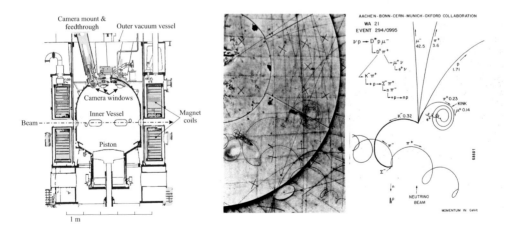

Figure 6.4: Big European Bubble Chamber (BEBC). Left: Cross-section of the detector (modified from [171]). Middle and right: Event display with production of a D meson in neutrino interactions [170].

For the analysis the photographic images created in these bubble chambers had to be scanned by semi-automated scanning machines under control of human operators and the kinematic parameters of the tracks extracted. To obtain a full three-dimensional reconstruction several images

These large bubble chamber experiments also were important in the development of organisational patterns that are followed also today: Large, complex and technologically challenging experiments at a central facility, with reconstruction and analysis performed at collaborating institutes.

from different angles were taken. A major limitation of the photographic imaging is the limited focal length of the cameras. To overcome this, the conventional photographic images have been supplemented with holographic images as they ideally have infinite focal length [287].

A modern application for bubble chambers is in searches for WIMPs. In these experiments there is no correlation with an accelerator duty cycle, and the goal is to maximise the time during which the chamber stays active. Hence, the detectors are optimised to delay spurious boiling, in particular on surfaces. This is achieved by enclosing the working fluid (for example, CF_3I) in a container of synthetic fused silica with high surface smoothness. One approach to prevent the working fluid from contacting any remaining metallic parts in the drive mechanism (piston) was to cover the working fluid with a layer of ultra-pure water as a buffer medium. With such a setup the chamber can be kept in a stable superheated state for several hundred seconds.

A recent example is the PICO experiment [56].

More recent versions of this experiment have eliminated the layer of water because of the acoustic signatures this layer created. The separation of the superheated working fluid is now achieved by a temperature gradient in the fluid, where the part of the liquid in contact with metallic parts is kept cold below the saturation point [165].

A dark matter signal would be observed in such an experiment as a nuclear recoil from a neutral particle. Such an interaction leaves only a single bubble at its interaction point, differently than the tracking bubble chambers used historically, which recorded the tracks from charged particles. Bubble chambers employed in the search or WIMPs are operated at combinations of temperature and pressure that make the detector insensitive to minimum-ionising radiation.

Another way to avoid the contact of the superheated liquid with any solid surface is by dispersing droplets of the working liquid in an insoluble gel or viscoelastic medium. Such a detector is called a Superheated Drop Detector (SDD) [61]. The energy required to make a bubble grow when a charged particle is passing is mechanical energy stored in the liquid, hence this detector does not need any power, once its pressure is set. Hence SDDs can be used for dosimetry, in particular for neutrons ('bubble detector') [436].

CLOUD CHAMBERS

A cloud chamber is filled with a gas (typically air), which is mixed with another substance in the saturated vapour state. If the temperature is lowered, the vapour will become super-saturated and condensation commences [218]. However, due to the surface tension, the saturation vapour pressure over a curved surface with radius r is larger than over a planar surface and given by

$$\ln\left(\frac{p_{\mathrm{svp}}(r)}{p_{\mathrm{svp}}(\infty)}\right) = \frac{2\gamma}{r}\frac{1}{nRT}, \qquad (6.1)$$

where γ is the surface tension, R the gas constant, n the molar density and T the absolute temperature. If the droplet has a radius less than $2\gamma/(nRT)$ (for water this is about 1 nm), the saturation vapour pressure above the droplet is significantly higher than for a flat surface, and thus for reasonable levels of supersaturation such a small droplet would immediately evaporate, rather than grow. However, if the droplet contains an electric charge e, for example from the ionisation due to the passage of a charged particle, eq. (6.1) is modified due to the electrostatic repulsion of the charge density,

$$\ln\left(\frac{p_{\mathrm{svp}}(r)}{p_{\mathrm{svp}}(\infty)}\right) = \frac{1}{nRT}\left(\frac{2\gamma}{r} - \frac{e^2}{32\pi^2\varepsilon_0 r^4}\right). \tag{6.2}$$

The potential energy of the drop increases as it grows in size due to the surface tension, but at the same time it decreases due to the electrostatic repulsion. The result is that the supersaturation required to make a small charged droplet grow is significantly reduced, which allows small droplets to form. If the supersaturation increases, these droplets can grow, until a critical radius r_{c} is reached. Beyond this level of supersaturation (for water $S \simeq 4$, corresponding to an expansion ratio of 1.25) the droplet dimension becomes unstable and the droplet grows rapidly to visible size. This is the principle of the cloud chamber.

Two types of cloud chambers exist: In expansion chambers (also called 'Wilson chambers'), sub-cooling is achieved by lowering the pressure in the active volume of the detector. The vapour expands adiabatically and thus cools down, until it becomes super-saturated. The chamber is sensitive until the new equilibrium is reached. The main drawback of expansion cloud chambers is the long time such a chamber usually needs until the original equilibrium is reinstated after a complete expansion/compression cycle. A further improvement to the Wilson cloud chamber was the use of Geiger-Müller counters to trigger the expansion of the chamber and the operation of the cameras photographing it [146].

In diffusion cloud chambers there is a vertical temperature gradient, and the warm vapour either rises, or, more commonly, falls towards the cold region of the chamber, where it becomes super-saturated [388]. A diffusion cloud chamber can only be operated stably if the total density of the gas-vapour mixture increases towards the bottom to prevent convection. Diffusion cloud chambers usually use alcohol instead of water vapour, because of its lower freezing point. The main advantage of diffusion chambers is that they are continuously sensitive (as long as the required temperature gradient can be maintained), but the vertical extent of the sensitive area is limited.

Today, cloud chambers have little scientific value, but they are often used for demonstration purposes, because of their continuous sensitivity, the absence of the need for electrical power, and the aesthetically pleasing and intuitive presentation of the charged particle tracks.

However, a large cloud chamber experiment with a volume of 26.3 m^3, CLOUD, is currently in operation at CERN, to replicate and study atmospheric cloud formation. The contribution of ionising cosmic rays to droplet growth is part of these studies [307].

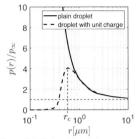

Figure 6.5: Ratio of saturation vapour pressure over a drop and saturation vapour pressure over a plane as a function of the droplet radius for water vapour (data from [218]).

C.T.R. Wilson developed the expansion chamber in 1911, and with it was the first to make individual charged particle tracks visible [499]. For this work he was awarded the 1927 Nobel Prize in Physics.

The Wilson cloud chamber was also the first detector used in human radiotracer studies that measured the transit time of a radioactive tracer in blood from one arm to the opposite arm in the human body [149].

See for example [502].

> *Key lessons from this chapter*
>
> - The energy transfer from a passing charged particle can create local changes to the chemical or physical state of a suitably prepared detector medium. The location of these changes reflects the trajectory of the incoming particle.
>
> - The local changes can be captured by optical imaging technologies, with subsequent manual or automatic scanning of the images for extraction of the kinematic properties.
>
> - Usually, the energy required to achieve the change of state is large and requires energy transfer in the tail of the charged particle energy loss distribution (δ-electrons).
>
> - These type of detectors have been historically important, but have still applications in neutrino physics and dark matter searches.

EXERCISES

1. *Bubble growth in a superheated liquid.*

    The surface tension of the liquid in a bubble chamber at temperature T is $\gamma(T)$.

    a) Show that a spherical bubble of radius r possesses an effective surface pressure $2\gamma(T)/r$. Hence show that a stable bubble will have a radius $r_c = 2\gamma(T)/\Delta p$, where Δp is the pressure difference between the pressure of the vapour inside the drop and the pressure in the liquid.

    b) The Gibbs free energy G to form a drop of radius r can be expressed as the difference of a surface term and a volume term. Find these terms and their dependence on r, and make a sketch of G as a function of r. At what value of r is the Gibbs free energy at a maximum? What will happen if the radius varies from this value?

    c) Show that the energy required to form a bubble with a radius r_c is

$$W = \frac{16\pi\gamma^3(T)}{3(\Delta p)^2}.$$

2. *Supersaturation due to expansion.*

    In a cloud chamber of Volume V_1 just before expansion, there is a non-condensible gas and vapour at a pressure p_1 at temperature T_1, with a mass M_1 in the volume and a molar mass m.

    a) First assume that just before expansion, there is non-condensible gas at a pressure p_g, and vapor at a pressure p_1 contained in a volume V_1 at a temperature T_1. Find the equation of state of the vapour, if M_1 is the total mass of vapour present in the volume V_1 and M the molecular weight of the vapour.

    b) Find the temperature T_2' and thus the equation of state after an adiabatic expansion to a volume V_2 (before any condensation takes place).

    c) However, this is an unstable state since at the lower temperature T_2' the amount of vapour that may be held in suspension is less than at the temperature T_1 and some of the vapour will condense, reducing the mass in the vapour from M_1 to M_2. After the condensation equilibrium is again reached and the vapour pressure falls to the saturation pressure p_2 at the lower temperature T_2. The equilibrium temperature T_2 is slightly higher than T_2', the temperature immediately after expansion, as some heat is liberated by condensation of vapour. Find the equation of state in the new equilibrium state.

d) The supersaturation S is the ratio of the density of vapour after expansion but before condensation, $\rho_2' = M_1/V_2$, to the saturation density at the lower temperature, $\rho_2 = M_2/V_2$, $S = \rho_2'/\rho_2$. Show that, if the slight difference between T_2 and T_2' is neglected,

$$S = \frac{p_1}{p_2}\left(\frac{1}{1+\varepsilon}\right)^{\gamma},$$

where $1 + \varepsilon = V_2/V_1$ is the expansion ratio, and $\gamma = c_p/c_v$, the ratio of the heat capacities at constant pressure and volume, respectively.

7 Gaseous detectors

When a charged particle passes through a gas, it will create ionisation charges as described in section 2.3. If an electrical field is applied, the charges will separate and drift towards the electrodes (see section 4.1), which induces electrical signals on the electrodes as discussed in section 3.1.

Because of the low density of gases, the amount of charge produced in a given path length is small, so that the charge typically needs to be collected over a larger volume, and internal amplification (see section 4.5) is usually required for gaseous detectors. On the other hand, the low density has the advantage that multiple scattering is less of a concern than in denser detector media. Gases do require containers, but there is flexibility in the shape of these containers as is in their size, and gases will readily fill out these shapes. Gases can also be easily replaced, so contamination and radiation damage effects can be reduced, as long as the damage products are not solidifying inside the detectors (see section 7.7).

In this section we will first discuss typical compositions of detector gases, and then discuss common geometries, starting with a simple planar geometry. The most common gaseous detector is the wire chamber, where the high fields required for internal amplification are produced close to a thin wire. The position resolution of simple wire chambers is given by the cell size and micro-pattern gas detectors (MPGDs) have been developed to reduce these.

A further refinement of the position information can be obtained when the finite time it takes the primary electrons to drift to the anode is taken into account, leading to the concept of the 'drift chamber'. Drift chambers are today the best tool for large area/volume, low-mass and affordable tracking detectors.

Finally, we will discuss the issues arising from the operation with high rates over extended times, and software that allows to predict and optimise the performance of gaseous detectors.

7.1 GASES FOR DETECTORS

To satisfy the different performance requirements for gaseous detectors it is common to use mixtures of gases. Typical components in these mixtures and the motivation for their use are:

- The main component is usually a heavy noble gas, because of the high charge yield and electron mobility they provide. Argon is the most affordable of these, thus is the most widely used main component of detector gases.

- In applications where a large primary ionisation density, and thus a large mass density, is required, heavy halocarbons are used as the main gas component, as they are dense and inert. An example is 1,1,1,2-tetrafluoroethane ($C_2F_4H_2$).

Halocarbons are compounds in which one or more carbon atoms are linked by covalent bonds with one or more halogen atoms.

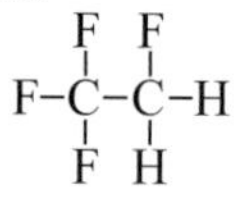

$$F-\overset{\overset{\displaystyle F}{|}}{\underset{\underset{\displaystyle F}{|}}{C}}-\overset{\overset{\displaystyle F}{|}}{\underset{\underset{\displaystyle H}{|}}{C}}-H$$

Figure 7.1: 1,1,1,2-tetrafluoroethane.

DOI: 10.1201/9781003287674-7

- A quencher is usually required to prevent photon feedback and discharges in the detector (see section 4.5). Good quencher gases (for example CH_4 or $i-C_4H_{10}$) are often flammable, non-flammable alternatives are CO_2 or CF_4.

- Cold gases like CO_2 might be used if reduction of diffusion is desired. Dimethylether (DME) is another gas if low diffusion and low drift velocity is required.

- CF_4 is also added sometimes because it can remove surface contamination from polymerisation of the detector gas (see section 7.7).

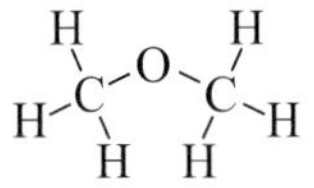

- If the detector should be sensitive for X-rays, then the gas should contain atoms with high Z, like for example xenon. If optical or near-UV photons need to be detected (for example in RICH detectors), TEA or TMP can be added to the gas (see section 7.6).

Isobutane ($i-C_4H_{10}$) is an isomer of butane.

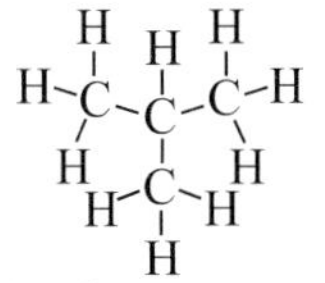

Figure 7.2: Isobutane.

Figure 7.3: Dimethylether.

All components are usually non-polar molecules, to avoid loss of electrons due to attachment, and they are chosen to be chemically inert and stable, and not strongly affected by radiation. Further concerns for the application of a specific gas can be safety (flammability), and the global warming potential (which is in particular a concern for halocarbons).

While in principle more than two components in the mixture are possible and tertiary mixtures are commonly used, control of the mixing ratios and thus achievement of stable drift and amplification properties becomes more challenging if more gases are used in the mixture.

Gas	Density at STP $[10^{-3}$ g/cm$^3]$	$(dE/dx)_{mip}$ [keV/cm]	E_I [eV]	$W_{e/ion}$ [eV]	N_p [cm^{-1}]	N_{total} [cm^{-1}]
He	0.18	0.32	24.6	41.3	3.5	8
Ne	0.90	1.45	21.6	37	13	40
Ar	1.78	2.53	15.7	26	25	97
Xe	5.89	6.87	12.1	22	41	312
CH_4	0.72	1.61	12.6	30	28	54
C_2H_6	1.28	2.91	11.5	26	48	112
$i-C_4H_{10}$	2.51	5.67	10.6	26	90	220
CO_2	1.96	3.35	13.8	34	35	100
CF_4	3.78	6.38	16.0	54	63	120
$C_2F_4H_2$	4.25	7.52	12.2		82	

Table 7.1: Properties of typical detector gases at NTP. $(dE/dx)_{mip}$: stopping power for minimum-ionising particle, E_I: ionisation energy, $W_{e/ion}$: average energy per electron/ion pair, N_p number of primary clusters (see section 2.5) and N_{total} total number of electron-ion pairs, for unit charge minimum-ionising particles (from [505, 443]).

7.2 RESISTIVE PLATE CHAMBERS

The simplest charge collection geometry is a parallel plate capacitor. However, the primary ionisation collected in a gas gap of reasonable thickness would be too small for direct detection, and thus avalanche amplification is required. In a practical design to achieve stability against discharge one or both of the HV electrodes are made of resistive material, so that the field collapses locally, once significant currents are generated by the avalanche. In that case the field decays exponentially with a time constant $\tau = \rho \varepsilon_r \varepsilon_0$, where ρ is the volume resistivity of the material (see exercise 6). Typical plate materials are glass (with a typical volume resistivity of $\rho = 10^{12}$ Ωcm and $\varepsilon_r \simeq 2.25$ resulting in $\tau \simeq 1$ s), or Bakelite (with $\rho \simeq 10^{10}$ Ωcm and $\varepsilon_r \simeq 10$, which gives $\tau \simeq 10$ ms).

'Too small' here means not distinguishable from the input noise of an electronic amplifier.

Such a detector is called a 'Resistive Plate Chamber' (RPC).

The RPC will be blind locally for this time.

Bakelite is a thermosetting phenol formaldehyde resin, and was the first type of plastic made from synthetic components (invented in 1907).

Typically, the main gas component in RPCs is 1,1,1,2-tetrafluoroethane ($C_2F_4H_2$), with a small fraction of i-C_4H_{10} as a quencher gas. RPCs are operated in avalanche mode or in streamer mode. In the latter case the output signals are much larger (between 10s of pC to a few nC), at the cost of rate capability (a few 100 Hz/cm^2) compared to avalanche mode RPCs, which can operate at rates a factor 10 higher. The addition of small amounts of SF_6, which has a high electron affinity, in particular for low electron energies, suppresses the development of streamers [175].

Tetrafluoroethane is used in the refrigeration industry, where it is known as R-134a.

See section 4.5.

RPCs are fast detectors, as the detectable signal is generated by avalanche amplification immediately after the primary ionisation charges are liberated by the incoming charged particle. Hence, in large particle physics experiments RPCs are used in two main roles, as timing detectors or as trigger detectors, and there are two main flavours of RPC designs to satisfy the respective requirements.

SF_6 is industrially used as electrical insulator and arc suppressant. It is also a potent greenhouse gas.

In trigger detectors the main goal is high efficiency. For this, they have larger gaps (typically a very few mm), with electric fields of a few ten kV/cm. A commonly used gas is $C_2F_4H_2$/i-C_4H_{10}/SF_6 96.7/3/0.3. Single and double gap geometries are being used. Efficiencies of 99% and time resolutions of the order of 1 ns are typically achieved.

Detectors with good time resolution can be used for time-of-flight (TOF) measurements (see section 12.2).

For timing applications, the gas gap of the RPC is small (250–300 μm) with very high fields (around 100 kV/cm). The high field demands extremely good surface quality of the resistive plates, leading to the choice of glass for the resistive electrodes, and also a gas mixture that is more heavily quenched and limiting streamer development (a common gas mixture is $C_2F_4H_2$/i-C_4H_{10}/SF_6 90/5/5). In a single gap the time resolution is limited by long tails in the time distribution caused by the stochastic fluctuations at the start of the avalanche process. For a time measurement that is robust against these tails, multi-gap geometries with typically five layers are used. The glass plates are stacked within conductive electrodes. No conductive connections to the intermediate plates are required, as they acquire the correct potential by electrostatic charge division. The time resolution achieved in such detectors is around 100 ps.

The time resolution in RPCs is caused by statistical fluctuations of the avalanche growth and is proportional to the effective Townsend coefficient $\alpha - \eta$ (see section 4.5), and the drift velocity [425].

About the time it takes light to cross the detector.

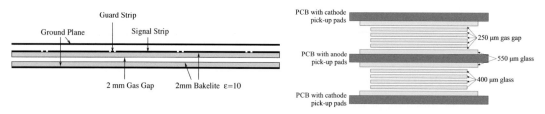

Figure 7.4: RPC geometries. Left: Typical trigger RPC [425]. Right: Typical multi-gap timing RPC (after [36]).

7.3   WIRE CHAMBERS

The classic geometry for gaseous detectors is the 'wire chamber', where the amplification field is generated close to thin anode wires ($E \propto 1/r$). The electrons from the primary ionisation first drift to the wire, where avalanche multiplication takes place in the high field close to the wire. The ions generated in the multiplication process then drift back from the wire towards the cathode. Because of the large drift time it is this movement of the ions that generates the largest part of the charge signal for typical wire chamber geometries. The electrostatics and signal generation in a cylindrical wire chamber are studied in exercise 3.

A simple implementation of a wire chamber for position detection over a larger area is the Multi-Wire Proportional Chamber (MWPC), which is a row or an array of cells with one anode sense wire each.

George Charpak received the 1992 Nobel Prize in Physics for his invention and development of the MWPC.

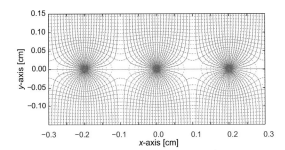

Figure 7.5: Field and equipotential lines in a cross-section through a planar MWPC (modified from [505]). The wires are orthogonal to the picture plane. For an analytical treatment of the fields in MWPCs see [234].

The spatial information in a MWPC is given by the binary information of the cell hit or not hit. Thus the position resolution of such a detector is $d/\sqrt{12}$, where d is the cell size. In practical detectors, the cell size is limited due to the effect of mechanical tolerances on the gas gain and due to the mechanical instability of the wires at high voltage.

In a MWPC geometry the anode wires are in an unstable mechanical equilibrium. Once one of the wires moves towards the cathodes, electrostatic repulsion will push the next wire to the opposite side and so on. This movement will lead to variations of the field and hence the gain, and in the worst case, to discharges that will render the chamber inoperable. This displacement is resisted by the wire tension.

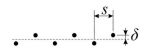

Figure 7.6: Wire displacement due to electrostatic repulsion.

See exercise 1.

It can be shown that the position of the wires is maintained as long as the tension of the wires is larger than

$$T_{\text{crit}} = \frac{1}{4\pi\varepsilon_0}\left(\frac{CV_0L}{s}\right)^2,$$

where C is the capacitance, V_0 the potential of the wire, L the length of the wire, and s the spacing between the anode wires [475].

The maximum tension that can be applied to a wire is limited by the properties of the wire, depending on the material and the radius, and thus, for a given length of wire a minimum spacing between wires must be maintained to prevent the displacement of the wires. Practically, wire spacings down to a very few mm are feasible, and thus the achievable position resolution of MWPCs is in the order of 1 mm.

To calculate the signal on the wire in a MWPC we can use the Ramo-Shockley theorem, eq.(3.1). Due to the proximity of the avalanche to the wire, the induced current due to the electron movement is very short, and contributes little to the induced charge on the electrodes. It is the motion of the ions that provides the bulk of the charge signal, but because of the long drift time of the ions the duration of this signal is usually truncated by an RC circuit.

See exercise 3.

To simplify the calculation of the signal we describe all electrodes as parallel wires or the superposition of these, which allows for a straightforward two-dimensional approach. In that case, the linear charge density on each wire can be found from the voltages of all wires V_n from $\lambda_n = \sum_m c_{nm} V_m$, where c_{nm} is the symmetric capacitance density matrix. The electric field close to wire n can then be found from the voltages

The capacitance density is the capacitance per unit length. c_{nm} is the capacitance density between wires n and m.

$$E_n(r_n) \simeq \frac{\lambda_n}{2\pi\varepsilon_0 r_n} = \frac{\sum_m c_{nm} V_m}{2\pi\varepsilon_0 r_n}.$$

The position of a drifting ion as a function of time in the vicinity of the wire, where the electric field is to good approximation cylindrically symmetric, is given by

See also exercise 3.

$$r(t) = a_n \sqrt{1 + \frac{t}{t_0}} \quad \text{with} \quad t_0 = \frac{a_n^2 \pi \varepsilon_0}{\mu \lambda_n},$$

where a_n is the radius of the wire. To find the induced current on the wire on which the avalanche took place we need the weighting field $E_w^n(r_n) = c_{nn}/(2\pi\varepsilon_0 r_n)$, and the induced current due to the movement of ions with a total charge Ne is

$$I_n(t) = -NeE_w^n(r_n(t))\frac{dr_n(t)}{dt} = -\frac{Nec_{nn}}{4\pi\varepsilon_0(t+t_0)}.$$

Similarly, for a different wire, the weighting field is $E_w^m(r_n) = c_{nm}/(2\pi\varepsilon_0 r_n)$, and the induced current is

$$I_m(t) = -NeE_w^m(r_n(t))\frac{dr_n(t)}{dt} = -\frac{Nec_{nm}}{4\pi\varepsilon_0(t+t_0)}.$$

The c_{nm} have the opposite sign to c_{nn}, and thus the induced signals on the wires without the avalanche have the opposite polarity as on the wire with the avalanche, as the ions move away from the wire with the avalanche and towards the other wires in the system.

There is also usually an additional capacitive cross-talk between the electrodes in the system.

The nature of the ions drifting back from the avalanche changes over time due to charge exchange reactions and due to ion clustering [298]. In the first kind of process ions from molecules with a high ionisation potential transfer their charge to other components in the gas that have a lower ionisation potential (for example $Ar^+ + CO_2 \rightarrow Ar + CO_2^+$). In the second type of process several molecules and ions form a charged molecular cluster (for example $Ar^+ + Ar \rightarrow Ar^+ \cdot Ar$ or $CO_2^+ + nCO_2 \rightarrow CO_2^+ \cdot (CO_2)_n$). The charge transfer occurs on a time scale of 1 ns, whereas the clustering happens within 10 ns. Both timescales are short compared to the usual drift time of the ions towards the cathode, so that most of the signal in a wire chamber is given by the drift of these ionisation clusters, and it is the drift velocity of these clusters that determines the signal induced.

To maintain energy/momentum conservation in these processes a third spectator molecule must be part of the reaction.

As discussed in section 4.1, the drift velocity is smaller for particles with higher mass.

7.4 MICRO-PATTERN GAS DETECTORS

One way to deal with the dimensional effects and the electrostatic instability issues of thin wires in MWPCs is to mount the conductors on an insulating substrate. Advances in micro-fabrication led to the development

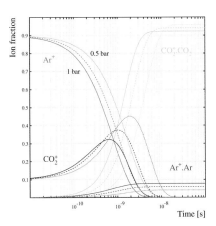

Figure 7.7: Computed evolution of the fraction of ions and clusters in Ar/CO$_2$ mixtures (initial Ar$^+$ fraction is 90%). The different lines are for different gas pressures (solid: 1 bar, dashed: 0.75 bar, dotted: 0.5 bar) [298].

of a class of gaseous detectors that comprises microelectronic structures with sub-millimetre distances between anode and cathode electrodes. Such detectors are called 'Micro-pattern gaseous detectors' (MPGDs).

The first representative of this class was the 'Micro-Strip Gas Chamber' (MSGC) [390], in which the electrodes are metal traces deposited on an insulating substrate like glass. However, this type of detector suffers from various instabilities ranging from time-dependent gain shifts to discharges between the electrodes, and charges generated during the avalanche multiplication that stick to the insulating substrate. Some of these charging effects can be remedied by using a substrate with controlled high resistance [159]. However, the problem of discharges, induced by extremely large primary ionisation events due to rare nuclear reactions or slow moving heavy ions associated to high charged particle fluxes, was never really resolved. These discharge events can have catastrophic consequences for the delicate conductive structures.

A whole range of other MPGDs have been proposed, which in addition to the reduction of dielectric material in the detector to alleviate charging issues also often use the capabilities of micro-fabrication to provide two-dimensional spatial information. A number of those still suffer from discharge and uniformity issues, and few have been developed beyond small bench prototypes. However, two amplification structures based on foils have emerged from these studies, which are now widely used.

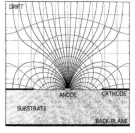

Figure 7.8: Field lines and equipotentials in an MSGC computed for a back-plane voltage close to the cathode voltage [454].

Figure 7.9: Discharge damage in a MSGC [272].

For an overview see [441].

MICROMEGAS

Micromegas (Micro-mesh gaseous structures) [262] consist of perforated metal foils, which are placed close (50–100 μm) to an anode plane, with a uniform high field in the gap (typically several 10 kV/cm).

Electrons from a drift volume enter the amplification gap through the holes in the micromesh, where a gain of up to 10^5 can be achieved. To maintain the correct distance of the foil to the anode plane spacers are located between them. Originally, the spacers were placed on the anode plane by lithography of a photoresistive film and the mesh was stretched and glued on a frame and then placed on top of the spacers. In an optimised production scheme the mesh is laminated on the photoresistive film, before the material between the spacers is etched away

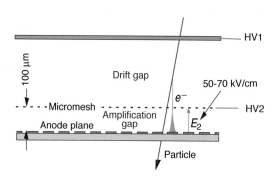

Figure 7.10: Schematics of a Micromegas detector [505].

('bulk micromegas' [263]). The thin gap and the high field result in a very fast signal from the ion drift (about 100 ns). The uniform field leads to uniform gain, and thus to good energy resolution for example for soft X-rays.

Again, the most demanding situation is the operation in a high-rate environment of charged particles, which creates discharges. Due to the absence of any fragile conductive structures damage to the micromesh is not significant enough to prevent continued operation of the device. However, the increase in current due to the discharge can exceed the limits of the HV supply, which requires a typical dead time of a few ms to recover.

Because of this, resistive micromegas detectors have been developed [40]. This design does not eliminate the chances of sparking but in the event of a spark being formed, it will strongly limit the current and hence avoid damage and reduce the droop of the high voltage.

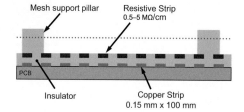

Figure 7.11: Schematic of a resistive Micromegas detector [40].

The induced currents on the readout electrodes can be read out with low noise amplifiers. Alternatively, a very powerful detector with high spatial resolution in orthogonal directions can be built using such amplification structures on top of a highly segmented silicon pixel readout chip (see below).

GAS ELECTRON MULTIPLIER

Gas Electron Multipliers (GEMs) [440, 442] are based on polyimide foils, clad with copper on both faces. First, a tight grid of holes is chemically etched in the copper. Then a different chemical is used to etch the polyimide through. The thickness of the polyimide and the diameter of the holes are $\mathcal{O}(100\ \mu\text{m})$.

For operation, a voltage is applied between the two faces of the foil, which generates a high electric field that is sufficient to induce impact ionisation when electrons pass the hole. A potential difference across the polyimide of a few hundred Volts will generate a sufficiently high local electric field to achieve charge amplification. To ensure high gains, the

Polyimides (PIs) are high-temperature engineering polymers. There are different flavours of polyimide with varying physical and chemical properties available from different manufacturers. A widely used polyimide film material is Kapton™.

Etching of copper-clad polyimide is a standard process in the PCB industry.

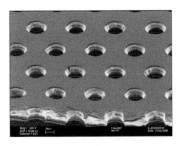

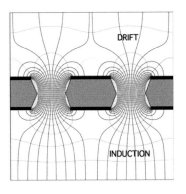

Figure 7.12: Left: Electron microscope picture of a section of typical GEM electrode, 50 μm thick. The hole pitch and diameter are 140 and 70 μm, respectively. Right: Field lines in a GEM [442].

optimum hole diameter should be of a similar size as the thickness of the foil. A fraction of the charge is lost because of attachment to the dielectric, and thus there is an effective gain that is smaller than the actual avalanche gain. A gas gain of 10^3 can be achieved with small GEM foils, but more typically effective gains are a very few 100 with one foil. Higher gains can be achieved by stacking several foils. The advantage of this approach is that a large overall gain can be achieved with lower voltages across the individual layers. The charge created in one hole in one layer spreads to several holes in the next, and thus the Raether limit (see section 4.5) can be avoided and the chance of electrical breakdown reduced.

Due to the high attainable gain GEMs can be used with a CsI coating to efficiently detect the single electrons released by ultra-violet photons for example in Cherenkov detectors.

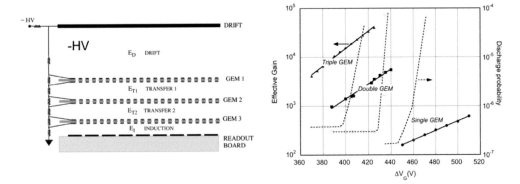

Figure 7.13: Triple-GEM. Left: Schematics. Right: Effective gain (full line) and discharge rates (dashed line) as a function of voltage [442].

Ultimately, the amplified charge is registered on a grid of readout electrodes on an insulating layer underneath the GEM foils. By using similar techniques as in the fabrication of the GEM foils, orthogonal strips can be created on the two sides of a copper-clad polyimide foil, which is then glued onto an insulating carrier. After chemical removal of the polyimide layer covering the lower strips, a grid of orthogonal readout strips is created that allows 2D reconstruction of the location of the hit [164].

PLANAR MPGDS WITH PIXELATED READOUT

Some of the most powerful gaseous detectors have been created by combining planar MGPDs, like Micromegas or GEMs, with high channel density readout ASICs that have been originally developed for the readout of silicon pixel detectors (a widely used readout chip is the Timepix chip [343]). The use of industrial post-processing techniques to build a Micromegas structure onto this readout ASIC led to the development of InGrid (integrated grid) structures [189]. Gaseous detectors with Ingrid structures have been called GridPix. A very low material GridPix detector with minimal material budget (gas gap 1 mm and thinned ASIC) for use in low-mass vertexing detectors is Gossip [179]. GridPix detectors are currently under investigation for use as active elements in future TPCs (see page 121). Due to the highly segmented readout the spatial resolution in these detectors is only limited by diffusion.

GEMs have also been integrated with Timepix chips, with the added benefit of increased mechanical robustness (GEMGrid) [147].

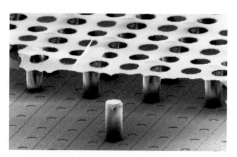

Figure 7.14: Pixelated MPGD readout. SEM image of an InGrid with partly removed Grid. The height of the pillars is 50 μm [348].

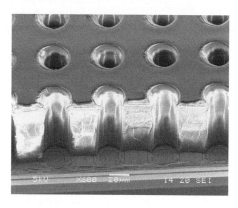

Figure 7.15: SEM picture of GEMgrid detector. The amplifying structure is 55 μm of SU-8 (a commonly used epoxy-based negative photoresist) with 30 μm diameter holes. The top metal layer is aluminium [147].

7.5 DRIFT CHAMBERS

A different approach to improve the position resolution of gaseous detectors is to exploit the finite speed, with which the electrons move through the gas. By measuring the time it takes the primary electrons to move from the point of the ionisation to the point where the actual electric signal is generated by the massive increase of charge in the avalanche, one can deduce the drift distance, if the drift velocity is known.

Such a detector is called a 'drift chamber'.

The start time for the drift time measurement can be provided by a fast external timing detector, or in experiments at accelerators from the known time structure of the beam. In both cases the time of flight and electric signal propagation speed in cables and active electronics, which are likely to vary between different channels, must be taken into account and calibrated. The start time can also be reconstructed from the data of the detector itself, if several cells of the drift chamber are hit.

This is often called the t_0.

In the simplest case the external timing signal could be from a plastic scintillator with a PMT (see section 5.1).

In experiments at accelerators this can also be used to identify the bunch-crossing from which the measured particle originated.

DRIFT CHAMBER CALIBRATION

For the reconstruction of the position of the incoming particle, the correlation of drift-time and drift-distance must be known. This relation depends on the local electric field, the drift velocity at that value of the field, and includes the increase of the drift-distance due to deflection in a magnetic field (if present) by the Lorentz angle. Particularly in magnetic fields, the drift-time/drift-distance relation can be asymmetric on the two sides of the wire. Generally, the relation will be non-linear, and is in many cases varying with the location of the drift cell within the experiment and/or time.

This is often called the r-t or the x-t relation.

While it is in principle possible to calculate the drift properties (see section 4.1), the quality of the results will be affected by the accuracy of the assumptions on the gas and field parameters made, small changes of which will have significant effects. Calibration of the drift-time/drift-distance relation using prototypes and reference tracking systems can also be compromised by the strong variations of the drift properties between an idealised set of operating parameters during the calibration and the actual operation of the detector.

In high rate/high gas gain applications, space charge from the slow ions created in the avalanches can also modify the electric field and thus the drift velocity.

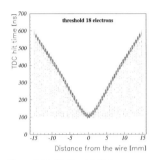

Figure 7.16: Digitised drift times versus distance of the muon track from the wire for the ATLAS muon chambers in a testbeam prototype (without magnetic field) [427].

The most accurate calibration is thus achieved using the data from the detector itself, taken during the actual operation. In 'autocalibration', the drift-time/drift-distance relation is found in an iterative process, where in a series of iterations an improved version of the relation is obtained using the position information reconstructed using the relation from the previous iteration. At the same time this method can also provide a calibration of the timing offset for each channel (the t_0). This method is capable of tracking slow changes of the operating parameters, and thus is the most accurate tool to calibrate a drift chamber response (see for example [77]).

The need for a stable drift-time/drift-distance relation puts an additional requirement on the drift chamber gas: the drift velocity should not depend strongly on the electric field, so that local differences of the field or variations in time do not change the drift-time/drift-distance relation significantly. A drift velocity that satisfies this requirement is called 'saturated'. It can be achieved for 'hot' drift gases for sufficiently high fields, but 'cold' gases typically display significant variations of the drift velocity with the electric field strength.

For the definition of hot and cold gases see section 4.1.

DRIFT-TIME MEASUREMENT AND POSITION RESOLUTION IN DRIFT CHAMBERS

As discussed in section 3.8 the drift-time of the ionisation electrons to the wire can be measured by time-to-digital converters (TDCs). The induced signal is a superposition from the signals generated by the movements of the ions created in the avalanches from several primary ionisation electrons. The time structure of this signal will reflect the arrival times of these electrons at the anode wire, where they start their avalanches. Due to the long duration of the signal produced by the drifting ions in a wire chamber, the signal from the individual primary electrons will pile up on top of each other with the difference in arrival times of the individual primary electrons being small compared to the overall signal length. Hence, by selecting a threshold for the comparator firing on the signal induced on the wire, one effectively sets a threshold on the number of primary electrons that have to arrive at the wire to stop the measurement of the drift-time. This threshold, in units of number of primary electrons, is given by $thr_{prim} = thr_{el}/g$, where g is the combination of avalanche gas gain and (pre-)amplifier, and thr_{el} the comparator electronics threshold (usually a voltage).

The position resolution in drift chambers as a function of the drift distance shows a distinctive pattern. At short drift distances the resolution is driven by the finite spacing between the primary ionisation clusters in the gas. It will be smaller for gases with a high ionisation density along the track. For larger drift distances the resolution becomes diffusion-dominated, and increases basically with the square root of the distance.

The resolution improves for increased gas pressures, as both the spacing between ionisation clusters and the diffusion in the gas are reduced.

Drift distance here is defined as the distance from the wire to the closest point on the track of the incoming charged particle.

These effects are also studied in exercise 4.

The price to pay is, of course, increased multiple scattering.

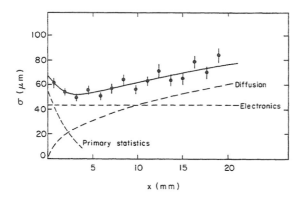

Figure 7.17: Typical drift distance dependence of the position resolution in a drift chamber [439].

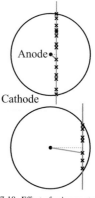

Figure 7.18: Effect of primary statistics on the drift distance measurement in a cylindrical wire chamber. Crosses: discrete primary ionisation clusters. Dashed lines: ideal distance. Full lines: actual distance to the first ionisation cluster.

By the choice of threshold the system can be optimised for improved resolution in either the region dominated by the primary ionisation statistics, or the region dominated by diffusion. A higher effective threshold (in number of primary electrons) improves the resolution at large distance, as a later electron will be closer to the centre of the diffusion distribution. A lower effective threshold (in primary electrons) will improve the resolution close to the wire, as the earlier electrons will be produced closer to the point of closest approach of the track.

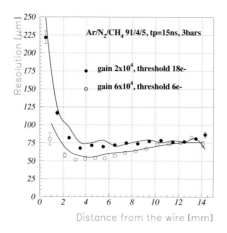

Figure 7.19: Measurements (dots) and simulation (lines) of the resolution in prototypes of the ATLAS MDT (cathode: aluminium tubes with 30 mm diameter, anode: wire diameter 50 μm) for different effective thresholds [422].

DRIFT CHAMBER GEOMETRIES

Typical drift chamber geometries are either individual tubular or square sections for integration into planar structures, or larger systems which are designed to fill the central cylindrical tracking cavity in a collider detector.

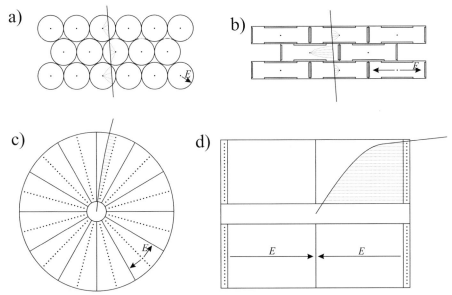

Figure 7.20: Drift chamber geometries. a) Planar made of tubes. b) Planar made of square cells. c) Jet chamber (endview). d) Time Projection Chamber (side view). Conceptual, not to scale.

A historically important drift chamber geometry for central tracking is the 'jet chamber', where radial layers of anode wires provide a large number of points (about 1 hit per cm in r) along tracks from the vertex. These type of detectors had good two-track resolution ($\sim$100 μm), good pattern recognition capabilities, and measurement of dE/dx. To compensate for the Lorentz angle the cells were often tilted.

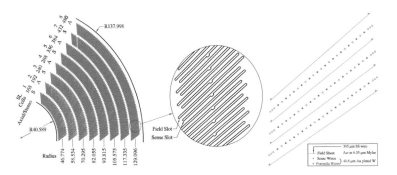

Figure 7.21: Example for a jet cell geometry: CDF Central Outer Tracker [25].

A gaseous drift chamber technology that provides 3D information over a large volume with minimum material is the Time Projection Chamber (TPC). It is a cylindrical gas-filled volume split into two halves by a central cathode disk at very high voltage (100 kV in the ALICE TPC). Primary electrons drift to the endplates, which are instrumented with a two-dimensional readout. In the case of existing and past TPCs this uses a grid of anode wires for avalanche multiplication. To prevent ions drifting back into the detector volume a gating grid is used, which closes their path electrostatically after an avalanche on the anode wires.

The position in the r-ϕ plane is determined from the ion signal induced on a plane of pads. The z information is obtained from the drift time of the electrons to the readout plane. The $r\phi$ information in this detector is maintained despite the very long drift distance because the transverse diffusion is reduced due to the parallel E and B fields (see section 4.1).

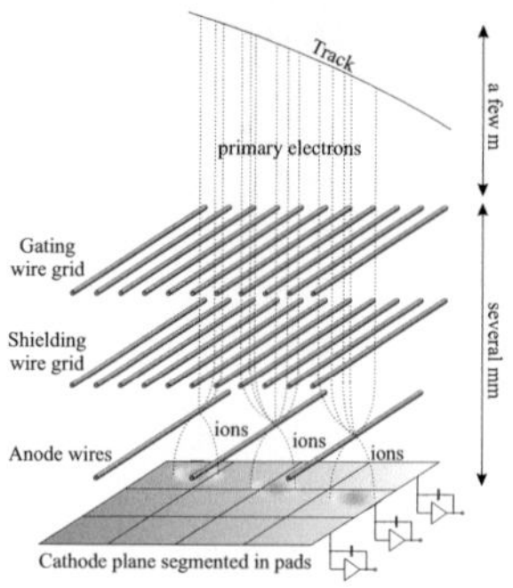

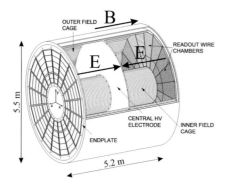

Figure 7.22: Schematics of a TPC with wire readout.

Figure 7.23: ALICE TPC (modified from [49]).

Because of the long drift-distance, calibration of the drift-time/drift-distance relation is particularly demanding. The long drift time makes this detector impractical for high luminosity hadron colliders, but they are used for heavy ion experiments (for example ALICE at the LHC) and are under consideration for future lepton colliders, where collision rates are lower. For future colliders, detector planes with better position resolution, for example based on MPGDs are being investigated [444]. One benefit of MPGDs like Micromegas or GEMs for the use in a TPC, in addition to their high position accuracy, is that their design limits ion feedback due to their amplification geometry.

TWO-DIMENSIONAL POSITION DETECTION

Typically, wire chambers with or without drift-time measurement most accurately record the position of the incoming track perpendicularly to the wire. However, sometimes position information along the wire is required, albeit usually with a much lower accuracy. Several approaches have been pursued for this.

Charge sharing

If a readout amplifier is connected to both ends of the wire, the charge signal induced on the wire at the location of the avalanche will be divided according to the ratio of resistances towards the two ends, independently of the overall length of the wire, and the capacitance or the inductance and their distribution. Position resolutions of less than 1% of the length of the wire are achievable [415].

Typical wire resistances are a few kΩ/m.

Delay line

Along the wire the signal propagates with a finite speed, close to the speed of light ($\sim$30 ps/cm). Again, both ends of the wire need to be instrumented, but this time the position along the length of the wire is found from the difference in arrival time at the two ends. Position resolutions of 1–2 cm, independent of the length of the wire, can be achieved in this way. It can be shown that the achievable position resolution for charge sharing and delay line readouts is comparable for a wire length that is about 250 times the gap between the cathodes surrounding the wire. Above this length the delay line measurement will yield the better result [153].

Stereo angle

Stereo angle is a method for measuring the second coordinate that can be employed if more than one detector layer is available. If the different layers are arranged so that the wires in one layer are tilted by a small angle (the 'stereo angle') with respect to the wires in the next, then the position along the wire can be found from the intersection of the positions reconstructed in the accurate direction in the two layers. The precision of the second dimension measurement can be improved by increasing the stereo angle, at the cost of a degradation of the combined position resolution in the accurate direction.

At high rates this method can lead to confusion, which creates spurious hits ('ghosts') in the reconstruction.

This approach is also used in silicon microstrip detectors.

Figure 7.24: Stereo angle: hit wires are indicated by arrows, reconstructed position by ellipsis.

Segmented cathode pads

The most accurate technique to establish the second coordinate of an avalanche in a wire chamber is by measuring the position of the induced charge on segmented cathodes.

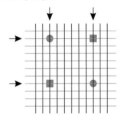

Figure 7.25: Confusion. Two actual hits (circles) and two ghosts (squares).

As we have seen, the dominant contribution to the induced charge in a wire chamber is coming from the drift of the ions. As well as on the anode, a charge in the chamber volume induces also a charge distribution on the cathode. As the ions drift towards it, the distribution on the cathode becomes narrower and narrower, until all the induced charge on the plane collects on the impact point of the drifting ion. Conversely, at a point on the cathode away from the impact point the charge density will first increase, and then decrease as the ion approaches the cathode. If the cathode is segmented into strips or pads the charge induced on these segments will be the integral of the charge density over their area and the overall behaviour will be similar.

For a discussion of a simplified charge distribution see exercise 5.

For practical applications the duration of the signal will again be limited by an appropriate high-pass *RC* circuit, which also limits the bandwidth for integration of random noise. For the cathode readout the signal shaping has the effect of capturing only the initial increase of the induced charge, so that typically more than one strip/pad shows a signal, and the centroid of the recorded signals gives a measurement of the position of the drifting ion, projected onto the cathode.

The weighting field for the signal induced on a cathode strip with infinitesimal width $2da$ by the movement of the ions in a MWPC with the anode plane centred between two cathodes spaced apart by $2D$ can be parameterised by the semi-empirical expression [259]

$$\frac{\mathrm{d}(rE_w(r))}{\mathrm{d}a/D} = K_1 \frac{1 - \tanh^2(K_2\lambda)}{1 + K_3 \tanh^2(K_2\lambda)}, \tag{7.1}$$

with $\lambda = x/D$, and K_1, K_2 and K_3 parameters that are found from the best fit to the data. The induced charge on strips of finite width can then be found from integration.

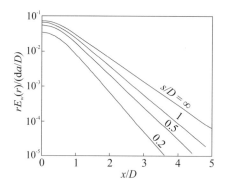

Figure 7.26: Cathode weighting field as a function of position for $D/R = 10000/15$ (modified from [259]). R is the anode wire radius and s the anode wire spacing.

Because of the symmetry of the induced charge distribution in the cathode strips/pads the position of its centre can be obtained from the centre-of-gravity of the measured signals in these segments,

$$x_{\text{centre}} = \frac{\sum_n x_n q_n}{\sum_n q_n}, \qquad (7.2)$$

In a detector with pad segmentation a similar expression would be used for the second, orthogonal direction.

where the sum goes over all the strips, x_n are the positions of the strip centres perpendicularly to their length, and q_n is the recorded charge on each strip.

In practice, only strips with a signal above a threshold given by the noise are considered in the sum, as the sum of many noisy pads would otherwise make the measurement meaningless. A common approach is to modify eq. (7.2),

This algorithm would fail in high track density environments. If separation of close tracks is needed, a more elaborate reconstruction is required, which compares the measured charge distribution with a standard charge distribution (for example using the parameterisation of eq. (7.1)). In this way a two-track separation of a few tenths of the strip width can be achieved [267].

$$x_{\text{centre}} = \frac{\sum_n x_n (q_n - B)}{\sum_n (q_n - B)}, \qquad (7.3)$$

with a bias level $B = b \sum q_n$, with $5 \times 10^{-3} \lesssim b \lesssim 25 \times 10^{-3}$, and only the elements are considered in the sum for which $q_n - B > 0$. However, the suppression of elements in the sum introduces a bias pattern of the measured position that is periodic with the strip width, and zero at the edges of the strip and the centre [408] (This is sometimes called an 's-shape' pattern.). The non-linearity can be improved by reducing the strip width. One way to achieve this without increasing the number of readout channels is by dividing the strips, and connecting only every second or third strip, with the intermediate strips floating, but tightly coupled capacitively to the readout strip [461].

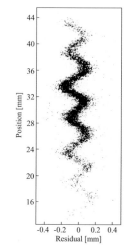

Figure 7.27: Reconstructed track position from cathode versus residual (measured track position minus reference position). The position has been determined using eq. (7.3) as described in the text with $b = 0.025$ (modified from [408]).

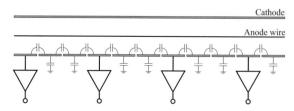

Figure 7.28: Floating cathode strips.

To obtain sensitivity to the avalanche position the sum in eq. (7.3) must include more than one element, and thus the strip width should be chosen so that a signal above the noise threshold is always obtained for at least two strips. A good rule of thumb is that the optimal strip width is about twice the distance from the wire to the cathode [259].

Position resolutions of a few 10 μm can be achieved in the direction of the wire using segmented cathode readout [188].

CATHODE STRIP CHAMBERS

Cathode strip chambers (CSCs) are large-area planar MWPCs that rely on the readout of the cathode pads to achieve good spatial accuracy. The gap is of the order of 5 to 10 mm, to allow for operation at high rates. Thinner gaps result in short electron drift times (up to a few 10 ns), and thus moderately good time resolutions. An advantage of the cathode strip readout is that the charge interpolation is a relative measurement of the signal in adjacent pads. Hence the chamber performance is not degraded by (modest) variations in environmental (temperature, pressure, magnetic field etc.) and operating (gas gain) conditions. However, calibration of the response of individual strips is required. This can conveniently be achieved by pulsing the anode wires, which will create a defined cathode charge by capacitive coupling.

THIN GAP CHAMBERS

A similar detector geometry, if very good time resolution is required, is the thin gap chamber (TGC), where the spacing between the anode and the cathode plane is reduced to order mm, thus reducing the maximum drift time in the gap. Because of the small scale, electrode non-uniformities become relevant. A uniform cathode surface is achieved by a layer of graphite, which has a sufficiently high electrical conductivity to allow for the removal of the induced charges after the drift, but is sufficiently transparent to allow for the induction of a signal on segmented readout electrodes on the backside of the cathode.

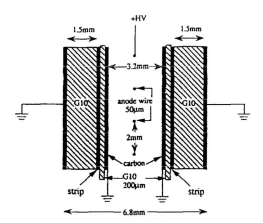

Figure 7.29: Schematics of the ATLAS muon TGCs [66].

7.6 PHOTON DETECTION WITH GASEOUS DETECTORS

The typical application of gaseous detectors is tracking of charged particles, where a sufficiently large signal size and hence efficiency can be achieved from gas amplification for detectors of moderate thickness (down to a few mm). However, there are applications where we want to observe and locate (single) photons. Two energy regimes are of primary interest: In Cherenkov detectors photons in the visible or UV range need to be detected, whereas for the detection of transition radiation it is X-ray photons for which the detector needs to be sensitive. In both cases the challenge for the use of gaseous detectors is the low photon absorption cross-section in these energy ranges (see section 2.1). While the energy transfer from the photon, in particular for X-rays, can be substantial, the small interaction probability due to the low density material in a gaseous detector limits the efficiency for photon detection. Typically, the detection efficiency for photons in a regular gaseous detector is of the order 10^{-2}.

GASEOUS DETECTORS FOR DETECTING CHERENKOV RADIATION

The challenge for Cherenkov detectors with gaseous photon detection is to achieve sensitivity of the gas, but transparency of the container (and the radiator). It requires the combination of a gaseous detector that has a low photo-ionisation threshold and photon entrance windows with a high transparency cut-off energy.

The figure of merit for the photon detection is the quantum efficiency (QE), which is defined as the fraction of photons creating ionisation, as a function of the photon energy. There are two approaches: The first is to add a chemical with low photo-ionisation threshold to the gas. Two compounds are commonly used: Triethylamine (TEA), with a threshold energy of about 7.5 eV, and tetrakis (dimethylamine) ethylene (TMAE), with a threshold energy of about 5.3 eV. With these thresholds, TEA requires windows made of CaF_2 (transparent up to about 9.5 eV), while for TMAE quartz windows provide a sufficiently large energy gap (transparent up to about 7.5 eV).

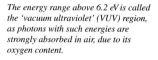

The energy range above 6.2 eV is called the 'vacuum ultraviolet' (VUV) region, as photons with such energies are strongly absorbed in air, due to its oxygen content.

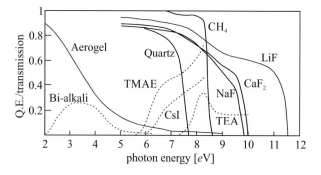

Figure 7.30: Quantum efficiency (dashed lines) and transmission (full lines) for common detectors and transparent materials in Cherenkov detectors (data from [386, 479]). The exact data may vary due to changes in the environmental parameters and purity of the gases.

One challenge with these additives in a gaseous detector is the increased potential for photon feedback, requiring heavily quenched gas mixtures. Typical quencher additives used are therefore pure CH_4 or CH_4/C_2H_6 75/25. TEA and TMAE are liquid at room temperature, and they are usually added to the detector gas by bubbling the main chamber gas through the liquid. The drawback of TMAE is its low vapour pressure

Another measure to mitigate photon feedback is to keep the gas gain in such detectors low. MPGDs like GEMs can be useful, as their geometry can prevent the propagation of photons [482].

at room temperature (0.67 mbar, compared to 96 mbar for TEA), which requires heating and tight temperature control for the bubbler.

The alternative approach to detect UV photons in gaseous detectors is to use a cathode coating with a low work function [373]. The material with the highest quantum efficiency for such purpose is CsI, because of its low work function, but also because it has a long electron absorption length (about 16 nm) [163]. Thin layers can be vacuum deposited either as a semi-transparent thin (about 100 μm) layer on the inside of the entrance window or on a metallic surface on the reverse cathode of the chamber. The quantum efficiency can be enhanced by heating the coating under vacuum after deposition. Contamination with oxygen or water must be avoided.

The quantum efficiency for CsI if used in gaseous detectors can be smaller than for emission into vacuum due to back-scattering of the photo-electrons into the electrode in elastic collisions on the molecules in the gas. Hydrocarbon or fluorocarbon gases have a higher probability for in-elastic collision due to rotational and vibrational levels and thus the back-scattering effect is strongly suppressed, with a photo-electron collection efficiency of about 90% of the vacuum value for CH_4 or $CH_4/i-C_4H_{10}$ mixtures [224]. In addition, for efficient extraction, the field at the location of the coating should be high. The quantum efficiency also can be increased by the adsorption of TMAE in the coating.

GASEOUS DETECTORS FOR DETECTING TRANSITION RADIATION

In principle, gaseous detectors are attractive for the active component of a transition radiation detector, as they provide an affordable solution for the large detection area usually needed in these detectors. For the detection of the high energies from transition radiation (typically a few keV), the strategy is to use a gas with high-Z atoms, as the absorption length for photoelectric effect scales with Z^m with m between 3 and 5 (see section 2.1). For xenon, for example, the absorption length in this energy range is about 10 mm. As usual, an appropriate quencher needs to be added for the use in an MWPC. CH_4 and CO_2 are widely used quenchers. It should be noted that due to the small emission angle for transition radiation photons (see section 2.9) the location of the energy deposition from TR photons will usually be indistinguishable from the charged energy loss along the track of the primary particle.

For further discussion of actual transition radiation detectors see section 12.5.

7.7 AGEING OF GASEOUS DETECTORS

In principle, the exchangeability of the gas in the detector would make radiation damage to the medium not a concern. However, after responding to a large number of incoming particles, polymers formed from the organic molecules in the gas and contaminants will begin to cover the electrodes in the detector as solid or liquid deposits, leading to performance degradation that ultimately will make the detector inoperable [296]. On the anode wires the deposits can be conductive (typically carbon) or a non-conductive polymer. In both cases the increase in radius lowers the field around the wire, whereas in the latter case the insulating layer will also trap electrons, further reducing the field. The net result is progressive gain

For a comprehensive summary of ageing in gaseous detectors see [474].

loss, which makes operation of the chamber more and more difficult, as the deposits are generally not uniform.

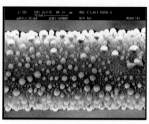

Figure 7.31: Examples of Si-deposits on anode wires. Left: Wire from the CDF central tracking chamber [143]. Right: Wire from CMS muon chamber prototypes [247].

On the cathode, the deposits lead to the Malter effect, where an insulating layer on the cathode traps ions on the surface, creating an electric field through the layer. The field can become strong enough to rip electrons out of the cathode, and accelerate them towards the positive charge layer. There not all of the electrons are captured, but some enter the gas volume, where they will be the source for another avalanche at the anode wire. The Malter effect thus leads to self-sustained dark currents, which decay only slowly when the high voltage is removed from the chamber.

Malter's original paper is [356], although the effect is described in an earlier paper [275].

The constituents for the polymerisation are free radicals dominantly formed from the organic molecules in the gas (in particular the quencher), but also contaminants. These radicals are created in abundance in the avalanche process. The radicals will then link to polymer structures that are often branched, making them particularly difficult to remove.

A free radical is an atom, molecule, or ion that has at least one unpaired valence electron. Radicals are typically highly chemically reactive.

A commonly used scale parameter to describe the effects of ageing in gaseous detectors is the charge accumulated over the lifetime per unit length of wire in a wire chamber, or per unit area for gaseous detectors with planar electrodes or MPGDs. The accumulated charge is given by the product of the rate of incoming particles, the primary ionisation per incoming particle, and the gas gain. Ageing effects are usually of a concern for wire chambers that accumulate 1 C/cm and above, although ageing effects can develop at significantly lower levels if care is not taken.

In typical detector gases it takes more than 10 eV to ionise a gas molecule, but only a few eV to break a chemical bond and create a radical.

It should be noted that the accumulated charge is not sufficient to describe ageing in gaseous detectors fully. Ageing does also depend on the rate of charge accumulation, with ageing effects more severe at lower rates. This is relevant when accelerated ageing tests are performed.

The chemistry of the polymerisation is not well understood, but some general trends are known, and lists of materials to avoid or to be used in contact with the gas have been compiled [182]. Of particular concern are silicon, which enters gaseous detectors as contaminant from silicon oils or grease or mould release agents, and phthalates, which are used as softeners in plastics. Both have a tendency to make their way into polymers accumulating on the anode wire.

The relevant field is plasma chemistry. However, there studies are usually performed at significantly lower pressures and in AC fields. Nevertheless, attempts have been made to carry over conclusions from there to wire chambers.

In general, radiation damage effects in gaseous detectors increase for lower gas flow, indicating that the flow helps removing active radicals, and the polymerisation is not constrained by the supply of a gas component from the input gas.

Some additives are known to reduce ageing, and even to sometimes reverse ageing effects. In particular, these are chemicals containing oxygen, which have a lower tendency to form polymer chains. Examples for such compounds are alcohols or methylal, but also water, which in addition is also believed to be beneficial in reducing Malter effect by increasing the conductivity of the insulating layer on the cathode, so that surface charges can discharge.

Figure 7.32: Methylal.

An additive that can have both beneficial and detrimental effects on ageing is CF_4. This ambiguity is a manifestation of the 'Competitive Ablation and Polymerisation' (CAP) principle [509], which states that ablation is creating reactive monomers from the neutral monomers entering the gas volume (detrimental), but also from the break-up of existing polymers (beneficial). These radicals can then be the precursor to long-chain polymerisation, but also to the formation of smaller stable molecules that are still volatile and are removed with the exchange of the detector gas. CF_4 in the presence of oxygenated chemicals will act as an etching agent, and is even capable of removing polymers containing silicon from anode wires, while CF_4 in conjunction with hydrogenated substances will tend to polymerise. No wire chamber has been able to operate at 20 C/cm and above without the addition of CF_4 to the chamber gas.

One other concern with CF_4 is that it can lead to the formation of hydrofluoric acid (HF), which is a potent acid and in particular etches glass.

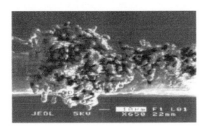

Figure 7.33: Left: Anode wire after an accumulated charge of 0.07 C/cm in Ar/CF_4/CH_4 67/30/3 [216]. Right: Anode wire after 20 C/cm in Xe/CF_4/CO_2 70/20/10 [35].

Ageing in MPGDs depends on the size of the electrodes. Detectors with small electrodes, like MSGCs, will suffer similar ageing effects as wire chambers (in addition to discharge damage as discussed in section 7.4) taking into account the smaller electrode size. Similarly as in wire chambers, high rate operation requires high purity of the gas and careful selection of detector materials. If these conditions are met, accumulated charges of 0.1 C/cm can be achieved [158].

Planar MPGDs like GEMs or Micromegas undergo the same ageing mechanisms, but the effects are less severe due to the increased electrode surface area. Operation of triple layer GEMs for accumulated charge densities in excess of 0.8 C/cm^2 without degradation has been achieved [241].

Ageing effects in RPCs have been observed in early applications. These could be traced to either significant amounts of water (2000 ppm), which together with tetrafuororethane in the chamber gas resulted in the formation of hydrofluoric acid, or the flawed application of linseed oil, which has been used to improve the surface quality of bakelite electrodes [481]. If these issues are avoided, RPCs demonstrate good ageing performance [261].

7.8 SIMULATION OF GASEOUS DETECTORS

We have seen in the previous chapters that in principle the physics of the various aspects of the operation of a gaseous detector is well understood. At the same time, the often complex stochastic nature of many of these processes makes an accurate analytic prediction of the detector performance impossible. However, Monte Carlo algorithms are capable of capturing these aspects and, in parallel to the development of gaseous detector designs, a suite of powerful programmes have been developed, which capture all aspects of the operation of a gaseous detector, from the generation of the primary ionisation charges to the electrical signal created [486]. These results allow for a detailed prediction of the detector performance for a given set of design and operating parameters, and thus are very valuable in the optimisation of gaseous detectors.

The central piece of this software is GARFIELD [487]. This programme provides facilities to calculate analytically the electric field in simple geometries, generates primary ionisation, drifts the charges in the field, including collision effects and diffusion, replicates avalanche multiplication and obtains signals from the drift of the final charges in the system. Some of its outputs are [487]

- field maps, contour plots and three-dimensional impressions;

- the wire sag that results from electrostatic and gravitational forces;

- optimum potential settings to achieve various conditions;

- plots of electron and ion drift lines;

- $x(t)$-relations, drift time tables and arrival time distributions;

- signals induced by charged particles traversing a chamber, taking both electron pulse and ion tail into account.

GARFIELD invokes inputs from several other pieces of code, some of which have originally been conceived as standalone programmes, but do have an interface with GARFIELD:

- HEED: This programme calculates primary ionisation yields in gases based on the PAIR model [460] (see section 2.4).

- MagBoltz: This code [138] calculates drift properties for electrons in gas mixtures under the influence of electric and magnetic fields, using the Monte Carlo approach described in section 4.1 [137].

  A python version of MagBoltz, PyBoltz [38], is available and can be found at [411].

- Garfield allows for the input of field maps from commercial finite element analysis (FEA) software, like Ansys, Maxwell, Tosca, QuickField and FEMLAB, which can compute approximate fields in nearly arbitrary three-dimensional configurations with dielectrics and conductors.

- FEA programs perform reasonably well in the calculation of potentials, but they are not well suited to the calculation of fields in typical gaseous detector geometries, as the quadratic shape functions used in FEA programs do not describe $1/r$ or $1/r^2$ fields well, and because electric fields can be discontinuous at boundaries. A different approach is the Boundary Element Method, where the elements are on the boundaries, not inside the problem domain and charges are computed for the boundary elements. A practical implementation that interfaces with GARFIELD is neBEM (nearly exact Boundary Element Method) [383, 382].

Contrary to popular belief, the original namesake for the programme's name GARFIELD is not a lasagne-devouring feline, but the 20th president of the USA, J.A.Garfield (1831–1881). In 1881 President Garfield was shot by an erstwhile follower, who felt betrayed for not receiving a political office in return for his support, and the bullet lodged in his body. Graham Bell then successfully developed the first metal detector to locate the bullet. However, despite working properly during the development, the detector failed on President Garfield. Only later on Bell realised that the metal detector was confused by the iron bed springs within the bed the wounded President was lying on [126]. In the end President Garfield did not die from the bullet, but two months later from infections caused by the doctors searching with their fingers for the bullet in his body without proper sterilisation. The lesson is: For a successful technology you need to know your physics and apply it correctly in the required environment.

Originally, this programme calculated the transport parameters by solving the Boltzmann transport equation, eq. (4.9) [136], which was the motivation for its name.

- Finally, the calculated signals from Garfield can be used as input to publicly available commercial or free electronic circuit simulation software like SPICE and its derivatives.

All this software is constantly maintained and improved to meet the challenges of new detector technologies and design ideas. Originally, GARFIELD has been written in FORTRAN, but a C++ version, GARFIELD++ [446], is now available. This currently represents the most accessible version of the code.

GARFIELD++ also can simulate the signals generated in semiconductor detectors.

Some examples for the use of this software for end-to-end simulation projects can be found in [422] (ATLAS MDT), [425] (RPCs), or [155] (GEMs).

A different set of detector simulation software is GEANT (GEometry ANd Tracking) [260], where the emphasis lies less on the prediction of the position measurement performance for single charged particles in a gaseous detector, but on the prediction of the physics-relevant response (momentum or energy resolution) of all components of a detector system in the presence of all the matter in the system. This includes modelling of energy loss of charged particles, including radiative energy loss, and multiple scattering [29, 45].

For a complete simulation of gaseous detector it is possible to connect the GEANT framework with GARFIELD++ [261].

Key lessons from this chapter

- The low density of the active material in gaseous detectors limits the primary ionisation charge from the passage of a charged particle. To still obtain detectable electrical signal internal avalanche amplification is required.

- A common way to achieve the fields for avalanche multiplication is to use thin wires. In a wire chamber the induced signal is dominated by the drift of the ions. Due to their limited speed the time to collect all the charge is often prohibitively long, but it can be truncated by appropriate electrical filtering, as most of the signal is induced early on.

- To suppress photon feedback resistive cathodes and/or the addition of a quencher gas are required.

- The gases used are mixtures designed to achieve various requirements (high ionisation, high gas gain, no photon-feedback, saturated drift velocity, low diffusion, high-rate operation, long lifetime). Several of these requirements cannot be satisfied at the same time, thus there is no gas that will work for all applications. Careful optimisation of gas mixture and geometric design for the specific application is required.

- This optimisation is helped by the availability of a suite of powerful software simulation tools (Garfield/HEED/MagBoltz).

- Even though gaseous detectors generally have a limited position measurement and high-rate performance they are still an effective and affordable way to instrument large area detectors like muon detectors.

- A TPC is a large-volume, low-material tracking detector which provides a large number of space points (with moderate precision), and the capability to measure dE/dx.

- The active development of MPGDs is pushing performance parameters in gaseous detectors to new limits.

EXERCISES

1. *Stability of wires in a MWPC.*

   Consider displacement of the anode wires in a MWPCs as shown in Figure 7.6.

   a) Show that the force per unit length between two parallel wires at the same potential V_0 and with a capacitance per unit length C_1 at a distance r is given by $F_1 = (C_1 V_0)^2/(2\pi\varepsilon_0 r)$.

   b) Assume that the wires are displaced alternately by a distance $\delta \ll s$ perpendicular to the central plane as shown in Figure 7.6, where s is the spacing between wires. Use superposition and the symmetry in the problem to show that the force on one wire from all the other wires in the system is $F = (\pi(CV_0)^2/(4\varepsilon_0))(\delta/s^2)$.

   c) To counter this displacement, the wire is held under tension. In equilibrium, the restoring force due to the tension must equal the displacing force caused by the interaction with the other wires. Find a solution for the displacement of the wire as a function of the position along the length of the wire. Use the boundary conditions to find a valid solution for the wire tension.

      If the wire tension exceeds this value, no solutions are possible, and the displacement δ must remain zero along the wire.

2. *Gas gain in a wire chamber.*

   What are the effects of the following changes to the gas gain in a proportional wire chamber?

   a) Increasing the diameter of the anode wire.

   b) Increasing the gas pressure.

   c) Increasing the fraction of quencher gas.

3. *Time response of the signal from a proportional chamber.*

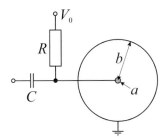

   a) A particular chamber configuration has an anode wire radius of $a = 10\ \mu$m and a potential on the wire of $V_0 = +3$ kV. The cathode is a cylinder of radius $b = 10$ mm at zero potential, concentric with the wire. Assume that the avalanching occurs at a distance $\delta = 1\ \mu$m from the anode. Show, using Ramo's theorem, that the induced charge signal from a charge Q displaced through a distance dr in the chamber is

   $$dq = \frac{Q}{V_0}\frac{dV}{dr}\,dr,$$

   calculate the relative contributions to the charge pulse from the proportional chamber by electrons moving to the wire, and by the positive ions moving away. Show that the contribution of the electrons to the signal is negligible.

   b) The velocity of the ions is a linear function of the electric field, E, in the tube:

   $$\frac{dr}{dt} = \mu E.$$

   The mobility μ is equal to 1.7 cm^{2}s^{-1}V^{-1} for this chamber. Using this relation and the expression for dq given above, show that the signal current for the positive ions (as a function of time) is

   $$i(t) = \frac{QC}{4\pi\varepsilon_0}\frac{1}{t + \frac{\pi\varepsilon_0(a+\delta)^2}{\mu V_0 C}},$$

   and find the total drift time (T) of the ions. Sketch the time development of the current pulse (from ions) due to the passage of this particle. Show further that the total charge (from the positive ions) collected (after T) is:

   $$q(T) = Q\frac{\ln\left(\frac{b}{a+\delta}\right)}{\ln\left(\frac{b}{a}\right)}.$$

   What is the charge collected from the movement of ions for the parameters given above?

4. *Position resolution in a drift chamber.*

   Consider a cylindrical drift chamber. A minimum-ionising particle crossing the chamber creates on average N_c electron-ion clusters per unit length. The closest cluster to the anode wire is used to estimate the perpendicular distance from the track to the anode, d_c. The radial distance from this cluster to the anode is r.

   a) Estimate the resolution in d_c as a function of r.

   b) Explain qualitatively the effect of diffusion on the resolution for d_c due to diffusion at large values of r.

   c) Why is the contribution to the resolution from the precision of the arrival time measurement independent of r?

5. *Charge induced on cathode planes.*

   A charge located between two cathode planes induces a charge density on the planes, which can be found using the method of mirror charges. For simplicity we assume here that the charge is central at $z = d/2$ between two infinite cathode planes at a distance d.

   a) Show that the electric field induced by the charge on the surface of the plane is given by

   $$E_z(r) = -\frac{\lambda}{\pi \varepsilon_0} \sum_{k=0}^{\infty} (-1)^k \frac{(2k+1)\frac{d}{2}}{(2k+1)^2 \frac{d^2}{4} + r^2} = -\frac{\lambda}{2\varepsilon_0 d} \frac{1}{\cosh\left(\frac{\pi r}{d}\right)},$$

   where r is the distance on the plane from the point of projection of the charge onto the plane.

   b) Find the charge density on the cathode from the field using Gauss's law.

6. *Charge decay.*

   Consider a material with a conductivity of σ and a relative permittivity ε_r. Show that any charge buildup will decay exponentially with a time constant $\tau = \varepsilon_0 \varepsilon_r / \sigma$. (Hint: use the appropriate Maxwell equation in 3D and the microscopic version of Ohm's law.)

   Evaluate the time constant for copper using $\sigma = 1.58 \times 10^{-8} \ \Omega m$ and assuming $\varepsilon_r = 1$.

8 Liquid detectors

Liquid detection media are attractive when a detector with large stopping power and full sensitivity over a large detector volume is required. Hence the main applications for detectors with a liquid detection medium are in calorimeters, in neutrino physics and in dark matter searches. The high density compared to gases entails large interaction probabilities and high energy loss per track length for charged particles. At the same time diffusion for drifting ionisation charges is low.

The high density makes liquids less well suited for tracking detectors due to multiple scattering, while the liquid lacks the fine segmentation of the readout that semiconductors can provide.

The largest particle detectors today use the most abundant liquid on earth, water, exploiting its transparency for the visible photons from Cherenkov radiation. We will discuss water Cherenkov detectors in more detail in section 12.4.

For detectors that are not relying on Cherenkov radiation, the most widely used liquids in particle detectors are noble liquids because of the low intrinsic attachment electron coefficient and/or self-absorption for scintillation photons. Of the noble liquids argon is the cheapest and hence liquid argon (LAr) has found wide use in particle detectors.

For review articles on liquid noble gas detectors see [190], for liquid argon [358, 355], and for liquid xenon [64].

Liquid noble gases are industrially obtained from fractional distillation of liquid air. Argon is the third-most abundant component of air (almost 1%), whereas the fraction of xenon in air is only around 90 ppb.

Liquid xenon (LXe) has a higher atomic mass and higher density, and thus a higher stopping power than other noble liquids, and low diffusion. In addition, it has properties that make it particularly interesting for detectors for dark matter searches: It would have a high interaction cross section with weakly interacting massive particles (WIMPs). The only common long-lived radioactive contaminant, ^{85}Kr, can be easily removed, and natural xenon includes odd-neutron isotopes with non-zero nuclear spin, which gives sensitivity also to a spin-dependent WIMP interaction component.

A liquid detection medium can be circulated, which allows for continuous purification of the medium. This is usually a necessity to achieve charge or photon transport over the distances required in large detectors, but it also can help to achieve radiopurity, if the radioactive contaminant can be removed.

Room-temperature liquids for particle detectors have challenging purity requirements, but for example tetramethylsilane (TMS) is being used in the KASCADE experiment [409].

These are often referred to as 'warm liquids'.

8.1 CHARGE READOUT

In the simplest geometry, liquid argon is used to fill the gap in an ionisation chamber, and an electrical signal is created by the movement of the ionisation charges created in the volume towards the collecting electrodes. This is a geometry often used in calorimeters (see chapter 11) with a typical drift distance of a few mm to a very few cm.

This geometry is appropriate for the energy measurement of high-energy showers, where the primary goal is the collection of a large fraction of the created ionisation charge, and not the exact localisation of the charge. In neutrino physics experiments we do want to reconstruct the tracks of the leptons and any charged hadrons recoiling from the

DOI: 10.1201/9781003287674-8

interaction of the neutrino with the detector medium, and thus a finer position resolution is required.

The obvious design choice is to increase the segmentation of the collection electrodes and to increase the drift length, so that a projection of the primary ionisation charge is recorded in the plane of the collecting electrodes, where a localised signal is induced due to the strong increase of the weighting field around the small collection electrodes. If the drift time of the primary electrons towards the collection electrode is measured, this can be used to determined the distance of the creation of the primary ionisation to the detection plane. This is the principle of the liquid TPC. Compared to a gaseous TPC, liquid TPCs produce significantly more primary ionisation charges, and the diffusion is lower.

This benefit can only be realised if the liquid is pure enough to allow for large drift distances.

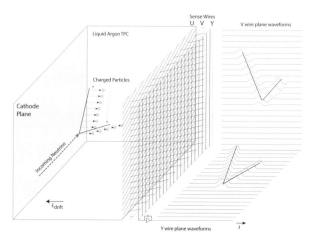

Figure 8.1: Schematic drawing of a single phase liquid TPC (left) and signals in the position/time domain (right) [14]. The ionisation electrons drift in a large electric field towards the anode plane. The relative voltage between the wire grids is such that the drifting electrons pass through the first two planes and are collected on the third plane.

A common detection geometry uses wires to collect the ionisation electrons. To provide two-dimensional position information, two planes of wires are needed, but to resolve ambiguities a third plane is usually added [258]. The plane farthest from the drift volume is kept at the most positive potential (usually ground or slightly above), and is called the 'collection' plane. The other wire planes are at a slightly (by a very few hundred V) lower potential and are referred to as 'induction' planes. As the electrons are drifting towards the collection wires, they pass through the induction planes, where they induce a bipolar signal on the induction wires, as they are drifting first with and then against the weighting field for the induction wires. On the collection wires the signal is unipolar. The weighting field towards the drift volume is small due to the large distance to the drift cathode, and thus the leading lobe of the signal on the first induction plane will be very small. Therefore sometimes an additional, uninstrumented, induction plane is added before the readout planes [110]. The scale of the duration of the recorded signals is given by the drift time between the wire planes and is typically μs, compared to ms for the drift time in the drift volume (with a typical drift field of a few hundred V/cm).

This is referred to as a 'gate' by analogy to the wire layer in similar location in a gaseous TPC, although it does not fulfil a gating function (prevention of ion back drift into the drift volume), as there is no avalanche multiplication.

The mobility for ions is significantly lower, so that it can take minutes until the ions are removed from a large ($\mathscr{O}(m)$) LAr TPC.

The LAr TPC with the longest continuous operation is part of the MicroBooNE experiment [14]. Its drift volume is $2.3 \times 2.6 \times 10.4$ m^3 filled with 90 t of LAr and a drift field of 500 V/cm for a maximum drift time of 1.6 ms. It is read out electrically with 8256 wires in three planes at

$+60°/-60°/0°$ at $-200/0/+440$ V with a wire pitch of 3 mm, and optically with 32 PMTs with 200 mm diameter.

Optical readout of scintillation in the liquid will be discussed in the next section.

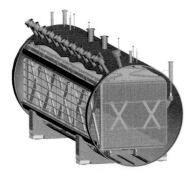

Figure 8.2: Left: The MicroBooNE LAr TPC in its cryostat [14]. The drift field is vertical with the readout planes on the left, and the neutrino beam running along the length of the chamber, entering from the rear face. Right: Zoomed event display from MicroBooNE with an electromagnetic shower [372]. The colour indicates the magnitude of the ionisation signal.

The choice of wires as the electrode in LAr TPCs is driven by the simplicity with which such electrodes can be assembled into a regular readout grid, and the straightforward connection to a readout amplifier, but not by the aim to achieve an electric field sufficiently high to cause avalanche multiplication. Stable high-gain (>10) avalanche multiplication close to a wire in pure noble gases at cryogenic temperatures has proven to be an elusive task. However, internal amplification is desirable, and has been achieved using similar amplification structures as in MPGDs [173].

One geometry making use of such amplification structures is the dual-phase TPC, where the drift field is vertical, bringing the electrons close to a liquid-vapour interface close to the top of the TPC. An extraction field created by an extraction grid within the liquid propels the electrons across the liquid-vapour interface. In the gaseous phase the electrons undergo avalanche multiplication in planar amplification structures called 'large electron multipliers' (LEMs) that are made of metalised and perforated printed-circuit board (PCB) plates and achieve avalanche multiplication inside the holes, similar to a GEM (see section 7.4).

Figure 8.3: Gain in a triple-GEM at cryogenic temperatures for different noble liquids as a function of the voltage applied across each GEM [154].

Also sometimes called 'thick gaseous electron multipliers' (THGEM).

PCBs are used instead of polyimide as in GEMs, because their stiffness allows for the simple construction of large-area amplification structures, and multiple scattering material is not a concern.

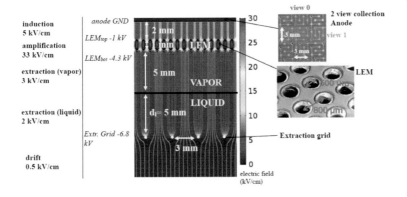

Figure 8.4: Vertical cross-section of the dual-phase detector as originally proposed for the DUNE experiment [187].

8.2   PHOTON DETECTION

As discussed in section 5.4, the passage of charged particles through cryogenic liquids also results in the emission of long-range scintillation photons in the VUV. These photons are prompt as the scintillation usually has

VUV is the acronym for vacuum ultraviolet, which covers wavelengths below 200 nm, where oxygen in the air is strongly absorbing.

a fast (a few ns) component, and they can thus be used to determine the time of the ionisation event. In neutrino experiments this is how the time of the interaction in the detector can be determined, which gives the start of the drift time of the electrons from the ionisation location to the detection plane, which in turn is used to obtain the drift distance.

Because of the size of typical neutrino experiments an affordable way to provide the large detection coverage required is by using PMTs. The necessary wavelength shift from the VUV to the visible where the PMT sensitivity peaks is often achieved by coating the PMT with tetraphenyl-butadiene (TPB).

An alternative method to achieve large area coverage for the detection of scintillation from liquid argon is the ARAPUCA (Argon R&D Advanced Program at UniCAmp) device [354]. In this device photons in the visible wavelength regime are detected by a silicon photomultiplier (SiPM). However, a typical SiPM is relatively small (a very few cm^2) and to achieve a large acceptance the photons from the scintillation need to be guided to this detector. This is achieved by trapping the photons in a box with highly reflective walls where they can bounce off several times until they hit the detector. The photon is trapped in the box by a dichroic entrance window. The window is coated with two wavelength shifters. On the outside, a wavelength shifter is used that shifts the VUV photon from the argon scintillation to a wavelength just below the threshold for transmission of the dichroic window. The photon can pass the window, where the wavelength is shifted by the inner wavelength shifter to a wavelength above the window threshold, so that the photon is trapped in the box. The wall reflection and the window transmission coefficients can be made high enough to obtain detection efficiencies of a few % for devices with a coverage of a few thousand cm^2.

In a dual-phase TPC a second source of photons is electroluminescence caused by the ionisation electrons in the high electric extraction field in the gaseous phase above the liquid. At low pressures ($\mathscr{O}(\text{mbar}_a)$ and less) the electrons collide with atoms in the gas, which get excited into a P state. In the subsequent de-excitation the atom emits VUV photons with wavelengths of about 105 nm and 107 nm in argon, and 130 nm and 147 nm in xenon for transitions from the excited singlet and triplet states, respectively [190]. At higher pressures formation of diatomic excimers becomes more probable and dominates at 1 bar$_a$. These excimers emit scintillation similarly to the primary scintillation process in the liquid, with similar wavelengths. The secondary photon yield is proportional to the drift length of the ionisation electron and, depending on E/p (the ratio of electric field to pressure), the yield is about several hundred photons per cm. As both emissions are in the VUV the same photon detectors can be used.

Readout of secondary scintillation is particularly attractive for dark matter searches because the ratio of the primary (from excitation in the liquid by the original particle) and secondary (electroluminescence from avalanches at the amplification structure in the gas) scintillation response is very different for electrons and nuclear recoils, allowing to reduce radioactive backgrounds. Further reduction of background from the walls of the chambers is achieved by rejecting events in the vicinity of the walls.

Silicon photomultipliers will be discussed in section 9.9.

Lightguides would be too bulky and not be efficient because of Liousville's theorem.

A dichroic window is only transparent at wavelengths below a certain threshold.

For example, p-terphenyl (PTP) with an emission wavelength of 350 nm for a window threshold of 400 nm. For example, TPB with an emission wavelength of 430 nm.

An improved version is the X-ARAPUCA, which uses a light guide that is doped with the wavelength shifter inside the box, to improve the guidance to the photon detector [166].

Thus, for a fixed drift length in the extraction field the secondary scintillation signal is proportional to the number of ionisation electrons.

In neutrino detectors it is usually tracking performance over a large volume that is required, and thus electronic readout is the standard.

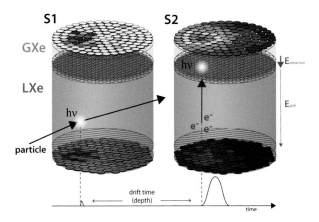

Figure 8.5: Schematic view of the working principle of a dual-phase TPC with photon readout. 'S1' is the signal from the prompt scintillation photons, 'S2' is the signal from the electroluminence photons created when the ionisation charge has reached the high field region in the gas layer above the liquid [223].

Several dark matter search experiments of this design with increasing size have been built. One of them is the LUX-ZEPLIN (LZ) experiment [34]. It consists of a xenon volume with a maximum drift distance of 145.6 cm and a diameter of 145.6 cm with 7 t of LXe in the active volume. The vertical drift field in the liquid is 300 V/cm. In the extraction region, the electric field strength in the gas is in the range 10.1 kV/cm to 11.5 kV/cm, sufficient for electroluminescence. The time between the primary and the secondary scintillation signal is used as a measure for the vertical coordinate of the primary interaction. The horizontal location is determined by the pattern of PMT hits. The spatial resolution for events at the minimum detector threshold is better than 1 cm and improving with increasing energy.

The grid for the 'gate' electrode is in the liquid and held at a voltage of −5.75 kV and the grid for the anode is at +5.75 kV. A woven grid with very fine pitch and very small diameter wires are used to ensure good transparency for the scintillation light.

For the primary scintillation response, the goal is to maximise the number of detected VUV photons, as this defines the threshold of the detector. This is achieved by the use of high quantum efficiency (QE) PMTs optimised for the wavelengths of the VUV scintillation photons, and materials in the chamber with low photon extinction. For the secondary scintillation response, the gain of the electroluminescence process makes it easier to collect enough photons even at the lowest energies, and the main design driver is instead to optimise the spatial resolution, especially for peripheral interactions.

Peripheral interactions are interactions close to the wall, which are rejected to reduce backgrounds.

COMBINED READOUT

The energy resolution in LXe detectors from the scintillation or the ionisation signal is worse than expected from Poisson statistics due to fluctuations in the electron-ion pair recombination rate. Since recombination results in the creation of excited xenon atoms, which subsequently tend to form excimers, which decay under emission of a VUV scintillation photon, the ionisation and the scintillation signal will be complementary and the combination of the two can improve energy resolution. This has been demonstrated for 662 keV γs from the decay of ^{137}Cs [65].

8.3 PURITY OF LIQUID DETECTION MEDIA

The key requirement for the use of liquids as particle detector media is their purity, which is primarily required to allow for long-distance

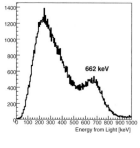

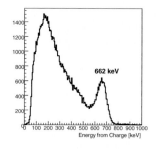

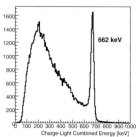

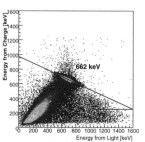

Figure 8.6: Energy spectra of 662 keV γs from ^{137}Cs at 1 kV/cm drift field in liquid xenon [65]. The top two plots are from scintillation (read out with PMTs) and ionisation (read out with charge-sensitive electronics in a gridded ionisation chamber), respectively. The charge-light anti-correlation is shown in the bottom-right plot. The straight line indicates the charge-light correlation angle. The left lower plot shows the improved energy resolution of the combined measurement.

transport of either ionisation charges or scintillation photons. The typical drift time of ms in a LAr TPC for neutrino experiments sets the scale for the required free electron lifetime in the liquid. Drifting electrons can be captured by contamination with high electron affinity like O_2, H_2O or CO_2. Photons are also captured by N_2. Impurity concentrations less than a few ten parts per trillion are required to achieve drift paths of several meters.

The use of a fluid medium allows for circulation of the fluid, with the possibility to continuously purify the medium. To remove oxygen, the fluid is flushed through Oxisorb cartridges and molecular sieves. Activated copper-coated alumina granules have also been used, which can be regenerated in an atmosphere containing hydrogen at elevated temperatures. With this purification electron lifetimes of 7 ms have been achieved [490].

Materials in the detector volumes have to be carefully selected and cleaned so that they do not introduce any contaminants. No effect on the electron lifetime might be observed when the material is immersed in LAr, but when it is positioned in the warmer region of the vapour phase (at ~200 K), the electron lifetime can strongly decrease.

Smaller cryostats (several m^3) can be evacuated before filling, but that will be impossible for large future neutrino experiments. There the volume will be filled first from the bottom with the dense argon gas replacing the air, followed by circulation of argon gas at 50 °C, and finally filling and circulation with LAr.

An empirical relation for the free electron lifetime τ as a function of oxygen-like contamination is $\tau/\text{ms} \sim 0.3/(N/\text{ppb})$, where N is the concentration of oxygen equivalent impurities.

Oxisorb is a commercial purification system that uses chemical absorption of oxygen by a silica gel with chromium oxide. Moisture is also absorbed physically.

> ### *Key lessons from this chapter*
>
> - Detectors based on liquids provide large uniform detection volumes, high stopping power, and low diffusion for drifting electrons.
>
> - The most common liquid detector media not relying on Cherenkov radiation detection are liquid noble gases. Of these argon is the most affordable, whereas xenon has a higher atomic mass and density, and additional features that make it particularly suited for dark matter search experiments, but also make it attractive for use in very high resolution electromagnetic calorimeters.
>
> - A 3D position information can be measured in a TPC configuration from the signals induced on a stack of anode wire planes (for 2-dimensional location in the plane), together with a measurement of the drift time for the 3rd coordinate. To start the drift time measurement the prompt scintillation signal in the liquid can be used. The scintillation signal is in the VUV.
>
> - While stable avalanche multiplication around wires is not possible in pure cryogenic liquids, ionisation electrons can be extracted from the liquid by high electric fields, and create secondary scintillation photons in the VUV (electroluminiscence).
>
> - One major challenge in these detectors is purity of the liquid, to prevent loss of ionisation or scintillation signal before detection. The fluidity of the detection medium allows for continuous purification. The required purity can also be beneficial in experiments where contamination can introduce radio-backgrounds.

EXERCISES

1. *Energy loss in cryogenic liquids.*

   Estimate the energy loss dE/dx for a minimum-ionising particle in liquid argon ($\rho = 1.4$ g/cm^3) and for liquid xenon ($\rho = 3.1$ g/cm^3). Estimate (for example from Figure 2.24) the range of a 1 MeV muon in LAr and LXe.

2. *Event display in a liquid detector.*

   Explain the features of the event display in Figure 8.2. How does the energy deposition along the charged tracks vary? Can you estimate the energy of the charged particles? Can you estimate the energy of the particle causing the electromagnetic shower?

3. *Dual-phase TPC.*

   A simple dual-phase argon TPC has a geometry as shown in the figure.

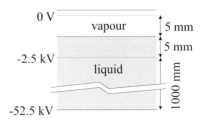

a) Calculate the electric field strength in the different regions of the TPC.

b) What is the polarisation charge density at the liquid-vapour interface? (The relative permittivities in the vapour and the liquid are $\varepsilon_r^{vap} \simeq 1$ and $\varepsilon_r^{liq} \simeq 1.5$, respectively.)

c) Estimate, using Figure 4.8, the maximum drift time for electrons in this TPC.

d) The speed of sound in liquid argon at $-186\,°C$ is about 810 m/s. Compare this to the electron drift velocity in this TPC.

9 Semiconductor detectors

In solids the stopping power for a charged particle is much higher than in gases due to the larger density. Hence, already thin layers of detector give sufficiently large signals to detect the impact of a charged particle. As a consequence, neither the spacing of primary ionisation clusters, nor diffusion of the charge carriers poses a practical limitation for the determination of the position in a solid detector, and δ-electrons are not a concern, because of their shorter range. Another advantage is that the sensor material is itself mechanically stable and does not need a container.

Diffusion can actually be useful, spreading the signal over several individual sense elements, which allows improved position resolution from the reconstruction of the charge centroid.

The key challenge in solid state detectors is the efficient collection of the ionisation signal. First and foremost, this requires high purity of the material, so that the charges are not trapped locally. Second, the material must be chosen so that the charges from the interaction of the incoming particle are mobile and thus can create a signal on collecting electrodes according to Ramo's theorem, while at the same time movement of other charges, not associated with the passage of a particle, is suppressed. As discussed in section 4.3, semiconductors are well suited for this task.

In this chapter we will focus on silicon detectors, as the most common semiconductor detector material. We will first discuss the principles of operation of these type of detectors, and discuss the two most common implementations, as strip or pixel detectors, which are based on the principle of a reverse-biased *pn* diode junction. One concern for the operation of these type of detectors is the leakage current and its implication for the thermal stability of silicon detector systems.

We will then discuss more advanced silicon detector technologies that rely on charge carrier drift, internal amplification, or the capabilities of modern semiconductor fabrication technologies. Finally, we will investigate the effects of high-intensity irradiation on silicon sensors and discuss alternative materials for solid detectors.

9.1 PRINCIPLE OF OPERATION

INTRINSIC CHARGE CARRIER DENSITY

While there is a finite gap between the valence band and the conduction band in semiconductors (see section 4.3), it is small enough that some electrons can be promoted to the conduction band due to thermal motion. At the same time electrons promoted to the conduction band leave a hole in the valence band. The volume number density of intrinsic carriers of energy E is given by $n(E) = \rho(E)F(E)$, where $\rho(E)$ is the density of states and $F(E)$ is the occupation probability. As electrons are fermions this is in principle given by the Fermi-Dirac distribution for electrons,

Intrinsic charge carriers are the charge carriers present in the conduction band in pure silicon.

$$F(E) = \frac{1}{\exp\left(\frac{E-E_F}{k_B T}\right) + 1}.$$

The Fermi energy E_F is defined as the energy at which the occupation probability is 50%. This is approximately half way between the conduction and valence bands.

However, at room temperature $k_B T \simeq 26$ meV, and as the Fermi energy is in the middle of the band gap, $E - E_F$ is a about two orders of magnitude

DOI: 10.1201/9781003287674-9

larger than $k_B T$, so that we can describe the occupation probability with a Maxwell-Boltzmann distribution, $F(E) \simeq \exp[-(E - E_F)/(k_B T)]$.

The integrated (over energy) negative and positive (holes) charge volume density are given by (see exercise 3)

$$n_n = 2 \left(\frac{m_n^\star k_B T}{2\pi\hbar^2} \right)^{3/2} \exp\left(-\frac{E_c - E_F}{k_B T} \right),$$
$$n_p = 2 \left(\frac{m_p^\star k_B T}{2\pi\hbar^2} \right)^{3/2} \exp\left(-\frac{E_F - E_v}{k_B T} \right),$$

(9.1)

The motion of an electron (hole) in electric and magnetic fields in a semiconductor can be treated as for a classical free particle but with an effective mass $m_n^\star$ ($m_p^\star$).

where E_c, E_F, E_v are the energies at the bottom of the conduction band, the Fermi energy and the top of the valence band, respectively, and $m_n^\star$ and $m_p^\star$ the effective masses for electrons and holes.

For *intrinsic* semiconductors the density of electrons and holes are equal, $n_n = n_p \equiv n_i$, where we have introduced n_i, the intrinsic carrier concentration, given by

Intrinsic semiconductors are chemically pure.

$$n_i = (m_n^\star m_p^\star)^{3/4} \left(\frac{k_B T}{2\pi\hbar^2} \right)^{3/2} \exp\left(-\frac{E_c - E_v}{2k_B T} \right).$$

(9.2)

For pure silicon at room temperature $n_i \simeq 10^{10}$ cm^{-3}.

THE *pn* JUNCTION

At room temperature the intrinsic charge carrier concentration is so high that the resulting leakage current will tend to be much larger than the signal current created by the passage of a charged particle (see exercise 1). Thus, to make a viable silicon detector, we need to remove the mobile carriers from a region of the silicon. This can be achieved by joining *n*- and *p*-doped silicon.

In *n*-doping, small amounts of a period 5 element (e.g. P, As or Sb) are used to replace silicon (period 4) atoms in the lattice (the period 5 atom is called a 'donor'). This leaves an unpaired electron that is only very loosely bound, with an energy just below the conduction band. At room temperature, these electrons will tend to be promoted to the conduction band and thus they are effectively free to move around the lattice if there is an external electric field.

In *p*-doping, small amounts of a period 3 element (e.g. B, Al, Ga, In) are used to replace silicon atoms (such an atom in the lattice is called an 'acceptor'). The period 3 element only bonds with three electrons to neighbouring silicon atoms in the lattice. Therefore a fourth electron can be 'borrowed' from a nearby silicon atom. This results in a less strongly bound state that is just above the top of the valence band. In a similar way as for *n*-doping, at room temperature electrons from the valence band will be promoted to this impurity level, resulting in a vacant electron state in the valence band. This vacancy can also move under the application of an external electric field. We simplify the picture of many electrons moving to fill vacancies by interpreting the charged vacancy as a positive 'hole', i.e. it is effectively a free positive charge.

Mathematically, doping changes the Fermi levels in the two regions by

As we will see, charge carrier removal at a pn junction works for thin detectors. However, if we want to use a semiconductor detector to detect gamma rays (~ 1 MeV) with high efficiency we need a large active volume because of the long absorption length. There, the only way to reduce intrinsic charge carrier density is low temperature. The finest gamma spectroscopy detectors are germanium detectors ($E_g = 0.67$ eV) cooled by liquid nitrogen (see section 11.1).

It should be remembered that while conceptually appealing, the electrons or holes here are not real, standalone particles, but really describe a state of the lattice (similar to an electron in a shell of an atom).

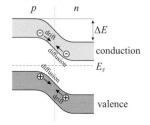

$$\Delta E_F^{n,p} = k_B T \ln\left(\frac{n_{a,d}}{n_i} \right),$$

(9.3)

Figure 9.1: Band structure at a pn junction.

where $n_{a,d}$ is the acceptor or donor density, respectively, and n_i the intrinsic carrier density. p doping brings the Fermi energy closer to the conduction band, n doping brings it closer to the valence band.

At the junction of n- and p-doped silicon the surplus charge carriers in each doping region diffuse into the other doping region and neutralise until thermal equilibrium is reached. As a result, the area around the junction gets depleted of charge carriers. The ions remain and create a space charge and an electric field. In equilibrium the resulting drift currents cancel the diffusion current.

At this point the Fermi energy in the two regions becomes equal.

The depth of the depletion zone can be found from integrating the Poisson equation, $d^2\Phi/dx^2 = -\rho/\varepsilon$, where $\varepsilon = \varepsilon_r\varepsilon_0$ is the permittivity of silicon. In a simple model the number of free charge carriers is assumed to vanish over the depth of the depletion zone $d = d_n + d_p$ where the $d_{p,n}$ are the depletion depths in the two doped materials, but to be constant outside. As a result of the first integration and assuming overall charge neutrality, we find $d_p/d_n = N_d/N_a$, where $N_a(N_d)$ is the acceptor (donor) volume carrier density. The second integration then yields

The relative permittivity of silicon is independent of doping, $\varepsilon_r = 11.7$.

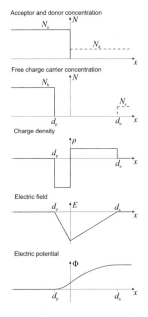

$$d = \sqrt{\frac{2\varepsilon}{e}\frac{n_a+n_d}{n_an_d}\Delta V}. \qquad (9.4)$$

For an unbiased pn junction, ΔV is the difference of the Fermi levels in the two doping regions (typically a few hundred mV). However, the depth of the depletion zone can be increased by applying a DC bias voltage (negative pole connected to the p doped, and positive pole to the n doped region). This configuration is known as a reverse-biased diode. In that case $\Delta V = \Delta E_F/e + V_{bias} \simeq V_{bias}$, i.e. a bias of 100 V will increase the depletion depth by more than a factor 10.

The simplest silicon micro-strip detector geometry consists of a heavily p-doped (p^+) region on a lightly n-doped bulk material. For example, $n_a \simeq 10^{15}$ cm^{-3} and $n_d \simeq 10^{12}$ cm^{-3}, so that $\Delta E_F \simeq 0.42$ eV and $d_p = 0.02$ μm and $d_n = 23$ μm without bias. With a bias voltage of 100 V this becomes $d_p = 0.4$ μm and $d_n = 363$ μm. In practice the intrinsic bias voltage is negligible compared to the applied bias voltage.

Figure 9.2: Simple pn junction model.

The notation n^+ (p^+) refers to very high doping concentration for n-type (p-type) material.

We usually want to collect a fast signal from the full detector depth, which requires a large electric field throughout the detector depth, i.e. the bias voltage must be at least as large as the depletion voltage. If the bias voltage is equal to the depletion voltage, there will be a region of low electric field and therefore the drift speeds of the charge carriers will be slow in these regions. This results in slower signals with 'tails' at long times. Therefore, in practice we want to operate the detectors with a larger bias voltage ('over-depleted') to increase the charge collection speed. The limited depletion depth of silicon detectors of the reverse-biased diode type limits the usable thickness of such silicon sensors to a few 100 μm.

The depletion voltage is defined as the minimum voltage required to bias the full depth of the detector.

This is studied in exercise 5.

9.2 FABRICATION OF SILICON DETECTORS

Nearly all the silicon detectors we will consider require HV to operate. We typically need a bias voltage of $\mathcal{O}(100$ V$)$. For a detector to work reliably, care must be taken to ensure that all components in the system are compatible with this high voltage, including the actual sensors. For

For extreme radiation levels like at the LHC we need higher values of the bias voltage, typically $\mathcal{O}(500$ V$)$.

example, the design of the sensors must minimise sharp edges where the electric field will be enhanced, which can result in local increases of the leakage current, called micro-discharges, that can, in the worst case, lead to electrical breakdown. The silicon detectors need to be rigorously tested to weed out devices with excess leakage current from HV breakdown.

Breakdown occurs when the leakage current becomes too large even for small voltages, and exceeds the capabilities of the power supply.

The production of silicon sensors requires similar steps to those used in the electronics industry for the fabrication of CMOS chips. However, the minimum feature sizes in silicon detectors are far larger than used in modern CMOS chips and the dopant concentration in the bulk has to be much lower. The other important difference is that for a sensor the full wafer has to be exposed at once as opposed to the CMOS production of chips in which a small area is exposed and a step and repeat process is used.

For a more detailed discussion of the process steps required, see [464, 316, 435].
The starting material for silicon detectors has to be high resistivity silicon (see exercise 11).

Silicon with the required purity is grown in cylindrical ingots, which are sliced into disk-shaped wafers. The surfaces are then lapped and polished. The silicon atoms at a bare silicon surface are missing one silicon atom and this creates electrically active interface traps ('dangling bonds'). These traps will reduce the carrier lifetimes and the charge collection efficiency, as well as create surface leakage current [76]. To prevent this, sensor manufacture starts with a thermal growth of a thin layer ($\mathcal{O}(1\ \mu\mathrm{m})$) of silicon dioxide ($SiO_2$) on the surface of the silicon. After this process oxygen atoms in this layer saturate most of the traps. A further reduction in the trap density can be achieved by adding hydrogen chloride to the oxygen [302].

Silicon detectors have been produced by specialist manufacturers. In section 9.10 we will also consider very different designs of silicon detectors based on CMOS imagers. CMOS imagers are used in cameras on nearly all mobile phones. They are widely used in industry and in medical applications. Their use in astrophysics is also growing.

This process is called 'passivation'.

As the silicon detectors are typically rectangular in shape they need to be cut out ('diced') from these wafers, for example using a diamond-tipped saw. Whichever technique is used, the cutting will inevitably result in ragged edges, potentially resulting in large electric fields that will make the edges conductive. Therefore, we need to prevent electric fields at the edge of the detectors. For p^+-in-n sensors we can achieve this with p^+ 'guard rings' at the edge of the detector (see also Figure 9.6), which decrease the field strength towards the edge. The outer guard ring has the same voltage as the 'backside' to minimise leakage current at the edges and prevent any conductive path from the high voltage side to the ground via the edge of the detector [122]. The drawback of this design is that there will be small regions at the edge of the detector which are not biased and therefore inefficient ('dead regions').

The surface leakage current of the sensors is very dependent on the quality of these processes. Even with passivation silicon sensors are very sensitive to mechanical damage. A surface scratch can result in electrical shorts. Shorts can also be created by chemical contamination. Bond pads need to be kept very clean for wire bonding. Therefore vacuum pickup tools are used and the sensors are kept in clean rooms.

The guard rings are typically floating, and acquire an appropriate potential by 'punch-through'. The potential of the floating implant adjusts so that it's just enough that the depletion zone from the adjacent (biased) implant extends to it and the potential difference is high enough to make charge carriers flow from the floating implant to the adjacent (biased) implant.

Strip detectors have readout strips that are usually AC coupled to the pre-amplifier. This is done with integrated capacitors using an insulating layer. For strip detectors a SiO_2 layer is often used to create the AC (capacitive) coupling of the p^+ implants to the readout electrodes to isolate the amplifiers from the leakage current.

The contacts are usually made of aluminium, which is connected to the readout amplifiers by wedge bonding. Similarly, aluminium is used for the bias voltage connection to the backplane, with a layer of n^+ implant to avoid a Schottky barrier and improve the ohmic contact.

The patterning to create the detector structures uses photolithography (as used for the fabrication of CMOS chips). The desired material is applied to the surface of the silicon and covered by a photoresist. The

A Schottky contact is a contact between a metal and a semiconductor. It results in a barrier for charge carriers due to the continuity of the Fermi levels between the two materials. The height of the barrier depends on the doping concentration in the semiconductor.

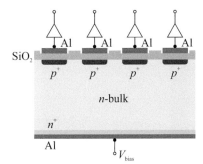

Figure 9.3: Conceptual cross-section (perpendicular to the length of the readout strips) through an AC-coupled p-in-n strip sensor (not to scale). The bias voltage is applied to the aluminium backplane.

photoresist is exposed by a UV source through a mask and then the exposed (or non-exposed) photoresist is removed chemically. Chemical etching is used to remove the material in the gaps of the mask, thus producing the desired structure. This process is used for

A photoresist is a light-sensitive polymer that becomes soluble when exposed to UV light.

- aluminium for surface electrodes used for connections by wedge bonding;

- polysilicon for creating resistors;

- silicon nitride (or silicon oxide) to create coupling capacitors;

- silicon nitride (or polyimide or silicon oxide) used to protect the sensors from mechanical or chemical damage.

Wedge bonding is a standard industrial process that typically uses 25 μm thick aluminium wire for the electrical connection. The wire is connected to the aluminium electrodes by applying a pulse of ultrasonic energy and pressure which results in a welded connection. Wedge bonding is done using standard machines from the semiconductor industry, which use a semi-automated process to achieve a high throughput.

For silicon detectors we need to introduce selective p or n-type doping into the high purity silicon. A common way to do this is to use electrostatic acceleration of suitable ions (e.g. B for p-type doping) to inject the ions into the silicon. The range of the ions in silicon varies with energy. Therefore the doping profile can be adjusted by varying the energy of the ions. The mask thickness for these steps is chosen to stop all the ions so that only the desired regions are doped [435].

Additional layers to the SiO_2 passivation layer are needed for practical sensors.

One disadvantage of the p^+-in-n approach is that most of the signal comes from drifting holes rather than electrons (see exercise 5). As discussed in section 4.3 the mobility of holes is lower than for electrons, while the carrier lifetimes are similar for electrons and holes. Hence, as holes spend more time drifting than electrons, they are more likely to be trapped during their passage through the bulk (see section 9.11). For n^+-in-n or n^+-in-p sensors most of the signal comes from the drifting electrons and these types of sensors are therefore more radiation tolerant than p^+-in-n sensors.

Ionising radiation will create electron-hole pairs in the oxide. The high mobility electrons will drift and be collected by a positively charged electrode. However, the low-mobility holes move only until they reach deep traps at the Si/SiO_2 junction. After further irradiation, this creates an electron accumulation layer. This causes one issue with n^+ implants in that the electron accumulation layer will result in the n^+ implants becoming electrically connected (at the level of kΩ). This can be avoided by introducing p-stops or p-spray between the n^+ implants. For p^+-in-n detectors, the accumulation will improve the inter-strip isolation and therefore the additional processing is not required.

The difference between p-stops and p-spray is the dose and depth of the doping. p-stops are made using ion implantation with a mask. p-spray is typically deposited at lower dose and does not require the use of an additional mask. It thus requires one less processing step.

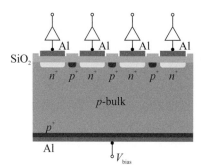

Figure 9.4: Conceptual cross-section (perpendicular to the readout strips) through an AC-coupled n-in-p strip sensor with p-stops (not to scale).

Again, to avoid excessive leakage currents or breakdown at the edges of the sensor guard rings are employed. For n^+-in-p they again need to be on the side of the readout n^+ implants, but for n^+-in-n they are also needed on the backplane, requiring additional processing steps for the sensor and thus increasing the cost [192].

For the first generation of the LHC silicon detectors the choice was to use the cheaper p^+-in-n sensors for the large area micro-strip detectors and the more radiation tolerant and expensive n^+-in-n for the smaller pixel detectors that need to survive higher fluences. However, for the upgrades for the HL-LHC, even the micro-strip detectors require better radiation tolerance and therefore, n^+-in-p sensors will be used for both the micro-strip and the pixel detectors there.

As discussed, n^+-in-p sensors have better radiation tolerance than p^+-in-n sensors but do not require double-sided processing and are therefore cheaper than n^+-in-n sensors.

9.3   SILICON STRIP DETECTORS

In silicon strip detectors the readout is segmented into strips that are typically of the order of 100 μm wide and several cm long. This design aims for good position resolution perpendicular to the strips while limiting the overall number of readout channels. The drawback of this geometry is that it does not give useful information about the location of hits along the direction of the strip.

Exercise 2 investigates how the position resolution in the direction of the strips can be improved.

In a strip sensor the aluminium readout strips are connected to the readout ASICs using wedge bonding. The width of these bonds makes it difficult to reduce the pitch below ∼50 μm on the width of the strips that can be used.

ASIC is the acronym for application-specific integrated circuit, a custom-designed electronic chip.

The connection of the bias voltage requires very good filtering close to the detector as we use single-ended amplifiers as discussed in section 3.3, which makes the readout very sensitive to noise on the high voltage (HV). For this type of single-sided detector the HV is connected directly to the backplane (i.e. the opposite side to the junction) of the detector.

This will typically be some type of RC filter.

For AC coupled detectors, a DC conductive path for the leakage current is required, which is provided by a bias resistor. As each strip needs such a resistor it is not feasible to use discrete components, so the resistor has to be built into the silicon sensor. A common way to do this is using polysilicon resistors which have sheet resistances of $\mathcal{O}(100 \text{ k}\Omega/\square)$[1]. To

The aluminium layer allows easy electrical connection to the detector (using either conductive glue or wedge bonding).

A typical silicon microstrip detector has $\mathcal{O}(1000)$ channels.

Polysilicon is a high purity polycrystalline form of silicon.

---

[1] Sheet resistance is the resistance of a square piece of a thin material with contacts made to two opposite sides of the square. It is common to use the unit $\Omega/\square$ or 'ohms per square', which is dimensionally equal to an ohm, but is exclusively used for sheet resistance. It is obtained when the resistivity of the material (in $\Omega\text{m}^2/\text{m}$) is divided by the thickness of the conductor (in m).

achieve good coupling of the signal into the amplifier the resistance needs to be much larger than the input impedance of the amplifier at signal frequencies. We also want to minimise the thermal noise in the resistor (see eq. (3.5)). However, too large a resistance would result in excess voltage drop and power consumption. Typically a resistance of $\mathcal{O}(1\ \mathrm{M\Omega})$ is used. This requires a very high ratio of length to width, which is achieved with a long meander.

It is difficult to achieve a very precise value of the resistance.

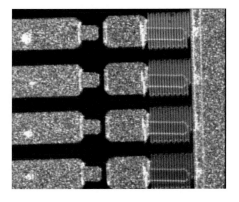

Figure 9.5: Photograph of a polysilicon resistor from a CMS sensor for the HL-LHC upgrade [319].

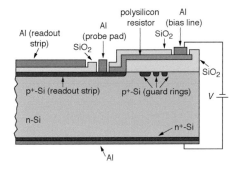

Figure 9.6: Schematic cross-section (along a readout strip) showing the edge region of one channel of a silicon microstrip detector [319]. The three regions of SiO$_2$ shown are grown at different stages of the fabrication.

SILICON DETECTOR SIGNALS

As the depleted region is not very deep, the duration of the signals is short (of order 10 ns), and the readout electronics is usually not fast enough to resolve the time structure of the signal.

One reason to keep the drifttime short is that this reduces the likelihood of charge trapping, and thus improves the charge collection efficiency.

The drift field in a silicon detector resembles the field in an ionisation detector as discussed in section 3.1 over most of the volume, apart from the region around the strip implant, where the density of field lines increases. The weighting field for one strip is more complex, as it extends into the volume depleted by the adjacent strips.

As a consequence, a charge induces a current signal on the strip on which it will arrive that is unipolar and that increases as the charge gets closer to the strip. On the other hand, a mobile charge (electron or hole) liberated close to the backplane in the volume covered by an adjacent strip will first create a similar signal, but as the charge is approaching the adjacent strip the polarity of the induced current signal will reverse, as the drift and the weighting fields become anti-parallel (see Figure 9.7). The net charge induced in the first case if the drifting charge is starting close to

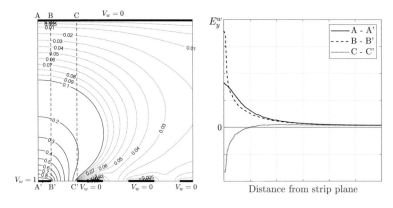
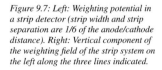

Figure 9.7: Left: Weighting potential in a strip detector (strip width and strip separation are 1/6 of the anode/cathode distance). Right: Vertical component of the weighting field of the strip system on the left along the three lines indicated.

the backplane will equal the drifting charge, whereas it will be zero in the latter.

As in an ionisation event electrons and holes are produced together both contribute to the signal. The dominant contribution comes from the type of charge carrier that drifts to the implant strip, because of the high value of the weighting field there, i.e. holes in the case of p^+ strips and electrons for n^+ implants. Therefore, the time structure is affected by the different mobilities for the two types of charge carriers.

An experimental technique to study properties of silicon detectors like the field distribution, the efficiency of charge collection, the drift velocity and the lifetime of charge carriers, is the transient current technique (TCT), which is the measurement of time-resolved current pulse shapes in silicon detectors induced by laser light pulses (with wavelengths of 660 or 1064 nm) of sub-nanosecond duration and a pulse power corresponding to from a few to about 100 minimum ionising particles (see for example [318]).

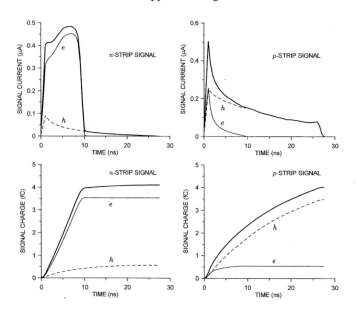

Figure 9.8: Strip detector signals for an p-in-n device with segmented strips on the p and n-sides of the sensor with 60 V depletion voltage operated at a bias voltage at 90 V for a uniform ionisation density in the sensor [465]. Shown are the electron (e) and hole contributions (h), and the total signal (thick lines).

The time response of a simple silicon pad detector to minimum-ionising particles for different bias voltages is considered in exercise 5.

ELECTRONIC READOUT

We have discussed the optimisation of the amplifier readout for low noise in section 3.6. As shown there, a cascode amplifier has the benefits of high gain and high bandwidth, and is therefore well suited for the pre-amplifier in a silicon system.

In silicon detectors it is common not to use a standard differential architecture of the amplifier chain, but single-ended inputs. The major advantage is lower power for the same noise performance. The fundamental noise for a differential amplifier is $\sqrt{2}$ times larger than for an equivalent single-ended amplifier (we assume that there is no correlation in the noise for the two inputs). Therefore, to achieve the equivalent noise performance we would need to increase the current for each transistor by a factor of two. The resulting differential amplifier would need 4 times the current as for the equivalent single-ended transistor.

The noise scales as $\sqrt{1/g_m}$ and the transconductance g_m scales like $\sqrt{I_d}$.

The drawback of the single-ended approach is that the amplifier is more susceptible to interference noise, such as noise from the power supply. This is quantified by the Power Supply Rejection Ratio (PSRR),

$$\text{PSRR} = 20\log\left(\frac{\text{Power Supply Variation}}{\text{Input Offset Voltage Variation}}\right). \qquad (9.5)$$

This expression gives the PSRR in dB.

The value of the PSRR quantifies how supply power fluctuations create a change in the input offset voltage. It is essential to use a suitably low-noise power supply and to have low-pass filtering on the power lines near the amplifier ASIC.

The input offset voltage of an amplifier is the voltage that must be applied between the two input terminals of the amplifier to obtain zero output voltage.

DATA TRANSFER

After suitable amplification there are three options for a silicon detector system for how to process the data on the detector and transmit it to off-detector data acquisition systems:

- *Binary:* For each detector element a digital signal with two possible states, logical '0' (no-hit) or '1' (hit), is produced.

  The output is typically the result of a comparison with a threshold.

- *Digital:* The signal from the output of each amplifier is digitised with more than one bit, and the digital output is transmitted.

  This architecture is used extensively in fields where amplitude information is required, e.g. in spectroscopy.

- *Analogue:* The amplitude of the pulse is converted to a proportional optical signal that is transmitted over optical fibre to the off-detector data acquisition system where it is digitised.

The first option is the simplest. In the most naïve version the output of the amplifier for each detector element (strip or pixel) is compared to a common discriminator level for all channels in one front-end ASIC. The problem is that there are usually significant channel-to-channel gain variations and these need correcting in order not to degrade the effective S/N.

Crucially, it is also the cheapest option and has been adopted by both ATLAS and CMS for the strip tracker upgrades for the HL-LHC.

For binary readout, the discriminator threshold is typically set by a Digital-to-Analogue Converter (DAC). It is not practical to have an independent DAC with the required number of bits (i.e. dynamic range) for each element. This problem can be solved by having a global DAC for each ASIC with the required number of bits and adding a 'Trim DAC' for every channel with a limited number of bits that are sufficient to compensate for the channel-to-channel variations.

Front-end refers to the electronics on the detector, in this case directly coupled to the silicon sensors. Back-end refers to the electronics in a 'counting room' which will be $\mathcal{O}(100\,m)$ away from the detector and not be subject to radiation.

As we have seen in chapter 3, there is a trade-off between speed, power and S/N. For the case of LHC detectors, there is an irreducible background of signals from additional proton-proton collisions in the triggered bunch crossing (called 'in-time pileup'). However, if the pulse

For example the readout ASIC for the ATLAS SemiConductor Tracker (SCT) has an 8-bit DAC and a 4-bit-trim DAC [177].

shaping is too slow, there will be genuine signals from the wrong bunch crossing ('out-of-time pileup').

One strategy used at LHC to minimise the power consumption and achieve the required performance is to allow the pulse-shaping to be such that signals extend over two bunch crossings. However, only strips which have a hit in the triggered bunch crossing and no hit in the preceding bunch crossing are read out into the buffer.

The second option, digital readout, requires a low power ADC for every channel. It enables the measurement of variations in the baseline level across a sensor/amplifier and thus slow drifts in time can be measured and corrected for. However, this option increases the power consumption and complexity for the ASIC compared to simple binary readout. The resulting increase in chip area and the higher bandwidth required for the data transmission to the counting room increase the costs.

The third option, analogue readout, which for example was used by CMS for the first generation of their silicon tracker, is attractive in that it minimises on-detector electronics and retains the possibility of correcting for baseline fluctuations in the off-detector receiver electronics. In the CMS case the analogue data is transmitted at 40 Msamples/s with an equivalent resolution of 8 bits. The data transmission requires analogue optical links with excellent linearity. While this system works well, the two key disadvantages with respect to the binary option are that the edge emitting lasers (EELs) used for the transmission are not as radiation-tolerant as the vertical cavity surface emitting lasers (VCSELs) commonly used for transmitting digital data and that it requires a much larger number of optical links.

Digital or analogue readout allow to calculate the charge-weighted centroid of a cluster of strips. This will result in an improved spatial resolution compared to that of the simple binary readout.

A buffer in this context means a memory used to temporarily store data before the data is transferred to another location. Therefore the use of the slower and lower power pre-amplifier does not increase the 'pile-up' from the previous bunch crossings. At high luminosity it will cause a small decrease in hit efficiency, $\mathcal{O}(1\%)$.

Good linearity enables the use of a simple calibration scheme. Electrical data links with this bandwidth would not be practical. This was achieved using EELs operating at 1310 nm and single-mode fibre.

VCSELs operate at 850 nm and are coupled to multi-mode fibres.

PIPELINE AND BUFFER

In applications with low channel counts and low rates each detector channel can simply be connected to a data transmission channel. However, at hadron colliders such as LHC the events are acquired at the beam bunch crossing rate of 40 MHz and it is generally not practical to read out the detectors at this rate. A trigger system is therefore required to reduce the data rate. The first trigger level, which is typically implemented in hardware, requires several μs to make a decision as to whether to accept or reject an event corresponding to a particular bunch crossing. Therefore the data are stored on the detector in a 'pipeline' memory while the trigger decision is being made.

A pipeline memory can be implemented as a circular memory. For the case of LHC detectors, data corresponding to each bunch crossing are written into a different cell, i.e. the write pointer advances by one cell every bunch crossing. If the depth of the pipeline is N, then after N bunch crossings new data is written into the cells. The external hardware trigger must decide whether to keep or reject the data from every bunch crossing. Clearly this decision must be made in less than a period equal to N times the spacing between bunch crossings.

For a discussion of triggers see chapter 13.

For the LHC bunch crossing frequency of 40 MHz a typical pipeline depth $N = 128$ gives 3.2 μs for a trigger decision.

If the data from a particular bunch crossing is accepted by the trigger, this bunch crossing is flagged and the data are not overwritten. The corresponding data are read out into a 'de-randomising' buffer for further processing and readout to the off-detector electronics. For the case of binary or digital data, this processing will involve 'sparsification', i.e. only reading out hit cells. In addition to further compress the data, neighbouring hits are gathered into clusters and the cluster information rather than the data from individual elements is read out. For a binary system this does not represent a loss of data, it is just a more efficient representation.

In general, arrival times of data at the processing facility are random. Without buffers, the processors remain idle after reading out one event while they wait for the next, and the system is inefficient. Adding buffers equalises the rate with which data arrives, with the result that the processors are kept busy nearly full-time, as if they were presented with a continuous stream of events matched to the processing rate ('de-randomised'). See also [295].

9.4 CALIBRATION OF SILICON STRIP DETECTORS

In large particle detector systems with millions of channels it is essential to determine optimised settings for the operational parameters (thresholds etc.) for each channel with an appropriate calibration system. These systems usually have too many channels to calibrate with sources or particle beams and therefore the calibration circuits are integrated into the readout ASICs.

The ATLAS and CMS silicon strip trackers have $\mathcal{O}(10^7)$ channels and the pixel detectors have $\mathcal{O}(10^8)$ channels.

Calibration is required to ensure the full functionality of all modules before installation in the overall detector. In addition, there will be variations in the performance of the modules over time, due to slow drifts in the electronics, for example due to changes in environmental conditions (typically temperature). If the detector is exposed to significant radiation (e.g. in the LHC detectors) the performance of the sensors and associated front-end electronics will be degraded over time. It is essential to track these changes and to adjust operational parameters. In principle this could also be achieved using data with and without beam present, but this would take far too long and would waste valuable beam time. Therefore built-in electronic calibration systems are used.

By a module we typically mean a silicon sensor and the associated readout circuitry.

The electronic calibrations can be performed in periods without beam and therefore do not waste any beam time.

An accurate calibration is required for all channels of a silicon detector to adjust for channel-to-channel variations in the sensor and electronics. Calibration systems are required to measure the gain and noise for all channels as well as, where possible, the response with no signals ('pedestals'). To know that a module is working and that the operational parameters are set correctly, we need to be able to measure the response to a known input charge to confirm a high efficiency for detection of charged particles. We also need to be be able to measure the ENC in absolute units (fC or number of electrons) to verify a satisfactory S/N. An accurate calibration is required for all channels of a silicon detector to adjust for channel-to-channel variations in the sensor and electronics.

We will discuss the calibration of analogue and binary systems. The calibration procedures are different for silicon strip systems with analogue or binary readout.

An example of a calibration procedure for a digital system is described in [193].

CALIBRATION OF ANALOGUE SYSTEMS

We will consider the calibration of the CMS silicon tracker [196] as an example of the calibration procedures required for an analogue system. The CMS silicon detectors use analogue optical links to transfer the data from the output of the front-end ASIC (APV25) to the back-end electronics in the counting room. There the signals are digitised at a rate of

40 Msamples/s, with an 8-bit resolution. Baseline levels with no signal (these are called 'pedestals') are measured for every strip using data with no beam. These pedestal values are subtracted from the measured data during physics operation for each channel. In addition, the median level of the ADC values for all the channels in each ASIC is measured and used to correct for common-mode noise.

Common-mode noise is coherent noise which has the same value for all channels in a given event (see section 3.2). It can vary from event to event and one of the attractive features of the analogue readout is that it allows for this common-mode noise correction.

The next aspect to the calibration is to determine the overall gain. This depends on the optical system as well as the APV25 and sensors. For this purpose, each event readout contains a 'tick mark' at the end of the analogue data that corresponds to a large and known charge, see Figure 9.9. The gain is measured using the ADC response to the tick mark after pedestal subtraction. The variation in gain calibration is further reduced using the measured pulse heights of clusters on tracks from collision data.

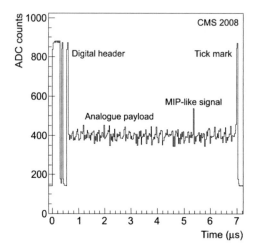

Figure 9.9: APV25 raw data in ADC counts, showing a digital header, analogue data and a tick mark at the end. The analogue data corresponds to the multiplexed data for one triggered event from two ASICs, with one channel carrying a signal from a minimum-ionising particle (MIP) [197].

This approach was validated at test beams using particles of known type and energy. The final validation uses in-situ data, for example cosmic ray muons (before the LHC started operation). The S/N of clusters of hits on reconstructed tracks is consistent with expectations.

By in-situ data we refer to data that is collected during operation of the full detector.

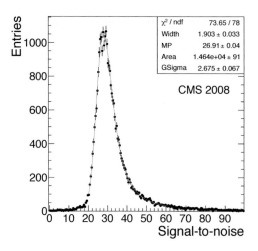

Figure 9.10: Distribution of S/N for clusters in the innermost barrel strip detector layer of the CMS tracker [197].

CALIBRATION OF BINARY SYSTEMS

Here we will consider the calibration system of the ATLAS ITk strip detector [404]. The calibration of binary readout systems is not as straightforward as for analogue systems. In a binary system, we can measure the ratio of the number of hit strips over the total number of strips (this is called the 'occupancy') for a given value of the discriminator threshold. If the noise has a Gaussian distribution, the variation of occupancy with threshold will be of the form of an error function. From a fit of the occupancy as a function of threshold we can determine two parameters: the value of the threshold at which the occupancy is equal to 50% (called $VT50$ in ATLAS) and the σ of the Gaussian noise distribution.

In a binary system we lack the data to determine a pedestal or the response to a known input charge from the measured ADC values.

The error function is the integral of the Gaussian distribution.

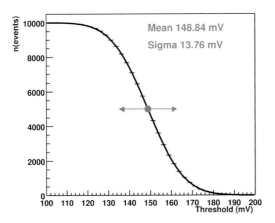

Mean 148.84 mV

Sigma 13.76 mV

Figure 9.11: Simulated scan of occupancy as a function of discriminator threshold and the error function fit for the ATLAS ITk strip system [404].

This scan is repeated for different values of known input charge, Q_i, and the corresponding values of $VT50_i$ and σ_i are determined from an error function fit for each of the scans. The gain at an input charge of q_i can then be determined as $G_i = VT50_i/Q_i$. In general this relation will not be linear.

The charge is injected by a fixed voltage over a capacitor of known value in the front-end ASIC.

The values of the Gaussian σ_i give an estimate of the noise at the output of the amplifier. With the gain obtained from the calibration, the input noise can be found from $\sigma_{\text{input},i} = \sigma_i/G_i$.

The amplifier is designed to give a linear response for typical signals expected from MIPs, but to have lower gain for larger charges. This results in a lower power consumption compared to an amplifier that is linear over the full range.

Once the gain and input noise are known, the discriminator threshold can be set low enough to give a high efficiency for MIPs, but large enough to be far enough above the noise.

The only way to be sure that these complex calibration procedures are correct is to study the response to charged particles. Therefore, the performance of modules is validated in test beams in which modules are exposed to particles of known energy. An example is shown for an ATLAS SCT module in Figure 9.12. The plots show the hit efficiency (for particle detection) and noise occupancy as a function of threshold for an unirradiated module [176]. The results show a wide range of threshold values which achieve a hit efficiency above 99% while maintaining a noise occupancy below 10^{-4}.

This calibration procedure needs to be performed on all channels. Therefore each channel has an adjustable discriminator threshold and its own charge injection circuitry.

This validates the intrinsic module performance and the calibration systems.

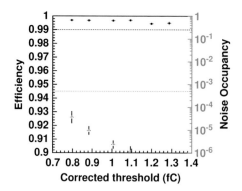

A similar study was performed for an irradiated module and that showed the expected degradation from radiation damage to the sensor and ASIC. However, it was still possible to find a suitable operating window (i.e. threshold setting) in order to achieve the required low noise and high hit efficiency [176].

Figure 9.12: Plots of efficiency versus threshold (upper data points and left hand axis) and occupancy versus threshold (lower data points and right hand axis) for an unirradiated SCT module [176].

9.5 LEAKAGE CURRENT

The signal in silicon detectors is the current induced by the charge carriers that have been promoted to the conduction band or created in the valence band by the interaction with the passing charged particle. However, even in the absence of such an interaction and even if the *pn* junction is fully depleted, a current will flow due to thermally created electron/hole pairs. Leakage current will be present in all silicon detectors, but it increases significantly after irradiation.

This current is called 'leakage current' or 'reverse bias current'.

We consider fully depleted detectors, i.e. we have used an electric field to remove the mobile carriers. The leakage current arises from thermal energy promoting electrons from the valence band to the conduction band, as any electrons in the conduction band will drift under the influence of the electric field.

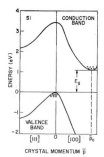

Again, we will focus the discussion on silicon. Crystalline silicon is an indirect band gap semiconductor. Hence transitions of electrons from the top of the valence band to the bottom of the conduction band are highly suppressed by the need to simultaneously conserve energy and momentum. Therefore the dominant mechanism for electrons to be promoted to the conduction band is via energy levels within the band gap. Such energy levels will exist because of defects in the silicon crystal but they are also created by radiation damage.

Figure 9.13: Energy band structure of silicon [349].

Such intermediate states are called 'traps'.

GENERATION-RECOMBINATION THEORY

In this section we will first consider the steady state achieved when charges are continuously injected into the conduction band.

For example, this could be done by shining laser light into the silicon. The theory was first worked out in [277] and [456]. Our discussion is based on the textbooks [316], [349] and [464].

There are four processes involving transitions from the conduction or valence band to/from trap centres (energy levels in the forbidden gap):

- *Emission* of *negative* carriers (electrons) from a trap centre to the conduction band with rate r_e^n;

- *Capture* of *negative* carriers (electrons) by a trap centre from the conduction band with rate r_c^n;

- *Emission* of *positive* carriers (holes) from a trap centre to the conduction band with rate r_e^p;

- *Capture* of *positive* carriers (holes) by a trap centre from the conduction band with rate r_c^p.

The rates for these four processes will be proportional to the number of available states. For capture of negative carriers by a trap centre, this involves the volume number density of negative charges, n_n, the volume number density of trap centres, n_t, and the occupation probability for the trap, f. The volume number density of the available trap centres is $n_t(1 - f)$.

Capture of positive carriers by a trap centre is equivalent to the emission of a negative carrier from the trap centre. The rate is therefore proportional to the number of occupied trap centres $n_t f$ and the volume number density of positive carriers, n_p. Therefore we can write down the capture rates per unit volume of negative and positive charges as

The occupation probability is given by a Fermi-Dirac distribution, however we can safely approximate this by the Maxwell-Boltzmann distribution as in section 9.1.

$$r_c^n = c_n n_n n_t (1 - f) \quad \text{and} \quad r_c^p = c_p n_p n_t f, \qquad (9.6)$$

where c_n and c_p are constants given by the product of the speed v and the appropriate capture cross-sections,

$$c_n = \int_0^\infty \frac{dn}{dv} \sigma_n \, dv, \qquad (9.7)$$

where dn/dv is the distribution of speeds, with the equivalent definition for c_p.

We can write down the equivalent formulae for the emission probabilities of negative or positive carriers

$$r_e^n = e_n n_n n_t f \quad \text{and} \quad r_e^p = e_p n_p n_t (1 - f), \qquad (9.8)$$

where the emission probabilities e_n and e_p are constants that we can determine by considering the situation without charge injection, i.e. in thermal equilibrium. In this case the rates of electron (hole) capture and electron (hole) emission must be identical.

This is the principle of detailed balance.

In *n*-doped silicon, we can write the negative carrier density in terms of the intrinsic carrier density n_i as

$$n = n_i \exp\left(-\frac{E_F - E_f}{k_B T}\right), \qquad (9.9)$$

where E_f (E_F) is the Fermi level for intrinsic (doped) silicon (see exercise 4). Combining this result with equations 9.6 and 9.8 gives relations between emission probability and capture coefficients,

$$
\begin{aligned}
e_n &= c_n n_i \exp\left(-\frac{E_f - E_t}{k_B T}\right), \\
e_p &= c_p n_i \exp\left(-\frac{E_t - E_f}{k_B T}\right).
\end{aligned}
\qquad (9.10)
$$

When we are injecting charge at a rate R_{inj} we no longer have thermal equilibrium but there is still a steady state, i.e. the rate at which electrons are transferred to the conduction band is equal to the rate at which they leave the conduction band (with an equivalent condition for the positive holes),

$$R_{\text{inj}} + r_e^n - r_c^n = 0,$$
$$R_{\text{inj}} + r_e^p - r_c^p = 0.$$
(9.11)

We can combine the two expressions in eq. (9.11) and substitute from equations (9.6) and (9.7) to solve for the occupation probability f in non-thermal equilibrium

$$c_n n_n n_t (1-f) - e_n n_n n_t f = c_p n_p n_t f - e_p n_p n_t (1-f),$$
$$f = \frac{e_p n_p + c_n n_n}{c_n n_n + c_p n_p + e_n n_n + e_p n_p}.$$
(9.12)

We can now calculate the net recombination rate, $G_t \equiv r_c^n - r_e^n$,

$$G_t / n_t = c_n n_n - f(c_n n_n + e_n n_n).$$
(9.13)

In thermal equilibrium the recombination rate is equal to the generation rate for negative and positive charges $r_c^n = r_e^n$ and $r_c^p = r_e^p$.

Substituting for f from eq. (9.12) gives

$$\frac{G_t}{n_t} = \frac{c_n n_n (c_n n_n + c_p n_p + e_n n_n + e_p n_p) - (e_p n_p + c_n n_n)(c_n n_n + e_n n_n)}{c_n n_n + c_p n_p + e_n n_n + e_p n_p} =$$
$$= \frac{c_p c_n n_n n_p - e_p e_n n_n n_p}{c_n n_n + c_p n_p + e_n n_n + e_p n_p}.$$
(9.14)

Eliminating e_n and e_p using eq. (9.10) yields

$$\frac{G_t}{n_t} = \frac{c_p c_n \left(n_n n_p - n_i^2\right)}{c_n \left(n_n + n_i \exp\left(\frac{E_t - E_f}{k_B T}\right)\right) + c_p \left(n_p + n_i \exp\left(\frac{E_f - E_t}{k_B T}\right)\right)}.$$
(9.15)

We can simplify this result if we make the approximation $c_n \simeq c_p$, and we can approximate the result for c_n from equation 9.7 as $c_n \simeq v\sigma$ where v is the mean thermal velocity and σ the effective cross-section,

$$G_t = \sigma v n_t \frac{n_n n_p - n_i^2}{n_n + n_p + 2 n_i \cosh\left(\frac{E_f - E_t}{k_B T}\right)}.$$
(9.16)

We can now combine the results from sections 9.1 and 9.5 to determine the variation of leakage current with temperature. We consider the situation of a region of the detector that is fully depleted of charge carriers (e.g. by applying a reverse bias voltage to a pn junction). Initially, $n_n = n_p = 0$ and we can calculate the rate of creation of carriers (electrons and holes will be created in equal numbers) per unit volume,

$$\frac{dn}{dt} = \sigma v n_t \frac{n_i}{2 \cosh\left(\frac{E_f - E_t}{k_B T}\right)}.$$
(9.17)

The leakage current in a volume V is

$$I_{\text{leak}} = \frac{dn}{dt} eV,$$
(9.18)

and substituting from eq. (9.17)

It is common to write $I_{leak} = e n_i V / \tau_g$, where τ_g is defined as the generation time constant. This is the exponential decay constant that describes the return to equilibrium $n_n = n_p = n_i$.

$$I_{\text{leak}} = \sigma v n_t \frac{n_i}{2 \cosh\left(\frac{E_f - E_t}{k_B T}\right)} eV. \tag{9.19}$$

We can see that the leakage current will be dominated by trap levels with an energy E_t close to the Fermi energy E_f. If we assume that $E_f \simeq E_t$, then the temperature dependence of the leakage current variation with temperature is given by the variation of n_i (eq. (9.2)) and v ($\propto \sqrt{T}$).

We neglect any variation of σ with v and the temperature dependence of the effective carrier masses.

With these approximations we find for the dependence of the leakage current (E_g is the band gap energy)

$$I_{\text{leak}} \simeq A V T^2 \exp\left(-\frac{E_g}{2 k_B T}\right). \tag{9.20}$$

A is a proportionality constant.

In case the trap levels are further away from the intrinsic Fermi level the cosh function in eq. (9.17) behaves like $\exp[-(E_f - E_t)/(k_B T)]$, and thus it is common to replace the band gap energy in eq. (9.20) with

$$E_{\text{eff}} = E_g + 2\Delta,$$

where $\Delta \simeq E_f - E_t$.

The band gap energy E_g is itself a function of the temperature and can be parameterised as

$$E_g(T) = E_0 - \alpha T.$$

For silicon, $E_0 = 1.206$ eV and $\alpha = 2.73 \times 10^{-4}$ eV/K. For the temperature dependence of the leakage current this means that the temperature dependence can be described by

$$I_{\text{leak}} \propto V T^2 \exp\left(-\frac{E_{\text{eff}}}{2 k_B T}\right) = V T^2 \exp\left(-\frac{T_A}{T}\right), \tag{9.21}$$

$T_A = E_{eff}/(2 k_B) \simeq 7000$ K is called the 'activation temperature'.

with $E_{\text{eff}} = 1.21$ eV [191].

It is common to quote the leakage current at a reference temperature (often $0\,°C$). It can then be scaled to the appropriate temperature using eq. (9.21). As a rule of thumb, the leakage current doubles for an increase in the temperature of about $7\,°C$.

The leakage current affects a silicon detector system in several ways:

- The readout electronics needs to accommodate the constant leakage current. For strip detectors the readout is typically capacitively coupled, so that the DC component of this current is disconnected from the readout electronics;

- Statistical fluctuations due to the discrete nature of the moving charge carriers lead to shot noise, which scales with $\sqrt{I_{\text{leak}}}$ (see section 3.2);

- The leakage current and the bias voltage across the *pn* junction result in ohmic heating of the silicon. This heat needs to be removed by some means of cooling. Otherwise the temperature of the junction will rise, and thus the leakage current will increase even further and the detector will become thermally unstable and go into 'thermal runaway'.

Practically, thermal runaway is limited by the ability of the power system to supply the current to maintain the bias voltage at the required sensor temperature.

The strong variation of leakage current with temperature implies that for irradiated detectors we must limit the effects of leakage current to a manageable level by operating the silicon detectors cold.

9.6 THERMAL MANAGEMENT OF SILICON DETECTORS

Because of the low mass in silicon tracking detector systems their heat capacity is small. At the same time, they typically include significant heat sources from on-detector front-end electronics and the temperature-dependent sensor leakage. The thermal management of silicon detector systems therefore has to achieve three requirements:

- Remove the heat generated by the front-end electronics and sensor leakage;

- Maintain sufficiently low temperatures to limit reverse annealing effects (see section 9.11), reduce shot noise due to sensor leakage current, and provide the required temperature for the front-end electronics components;

- Provide enough heat removal capacity at sufficiently low temperature, so that thermal stability is achieved.

Thermal stability means margin against thermal runaway.

These requirements are usually achieved by means of a local heat sink (typically a coolant) at an appropriate temperature, and a highly conductive heat path from the heat sources to the sink.

COOLING SYSTEMS

The typical power in silicon systems leaves only forced convection as a practical means to remove the heat from the silicon detector system over the large distance to an off-detector heat sink. Coolants can be in a single phase, or they can absorb the heat in a phase change. Any coolant must be chemically inert.

Monophase coolants can be liquid or gaseous. In the latter case the specific volumetric heat capacity is lower, limiting its use to systems with low front-end power, but the lower density is attractive if a low mass of the silicon system is one of the driving requirements. The heat transfer from the coolant to stationary parts of the heat path (e.g. the cooling tube wall) is higher for turbulent flow ($Re \gtrsim 2400$). In monophase cooling systems the temperature of the coolant increases along the contact to the heat source.

Not only is the density of the gas lower than for a liquid, but often gas-cooled systems do not require a container for the flow.

The Reynolds number Re is defined as $Re \equiv GD/\mu$ with G the mass velocity (in $kg/m^2 s$), D the diameter of the flow, and μ the dynamic viscosity of the coolant (in Pa).

Phase-changing cooling is attractive for the large latent heat typically associated with the phase transition. The most common type of phase-changing cooling in silicon detector systems are evaporative systems. Evaporative cooling systems can be pump-driven, compressor-driven or gravity-driven. In pump-driven systems (see for example [484]), the challenge is the subcooling required to prevent the pump from cavitating. In compressor-driven systems (see for example [105]) it is the need for oil-free compressors with a high compression ratio, as lubrication with oil risks contamination of the thin on-detector cooling pipes. Gravity-driven systems (see for example [118]) need a considerable height difference (order 100 m) to achieve the required drive pressure for typical coolants.

Cavitation is evaporation caused by local pressure drops close to the fast moving parts of the pump, which makes it impossible for the pump to generate pressure. The fluid needs to be cooled sufficiently below saturation (subcooled) to prevent this.

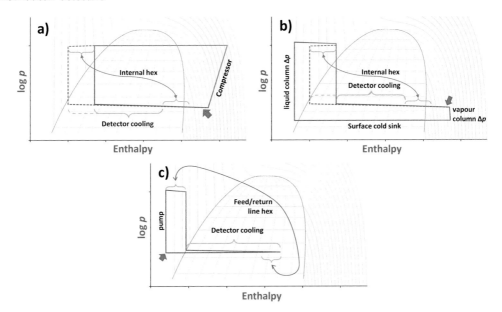

Figure 9.14: Conceptual pressure enthalpy diagrams for different cooling cycles [489]. a) compressor-driven. b) gravity-driven. c) pump-driven. The thin lines are isotherms for the coolant, and the bump is the saturation curve, separating, from low to high enthalpy, the liquid, two-phase and gaseous phase space, respectively. The loop direction is counter-clockwise in a) and clockwise in b) and c). The pressure control points in the three cycles are indicated by the thick arrows. The thin arrows indicate heat transfer via internal heat exchangers (HEXs), which can be used to increase enthalpy available for cooling (in a) and b)), or guarantee saturated conditions in the evaporator (in c)).
Pressure-enthalpy (or p-h) diagrams are useful to describe loop systems as the enthalpy ($h = u + pv$, where u is the internal energy) of the fluid stays constant in a thermally insulated flow process, and pressure drops are caused by flow restrictions (for example the pressure drop due to flow in a pipe). Exchange of heat with the environment changes the enthalpy of the fluid by the amount of heat exchanged.

In evaporative systems the evaporation temperature on the detector is controlled by the pressure in the return line on the outside of the detectors, either by a backpressure regulator or by a temperature-controlled accumulator in the cooling plant. In either case, the on-detector evaporation pressure is offset from the set value by the temperature drop corresponding to the pressure drop in the return lines from the detector to the pressure-setting element. Great care has therefore to be taken to reduce this pressure drop.

In evaporative systems the heat transfer depends on the physical properties of the coolant, but also on the geometry of the flow (the 'flow pattern'), which changes as the coolant absorbs heat. The best heat transfer is achieved for a thin liquid film along the tube wall, with a vapour core ('annular flow'). Once a significant fraction of the fluid has evaporated (typically 50% to 80%), the liquid film will lift from the wall, resulting in mist flow. At this point the heat transfer characteristics deteriorate significantly, which usually constitutes the limit of operation for evaporative cooling systems.

The physics of bi-phase flow is complex and no complete theory exists. Computational fluid dynamics methods are hampered by the difficulty to model the interfaces between the phases, and thus a wide range of semi-empirical correlations have been developed, with limited prediction power [284], some of which have been included into software to calculate pressure drops and heat transfer properties in evaporative silicon detector cooling systems (for example CoBra [488]).

The currently most commonly used evaporative cooling system design is the pump-driven, accumulator-controlled 2PACL (2-phase accumulator-controlled loop) design with CO_2 as cooling fluid [484]. The accumulator is a storage tank for 2-phase coolant in saturation, in which the pressure is set by controlling the temperature.

The feature of the line that most strongly affects the pressure drop is the inner diameter of the line, as well as its length, and to a lesser extend bends, abrupt changes in diameter, and surface quality.

This happens before complete evaporation ('dry out') of the liquid.

Because of the limitations of the correlations used it is mandatory to perform representative prototype tests to improve these predictions.

CONDUCTIVE PATH

Generally, heat conduction is a three-dimensional problem, but for benchmarking purposes the performance of the conductive thermal path can be characterised by a scalar thermal impedance. It includes the heat transfer at interfaces (including the interface from the coolant to stationary parts of the heat path).

Sometimes a 'thermal figure of merit' (TFOM) is used, which is the inverse of the thermal impedance. Attempts to normalise the TFOM by multiplying with the sensor area are usually not improving understanding, as this property does generally not scale simply by area.

The thermal impedance of a simple heat-conducting element can be described by $R = l/(\kappa A)$, where κ is the heat conductivity and l and A geometrical factors (the length and cross-section of the heat path). Low thermal impedance can therefore be achieved by materials with high thermal conductivity, and geometries with short heat path and large cross-section and contact areas.

A whole industry exists for the development of high-thermal-conductivity materials. In general, higher density materials will display better heat conductivity, but one class of materials that has high thermal conductivity, moderate density and excellent structural properties are carbonised materials (carbon fibre, but also carbonised foams). A special class of materials are thermal interface materials (TIMs), which are typically viscous substances used to improve the heat conductivity between two contacting solid surfaces by filling in microscopic features on the surfaces, which otherwise would result in a small insulating gap.

For a slightly more detailed discussion see [489].

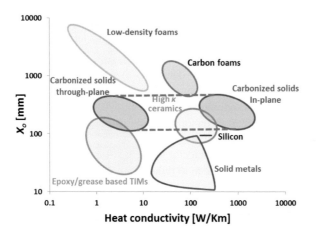

Figure 9.15: Radiation lengths versus heat conductivity for common groups of materials in silicon detectors [489].

Geometrically, the thermal impedance can be reduced by bringing the local heat sink as close as possible to the heat sources, and in particular the sensors. One approach used in modern pixel detector systems uses micro-channels, which are small-diameter cooling channels in direct contact with the sensor. These can be etched channels in a silicon substrate, which has the benefit of matching thermal expansion to the sensor. The challenge here is the number of connections required due to the limited size of the substrates, and the need to achieve leak-tight fluid connections at the required pressure to the silicon material.

See for example [220].

SENSOR TEMPERATURE AND THERMAL STABILITY

While the temperatures in a silicon detector system are generally determined by a three-dimensional heat flow, it is instructive to model the thermal behaviour using a simple linear network model using thermal impedances and heat sources. Temperatures then correspond to the potentials in the system. We will demonstrate this here for a system of a sensor (generating a thermal power Q) and associated front-end electronics generating Q_h. The first benefit of the linear network model is that the standard network rules (Kirchhoff's rules, Thevenin equivalents etc.) apply.

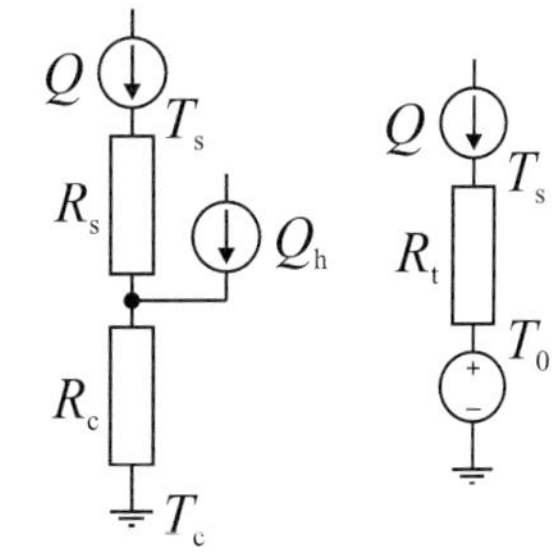

The system is thermally stable if the heat removal capability is sufficient to remove the sensor heat. Mathematically, this is achieved if a solution for the network equation

$$Q(T_s) = \frac{T_s - T_0}{R_t} \tag{9.22}$$

can be found, where $Q(T_s)$ is the sensor leakage power at the sensor temperature T_s, and R_t is defined as in figure 9.16.

Figure 9.16: Linear thermal network for a simple silicon detector system (left) and its Thevenin equivalent (right). R_s: thermal impedance for sensor, R_c: common thermal impedance, $R_t = R_s + R_c$: total thermal impedance, T_c: coolant temperature, T_s: sensor temperature.

$T_0 = T_c + R_c Q_h$.

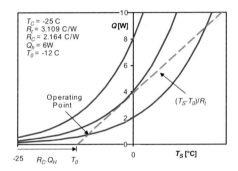

Figure 9.17: The balance of dissipated and conducted heat. The solid curves represent sensor heats of $Q(0°C) = 2$, 4 and 8 W. The dashed line represents conduction by the thermal path. Numerical values are appropriate to the ATLAS SCT barrel module [121].

For $Q(0°C) = 8$ W (the topmost curve) no solution to eq. (9.22) exists, the system is thermally unstable and thermally runs away.

As will be discussed in section 9.11, one of the consequences of radiation damage in silicon detectors is an increase in leakage current. It is therefore vital for the design of the thermal management in a high-rate environment to understand the thermal stability of the system as the leakage current increases over the lifetime of the experiment.

A relation between the leakage power at the reference temperature Q_{ref} and the sensor temperature T_s can be found using eq. (9.21),

$$Q_{ref} = \frac{T_s - T_0}{R_t} \left(\frac{T_{ref}}{T_s}\right)^2 \exp\left[T_A\left(\frac{1}{T_s} - \frac{1}{T_{ref}}\right)\right]. \tag{9.23}$$

It is standard practice to display this curve in its inverted form, $T_s = T_s(Q_{ref})$ (even though it cannot be inverted analytically).

The condition described in eq. (9.22) can be used to find an expression for the critical sensor temperature,

$$T_{s,crit} \simeq T_0 + \frac{T_0^2}{T_A}. \tag{9.24}$$

This is often wrongly called the 'thermal runaway curve'. In reality this is the stable sensor temperature curve. Runaway only occurs at the endpoint of this curve (where the slope becomes singular).

'Critical' here describes the value of a parameter at which the thermal behaviour becomes unstable.

See exercise 9.

Using the scaling law eq. (9.21) this gives the critical leakage power,

$$Q_{ref,crit} = \frac{T_A T_{ref}^2}{(T_A + T_0)^2} \exp\left[T_A\left(\frac{1}{T_{s,crit}} - \frac{1}{T_{ref}}\right)\right] / R_t.$$

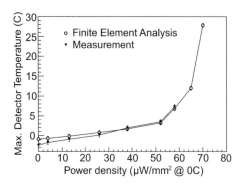

Figure 9.18: Silicon sensor temperature as a function of power density for a prototype silicon detector module for the ATLAS Semiconductor Tracker endcaps [194].

To find the critical parameters in terms of a given leakage power we start with the coolant sink temperature to achieve a sensor temperature T_s

$$T_c = T_s - R_c Q_h - R_t Q_h \left(\frac{T_s}{T_{ref}}\right)^2 \exp\left[T_A\left(\frac{1}{T_{ref}} - \frac{1}{T_s}\right)\right]. \qquad (9.25)$$

At the critical point (onset of runaway) $(dT_c/dT_s)|_{T_s = T_{s,crit}} = 0$. This yields for the critical sensor temperature

$$T_{s,crit} \simeq \frac{T_{ref}}{1 - \frac{T_{ref}}{T_A}\ln\left(\frac{T_{ref}^2}{R_t Q_{ref} T_A'}\right)}, \qquad (9.26)$$

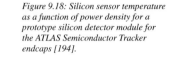
See exercise 9.

with $T_A' = T_A + 2T_s \simeq 7500$ K. Inserting this in eq. (9.25) gives for the critical cooling sink temperature

$$T_{c,crit} = T_{s,crit}\left(1 - \frac{T_{s,crit}}{T_A'}\right) - R_c Q_h.$$

Figure 9.19 summarises all the results in this section in one figure using the example of the ATLAS SCT module.

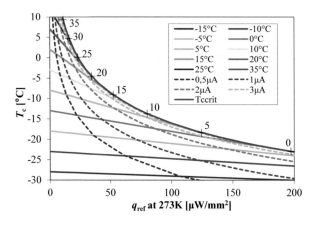

Figure 9.19: Sensor isotherms (full lines) and lines of equal strip current (dashed lines) for the ATLAS barrel SCT module from the minimal model ($R_s = 0.945$ K/W, $R_c = 2.164$ K/W, $Q_h = 6$ W, sensor area 163 cm², strip area 10.7 mm², $V_{bias} = 460$ V) as a function of leakage power density and coolant temperature. The area above the red line is excluded by thermal runaway. Crosses with numbers indicate sensor temperatures on the boundary line [121].

While we have discussed a specific linear thermal network in this section, the model is relatively generic and can be used for many thermal designs, and the conclusions can be used generally. It is straightforward to extend these concepts to create more complicated networks, which can also include other non-linear power components. At this point, the

A 3D solution using FEA is not possible for such problems anymore, and the network model is the only remaining calculation tool.

model does not necessarily have analytic solutions anymore, but can still be solved numerically with little effort (see for example [123]).

The numerical values for the thermal impedances in the model can be obtained from FEA, or from measurements of the sensor temperature for known leakage power configurations. For the best estimate of the thermal impedances in the network model the average of the temperatures over the silicon sensor should be used [123].

In both cases only as many FEA runs/measurements are needed as there are thermal impedances in the network.

The linear network equivalent can also be used to estimate the dynamic response of the thermal design. The time constant for a response can be estimated from $\tau = RC$, where R is the thermal impedance and C the heat capacity of the silicon module [121].

9.7 PIXEL DETECTORS

For many experiments at accelerators, to cope with the high track densities close to the interaction vertices and for improved resolution of their spatial reconstruction, smaller size detector elements are needed. Pixel sensors have a typical element size of 50 to a few 100 μm per side. The pixel cell is based on the same type of reverse biased junction as used for strip detectors. However, there are many additional technical challenges that have to be solved for pixel detectors, in particular for high-rate and high-radiation applications like at LHC.

Smaller pixel sizes can be achieved with CMOS sensors (see section 9.10).

Perhaps the most obvious challenge is how to connect the individual pixels to the cells of the readout electronic ASICs. As the pixel cells cover the full area of the sensor, they cannot be connected using the wedge bonding technique used by strip detectors. There is a well established industrial process for higher density interconnects than is possible with wedge bonding called 'Controlled Collapse Chip Connection' (C4) or flip-chip bonding. This process allows chips to be connected to Printed Circuit Boards (PCBs) or other chips. The steps required at the wafer level before dicing into individual chips are [435]:

- Under Bump Metallisation (UBM): A sequence of metal layers is deposited through a mask on to the implant pads to ensure good adhesion, and to provide a solder barrier and a very thin gold layer to prevent oxidation;

- Solder bumps: Solder is deposited on the UBM through a mask with typical dimensions of 100 μm $\times$ 100 μm;

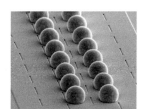

Figure 9.20: PbSn solder bumps on ATLAS read out circuit (25 μm size, 50 μm pitch) [503].

- Reflow: The wafer is heated to melt the solder and the surface tension of the liquid solder results in approximately spherical shaped solder balls;

- Dicing: The individual chips are then cut out of the wafer.

After similar solder bumps are created on the mating chip, one chip is 'flipped' so that the solder balls line up (hence the alternative name flip-chip) accurately and then heat is applied for a second reflow process. The solder balls melt and surface tension pulls them into accurate alignment.

The solder is lead-rich PbSn for the C4 process but eutectic PbSn (i.e. it melts at a lower temperature than either of its constituents) or Indium can be used for detectors.

The problem with the standard industrial process is that the minimum pitch is limited by the mask to about 250 μm, too coarse for the requirements of high resolution pixel detectors. This limitation has been overcome with the use of novel masks which allow the deposition of 25 μm diameter solder balls and thus allowing for a pitch as small as 50 μm. The two wafers need to be very accurately aligned and held very accurately parallel during this process to ensure a high yield. Failures that can occur are short circuits between neighbouring solder bumps and open circuits where the solder connection between two bumps failed.

Bump bonding is used to connect the silicon pixel sensor to the corresponding ASIC in the first step of the fabrication of a silicon pixel module.

Such a pixel detector is called a 'hybrid' pixel detector.

To cope with the effects of radiation damage in high-radiation environments, pixel detectors need to be operated cold, which creates additional stress from any mismatch in the coefficient of thermal expansion (CTE) between the bonded parts and can lead to solder connection failures. This makes the process very technically challenging and the cost of bump bonding is a significant fraction of the total cost of a pixel module.

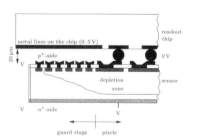

Figure 9.21: Schematics of a hybrid pixel detector with bump bonding. In a pixel module the readout chips will be connected to a flexible PCB by wire bonding.

As discussed in section 9.1, pixel detectors for high radiation environments use structures that have electrons drifting towards the pad implants, rather than holes, which improves the speed and therefore reduces the losses from charge trapping. This is achieved with n^+ implants in n^+-in-p or n^+-in-n detectors. For p^+-in-n sensors, after radiation-induced type inversion (see section 9.11), the naïve expectation is that the diode junction will grow from the backside. As most of the induced current comes from the motion of the electrons near the collection electrodes this would result in very small signals for sensors that are not fully depleted. This simple picture predicts that for n^+-in-n sensors, after radiation damage the junction would grow from the collection electrode side resulting in useful signals even for detectors that are not fully depleted. Therefore n^+-in-n sensors were used for pixel detectors in the first generation LHC detectors. The negative HV bias required for the n^+ implants is applied to the backplane of the detector so the resulting electric field accelerates the electrons towards the collection electrodes. As discussed in section 9.1, appropriate guard ring structures are required.

Near the collection electrodes the weighting and drift fields are high.

Further studies have revealed a more complex picture, in which the electric field peaks at the junction and backside of heavily irradiated sensors [233].

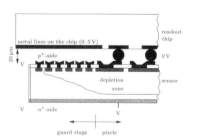
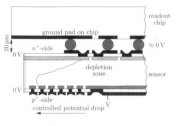

Figure 9.22: Schematic cross-sections of a p^+-in-n pixel module (left) compared with an n^+-in-n pixel module (right). This does not show the biasing structure or the inter-pixel isolation technique [435].

The biasing technique described for strip detectors using polysilicon resistors would use too much area for a pixel detector. The issue is how to connect the very large number of pixel sensor elements to a common voltage level. This can be done using the virtual earth of the preamplifier when the sensor is bonded to the readout ASIC.

It is essential to be able to test components like sensors and ASICs before bump bonding. To connect all the pixels, a 'punch-through' biasing technique is used. This technique provides a connection from a bias rail to individual pixels. As the bias voltage is increased, the depletion region around the bias rail increases until it connects to the implant region for the pixel. Thermionic emission allows current to flow from the bias rail to the pixel, thus establishing the required bias voltage on the pixel. Using this technique we can measure the current as a function of the bias voltage for unbonded sensors and reject failed sensors before further integration [316].

The key advantage of a pixel detector compared to strips apart from the obvious smaller size of the active element size is that the much smaller area of one pixel compared to a strip results in a lower capacitance and therefore lower noise (see eq. (3.11)). Another advantage of the smaller cell size is that leakage currents per channel are much smaller. Hence the shot noise contribution (eq. (3.4)) is also much reduced. In addition, the leakage currents are sufficiently low that it is not necessary to AC-couple the readout. The much lower noise implies that pixel sensors can operate successfully even after radiation damage has reduced the magnitude of the signal (see section 9.11).

The disadvantage of pixel detectors compared to strip detectors is that they will present more material per layer than strip detectors because of the bump bonds and the fact that the size of the readout ASICs is as large as that of the sensors. Pixel sensors also have a considerably higher cost per area. These advantages and disadvantages are some of the key factors that drive the choice that in present collider detectors pixel detectors are placed close to the beam, and strip detectors at larger radii.

The simple pipeline concept used for the readout of strip detectors would use up too much area to be practical for a pixel chip with a much larger channel count. However, we still need a way of storing pixel data during the latency period required by the trigger logic. One architecture transfers hit pixel address and time stamps to register the hits. When a trigger is received the hits with the correct time stamp are read out [2].

As the hit occupancy is much lower than for a strip detector, in a pixel system the signals from the preamplifiers can be slower. In that case in high intensity colliders the signal will in general be above threshold for several clock cycles. The number of clock cycles is a measure of the 'time over threshold' (TOT) and the TOT can be used as proxy for the charge deposited in the particular pixel.

SILICON DRIFT AND DEPFET DETECTORS

The strength of pixel detectors is also their biggest challenge: a very high channel density, requiring a large number of readout channels and connections between the sense elements and the electronics. For applications where high position resolution is needed, but at lower rate, alternative designs have been developed, where the charge carriers created by the incoming particle are transported within the silicon towards the edge of the

---

A virtual earth (sometimes called a virtual ground) is effectively held at ground without being directly connected. For example, consider an op-amp in a circuit using negative feedback with the non-inverting input connected to earth. The negative feedback will result in the inverting input being held at a voltage very close to earth even though the impedance between the two inputs will be very high, hence the name virtual earth.

The typical failure mode for silicon detectors is a rapid increase in detector leakage current at voltages below that required for full depletion.

Therefore one can use thinner detectors which produce a smaller signal, but have higher electric fields and faster signals as well as reduced material, which reduces multiple scattering.

The amplifiers then need leakage current compensation to prevent saturation and slow drifts as the leakage current changes.

This is called 'column drain'.

This is used to obtain higher precision centre of gravity measurements of pixel clusters than could be achieved with a simple binary approach. In addition, this allows for particle identification using the dE/dx method (see chapter 12.3) [401]. Note that this is only useful for PID at low momenta (below minimum-ionising) due to the density effect.

sensor or towards a single collection point, thus reducing the spatial dimensions of the readout by one. This information can be recovered from the arrival time of the signals.

If two planar *pn* junctions are placed back-to-back (with the p^+ layers on the outside) and connected as usual to a bias voltage through an n^+ contact, but this time at the edge of the silicon, the potential will form a potential trench across the thickness of the silicon when the bulk is fully depleted. Electrons liberated in the bulk will be contained by this potential and move towards the bias connection, drifting in the plane of the sensor to the edge. This is the principle of the silicon drift detector (SDD) [257].

This is referred to as 'sideward depletion'.

Practical silicon drift devices are made of a number of p^+ strips on both faces of the sensor at appropriate potentials to create the drift field, and a single anode n^+ strip for the readout. To obtain position information orthogonal to the drift direction the anode can be divided into pads, and the second coordinate obtained from the sharing of the charge induced in the pads.

A feature of silicon drift detectors is that the detector capacitance drops abruptly when the sensor becomes fully depleted. While in the undepleted case the capacitance is formed over the whole sensor area, in the depleted case it is given only by the capacitance from the anode to the adjacent electrodes, and thus is much smaller and becomes independent of the area. To further reduce the input capacitance to the readout amplifier, the first transistor can be implemented as a FET on the detector with the sensor anode providing the gate. The small capacitance results in a very fast signal and low noise.

Another type of detector with lateral charge carrier movement is the charge coupled device (CCD), where the individual pixels are made of CMOS capacitors. In the readout, the charge accumulated in each capacitor from the passage of ionising radiation is passed on to the next capacitor in a row controlled by a clock signal. The charge is thus moved towards the edge, where it gets amplified and digitised. This is done until the charge from all capacitors has been collected. For a long time CCDs have been the most important image sensor technology, and the have been used in the SLD vertex detector [363], but they are now mainly replaced by active CMOS sensors.

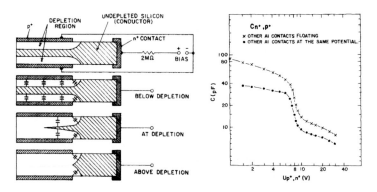

Figure 9.23: Depletion at different stages (left) and capacitance as a function of depletion voltage (right) for an SDD [257]. Shown is a bias connection at the edge. In practical detectors the p^+ electrodes are segmented into strips and the n^+ electrode is also placed on one of the facings of the sensor, but the basic principle remains.

In the SDD the cathode strips shield the anode from the drifting electrons, so that they only induce a signal on the anode when they are very close, and, like for gaseous drift chambers, the position can be inferred from the time it took the electrons to drift from the point of ionisation to the anode. Like in the case of gaseous drift chamber, calibration of the drift velocity is important, and can be accomplished by injecting charges at known locations.

The position resolution achieved in SDD position detectors is a few tens of μm. The challenge is that the material must be very pure to prevent trapping of the electrons during the long drift. Because of the low density of readout channels, and to allow for a charge sharing measurement, the readout is typically analogue.

The additional benefit of analogue readout is that the result can also be used in a measurement of the energy loss dE/dx.

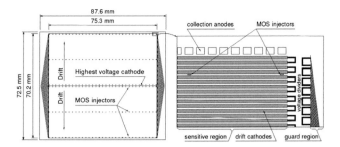

Figure 9.24: ALICE Silicon Drift Detector layout with a zoom of the top right corner [240].

Silicon drift detectors are attractive in energy-dispersive spectrometer (EDS) X-ray spectroscopy because they are fast and low-noise, achieve a good energy resolution at room temperature, and, as the devices are fully depleted and all the device thickness is sensitive for the absorption of X-rays, they have a high efficiency (for a 300 μm thick sensor the detection efficiency for 10 keV X-rays is 90%).

The EDS technique works by using a electron beam to excite electrons from inner shells. The electrons from higher energy levels will fill these holes and the energy of the resulting fluorescent X-rays can be used to identify individual chemical elements.

Such detectors are typically circular with a small central anode pad, again to minimise the detector capacitance [325]. The energy resolution and noise can be further improved by cooling the devices to a few tens of degrees below 0 °C, which can be conveniently achieved by Peltier coolers.

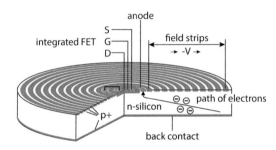

Figure 9.25: Cross-section of a cylindrical silicon drift detector for X-ray spectroscopy with integrated n-channel JFET. The gate of the transistor (G) is connected to the collecting anode. The radiation entrance window for the ionising radiation is the non-structured backside of the device [214].

The combination of a fully depleted bulk with a FET results in a sensor that produces a current signal without the movement of charge carriers, but by a static field effect [303]. Like in the silicon drift detector the ionisation charges are captured by the potential in a fully depleted *n*-type bulk but this time underneath a *p*-type FET, where they control the current between the drain and source terminals of the FET (they provide an 'internal gate' function).

Such a device is called a depleted FET (DEPFET).

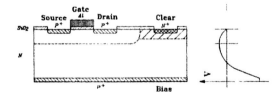

Figure 9.26: Schematics of a DEFPET detector (left) and potential within the bulk underneath the gate contact (right) [304].

The benefit of this is that the signal charge can be much lower, and thus the sensor thinner (about 50 μm) than in a conventional charge-collecting pixel sensor. As the current signal in a DEPFET is not created by the movement of charges, but just by their presence at the internal gate, the measurement is non-destructive and the device can store the charge.

The pixel detector of the Belle experiment [492] is using DEPFET detectors that are 75 μm thick in the active regions.

Only when a suitably high voltage is applied to a 'clear' contact (an n^+ electrode) the charges are drained away.

The charge storing capacity of the DEPFET and the non-destructive measurement allows for dual-FET devices, in which the signal charge can be repeatedly transferred between the two FETs, allowing for a statistically independent repetition of the charge measurement and thus a reduction of the noise to sub-electron level. This allows the resolution of the signal from single photons [351].

These are called repetitive non-destructive readout DEPFETs (RNDR-DEPFETs).

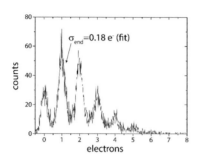

Figure 9.27: Single photon spectrum measured at low light intensity from an optical laser ($\lambda = 672$ nm) with a circular RNDR-DEPFET at a temperature of 55 °C, obtained with 300 readouts [351].

9.8 PHOTON DETECTION WITH SEMICONDUCTORS

The small band gap in semiconductor detectors makes them attractive for the detection of photons. The *pn* diodes we considered for charged particle detectors can also be used as photon detectors.

Improved photodiodes use a *p-i-n* structure, in which there is a thin intrinsic (*i*) layer between heavily *p* and *n*-doped regions. The *i* layer can be fully depleted with a low bias voltage because of the low carrier concentration and the small thickness of the *i* layer. The increased depletion thickness increases the sensitive volume for photon detection. If a photon is absorbed in the *i* layer it will create an electron/hole pair. The electron and the hole will drift in the high field of the depleted region. The high drift velocities combined with the small thickness of the *i* layer will result in a fast signal.

When silicon detectors are used for detecting charged particles they need to be in the dark to avoid large unwanted currents.

The photon absorption coefficient for intrinsic silicon as a function of wavelength is shown in Figure 9.28.

The intensity of light penetrating a distance x is given by $I = I_0 \exp(-\alpha x)$, where α is the absorption coefficient.

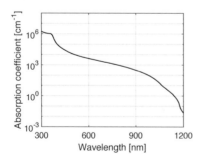

Figure 9.28: Attenuation constant for intrinsic silicon as a function of wavelength. Data from [269].

The band gap in silicon is 1.12 *eV*, which would correspond to a wavelength of ∼1100 nm. However, as silicon is an indirect bandgap semiconductor, the energy required for a transition without any change in momentum is ∼3.4 *eV*. At intermediate energy (i.e. in the range between

1.12 and 3.4 eV), transitions from the valence to the conduction band require phonons as well as the photon. This explains qualitatively why the attenuation constant is decreasing rapidly starting with wavelengths of ~400 nm. We can see that a very thin intrinsic layer can give a high probability for absorbing a photon for visible light and the near infrared. For longer wavelengths we need to use compound semiconductors like GaAs or InGaAs.

For example for a wavelength of 750 nm and a 10 μm thick layer, the probability of a photon being absorbed is 73%.

The intrinsic layer can be chosen to be just thick enough for there to be a high probability of photon absorption for the target wavelength. This allows for the photodiode to be fully depleted at very low bias voltages. It also results in very short transit times, which allows for very high data rates.

GaAs and InGaAs are examples of direct bandgap semiconductors and therefore have shorter attenuation lengths for energies just above the bandgap energy. These compound semiconductors are used for long distance telecommunications that use a wavelength of 1310 nm or 1550 nm.

One critical figure of merit for a photodiode is the responsivity R. The photocurrent I for a photon power incident on the photodiode P_γ is $I = RP_\gamma$. The responsivity can be written in terms of the external quantum efficiency η_{ex} and the photon frequency ν as $R = \eta_{ex}e/(h\nu)$. We can re-write this in terms of the reflectivity S of the surface and the internal quantum efficiency η_{in}

The external quantum efficiency is defined as the probability of a photon hitting the device being converted to an electron-hole pair. The internal quantum efficiency is the probability of a photon entering the device being converted to an electron-hole pair.

$$R = \frac{[1 - S(\lambda)]\,\eta_{in}e}{h\nu}.$$

The responsivity can be increased by an anti-reflective coating on the surface.

A typical value for a silicon photodiode operating at 850 nm is R = 0.5 A/W.

The other important parameter for a photodiode is speed. The intrinsic speed of a photodiode depends on the drift time and hence on the thickness of the active layer, the bias voltage and the carrier mobilities. However, we also need to match the speed of the amplifier, which depends indirectly on the capacitance and hence the area and depth. The absorption length in InGaAs is shorter than in Si. Therefore, thinner active regions can be used. In addition, the electron mobility is larger in InGaAs than Si. Therefore the maximum drift time for InGaAs photodiodes will be less than that of Si photodiodes. This provides an additional reason that the fastest diodes used in high speed telecommunications are based on compounds like GaAs or InGaAs.

For a given power, if there is an upper limit on the acceptable noise, this will place an upper limit on the bandwidth of the amplifier as for silicon particle sensors.

9.9 SILICON DETECTOR APPLICATIONS WITH INTERNAL GAIN

There are three main motivations for the use of internal amplification in semiconductor detectors.

- Compensation for the reduced ionisation charge in thinner silicon sensors (about 50 μm) to reduce material. This requires only moderate gain;

- Improved timing performance to tag hits belonging to primary vertices close in time (for example within the same bunch crossing at LHC, see section 9.9);

- Detection of photons in low-intensity electromagnetic radiation, in particular if single photons need to be detected.

In chapter 4 we have introduced the multiplication of charge carriers due to impact ionisation. In semiconductors fields high enough to cause impact ionisation can be achieved by appropriate doping and bias voltage. In section 9.1 we have discussed the concept of silicon detectors based on highly doped implant regions in lightly doped bulk material, either p^+-in-n^- or n^+-in-p^- or n^+-in-n^-. If an additional doping layer of n in the former case or p in the latter is introduced in between, a region of high field is created at the p^+/n or n^+/p interface, respectively. This field can be high enough to cause impact ionisation and create more ionisation in avalanche multiplication. The strength and size of this accelerating field can be controlled by the doping concentration and the depth of the additional doping, and the bias voltage.

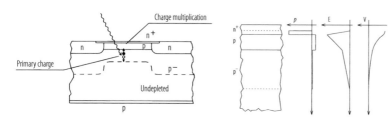

Figure 9.29: Schematic cross section of an $n^+/p/p^-$ avalanche photodiode (left) and doping profiles and fields (right) [350].

A range of device designs use this approach, but with different ranges of gain. Because the primary ionisation from charged particles in silicon sensors is already high, a small gain will cause a large signal.

- *Low gain avalanche diodes (LGADs)* have a gain of around 10, which can be useful to boost the signal from a charged particle, allowing for the operation with thinner silicon sensors, or, for the same thickness, providing better time resolution (see next section).

  These type of sensors allow precision measurements in space as well as in time and this can be used for 4D tracking as discussed in section 10.2.

  In a $p^+/n/n^-$ LGAD the secondary electrons only drift a short distance to the collection electrode and they get quickly removed from the induced current signal. Their contribution to the integrated signal is correspondingly very small. The secondary holes contribute most of the current and the integrated signal as they drift over the full depth of the detector. In order for the weighting field to be large over the full depth of the detector, the width of the collecting electrode needs to be comparable to the depth of the detector.

  This is more like the weighting field for a simple parallel plate capacitor, rather than a micro-strip.

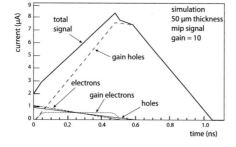

Figure 9.30: Simulated pulse from an LGAD [256], originally published in [184]. The curves labelled 'electrons' and 'holes' are due to the primary ionisation charges, whereas 'gain electrons' and 'gain holes' are the signals from charges that have been produced by impact ionisation.

One of the big challenges for LGADs is radiation damage. Apart from the effects discussed for silicon detectors in chapter 9.11, the acceptor creation in the p^+ layer will decrease the net carrier concentration and therefore lower the gain.

- *Avalanche photodiodes (APDs)* have a gain of about 50 to 500. They are in wide use for the efficient detection of photons in, or close to, the visible region. In this 'reach-through' structure, photons enter from the n^+ side and have a high probability of penetrating into the depleted layer where they are absorbed. The resulting electrons drift towards the highly doped 'metallurgical junction' formed by the p^+/n^+ doping. In this region the electric field is large enough to create impact ionisation and hence create a carrier gain M. For operation in the linear mode the peak electric field should be lower than the breakdown field.

For wavelengths $\lambda < {\sim}400$ nm the photons are typically absorbed before they reach the depleted region. Therefore APDs designed for shorter wavelengths use side illumination.

The avalanche process introduces additional fluctuations and the resulting shot noise is increased compared to a simple photodiode. This is accounted for by an excess noise factor f. An expression for the excess noise factor from a theoretical analysis is [362]

$$f = kM + (1-k)\left(2 - \frac{1}{M}\right), \qquad (9.27)$$

where k is the ratio of hole-to-electron ionisation rates. Silicon has a much lower k value (typically $k < 0.1$) than other semiconductors. In addition, typical leakage currents for silicon photodiodes are much lower than for photodiodes made with other semiconductors. Therefore the lowest noise APDs use silicon (unless the wavelength $\lambda > 1100$ nm, for which there is almost no absorption in silicon). For longer wavelengths other semiconductors like InGaAs are used. The excess noise in the simple *p-i-n* structures for these materials is avoided by the separation of the absorption and multiplication regions. These devices are called separate-absorption-multiplication (SAM) APDs [301].

If we had a noiseless amplifier, the noise from an APD would always be worse than that from an un-amplified photodiode. However, for a real amplifier we can minimise the overall noise for low light intensities. For a shaping time τ, leakage current I, an equivalent noise resistance R for the amplifier and a total capacitance of C, the noise is given by (see eq. (3.16))

We are ignoring any surface leakage current because it will not be amplified.

$$\sigma_q^2 = 2eIf\tau + \frac{4k_{\mathrm{B}}TRC^2}{M^2\tau}. \qquad (9.28)$$

Note that f in this expression is the excess noise factor defined above, and not the frequency.

Silicon APDs usually operate with $M > 20$, therefore the excess noise factor is given to a very good approximation by $f \simeq 2 + kM$, where k is defined in eq. (9.27). We can then determine the optimal multiplication by differentiating eq. (9.28),

A typical value for M for a silicon APD is ${\sim}100$.

$$M = \left(\frac{4k_{\mathrm{B}}RC^2}{ekI\tau^2}\right)^{1/3}. \qquad (9.29)$$

- *Single photon avalanche diodes (SPADs)* provide an even higher gain of 10^5 to 10^6. They operate in breakdown mode, which means that a single charge carrier injected into the depletion layer can trigger a self-sustaining avalanche in the device. Here the feedback involves the creation of holes, which at the high field lead to further

This can be a thermally produced charge carrier.

multiplication. The discharge is stopped by temporarily reducing the bias voltage ('quenching'), either by a resistor or by other means. Proportionality of the output signal with the input is lost.

Because a single electron, independent of its origin, can cause an avalanche in SPADs, they have a significant 'dark rate' that is caused by thermally generated electron-hole pairs.

As this is similar to the operation of a gaseous Geiger counter, SPADs are sometimes referred to as operating in Geiger mode, although the feedback mechanism is not involving long range photons.

HIGH PRECISION TIMING DETECTORS

The different types of silicon detector we have considered so far are typically optimised to obtain high spatial resolution. Therefore, these systems were optimised to maximise the S/N at the cost of limited timing precision ($\mathcal{O}(10\,\text{ns})$). For LHC detectors, the time resolution achieved this way was sufficient to enable hits to be uniquely associated with a given bunch crossing (every 25 ns). For HL-LHC there will be up to 200 interactions per bunch crossing. Therefore, disentangling the vertices will be much more challenging. The collisions occur over a range in distance along the beam and also in time. Therefore, if we can achieve good timing precision we can greatly enhance the ability to separate the different vertices.

This is called 4D tracking, see section 10.2.

For detectors not targeting timing measurements, the goal of the optimisation was to maximise the S/N, but if we want to measure time precisely we need a different optimisation. The precision of the timing measurement due to noise is

$$\sigma_t = \sigma_V \bigg/ \frac{\mathrm{d}V}{\mathrm{d}t} \,, \qquad (9.30)$$

where $V = V(t)$ is the output voltage as a function of time t, and σ_V is the voltage noise. The 'slew rate' $\mathrm{d}V/\mathrm{d}t \propto 1/t_\mathrm{r}$, where t_r is the rise time of the preamplifier and $t_\mathrm{r} \approx 1/\Delta f$ where Δf is the bandwidth. The voltage noise of the amplifier $\sigma_V \propto \sqrt{\Delta f}$. Therefore we need a higher bandwidth to achieve a good timing resolution than for the optimisation of the S/N.

As discussed in section 3.2, the voltage noise power density $\mathrm{d}v^2/\mathrm{d}f$ is constant.

However, there is no benefit in using a faster amplifier rise time than that of the signal in the sensor (t_s) and the optimum is when $t_\mathrm{r} = t_\mathrm{s}$ (see exercise 6). In order to further improve the timing precision it is necessary to increase the sensor signal, and sensors with internal signal amplification will be useful. An additional benefit of the amplification is that it allows for a reduction of the sensor thickness, which will reduce the length of the arrival time distribution, further improving the time resolution.

Already the limited gain provided by an LGAD is sufficient, and a resolution of 34 ps has been achieved with a 50 μm thick LGAD [185].

As usual for all precision timing measurements of analogue signals, time walk contributes to the timing resolution, and the use of a constant fraction discriminator is required for ultimate timing performance (see section 3.8).

Alternatively, the timing information can be corrected for signal height variations after measuring the signal height using time-over-threshold (TOT), or sampling the signal (at the cost of an increase of data).

It is possible to achieve even better timing resolution using SPADs. However, the 'dark rate' for these type of detectors in a high-radiation environment is extremely high, limiting their use.

A very large dark rate and a very large number of channels places challenging requirements for the readout bandwidth.

PHOTON DETECTORS WITH INTERNAL GAIN

For the readout of particle detectors like scintillators we usually need to achieve low noise for low photon intensities. For low intensities (low photon incidence rate), the signal becomes too small for detection at the desired data rate and power in the amplifier. The answer is again to employ an internal amplification process based on impact ionisation.

We have already introduced APDs and SPADs as semiconductor detectors with internal amplification that are optimised for photon detection. Silicon photomultipliers (SiPMs) are arrays of SPADs (up to 10000 per mm^2). Their typical gain is similar to conventional PMTs, but with the benefit of small pixel size and being unaffected by magnetic fields.

SiPMs are sometimes also called Multi Pixel Photon Counters (MPPCs).

The rise time of the pulse from a SiPM is determined by the avalanche process, so it is very fast ($t_r < 1$ ns).

There are two possible readout modes of a SiPM:

1. *Analogue*. The output of all the quench resistors are ganged together. Excellent timing resolution can be achieved [316] even for single photons ($\sim$100 ps). The speed of the falling edge is determined by the RC time constant, where R is the value of the quench resistor and C is the pixel capacitance. Typical values for the time constant are in the range 30 to 100 ns [316];

2. *Digital*. The output of each pixel goes to a discriminator and the number of 'fired' pixels are counted by digital logic. The quenching is performed using MOS switches rather than passive resistors. Noisy pixels can be switched off. One disadvantage of this mode is the higher power consumption.

The resistivity of polysilicon is low, so that a long, thin track is needed to create a passive resistor, which is typically larger than a transistor, so that this mode has the advantage of requiring less silicon area.

The advantages of SiPM over PMTs are a higher quantum efficiency, a very fast rise time and that they can operate in a strong magnetic field, which is a benefit in large particle experiments with their spectrometers, but also in Positron Emission Tomography – Magnetic Resonance Imaging (PET-MRI). PMTs on the other hand have a lower dark count and can have a large area photocathode. UV detection is possible for PMTs with custom photocathodes. APDs are usually cheaper than PMTs.

This comparison suggests that there are applications that benefit from a combination of these two technologies. An example for this is the Hybrid PMT (HPMT). An HPMT has a conventional photocathode like in a PMT. Instead of having a chain of dynodes there is one stage of electrostatic acceleration with a voltage of ~ 10 kV. The resulting electrons are detected in an APD. The high energy electrons create $\mathcal{O}(10^3)$ electron-hole pairs. The electrons create avalanches in the APD resulting in a further gain of $\mathcal{O}(10^2)$. Therefore HPDs can have very large gains like a PMT, but faster rise times which allow better timing resolution. As there is a much larger gain from the electrostatic stage, the pulse height resolution for single photons is better than for PMTs.

Another type of hybrid photodetector is the hybrid photodiode (HPD), which uses a conventional photocathode and a very high electric field to accelerate the electrons towards a pixel *pn* diode.

Photoelectrons emitted by the photocathode are accelerated by a potential of $\sim$20 kV towards a silicon pixel *pn* diode. The electric field is

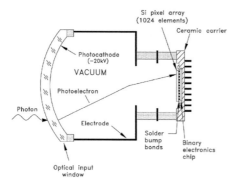

Figure 9.31: Schematic cross-section of an HPD [28].

also used to 'demagnify' and focus the photoelectrons on to the diode. This creates a signal of about 5000 electron-hole pairs in the silicon. The diode is read out by a custom ASIC which is bump-bonded to the diode. The very small pixel area corresponds to a very small capacitance and hence the readout speed can be high while the electronic noise can be very low. Therefore there is no need for the amplification of an APD and there is no excess noise factor. The resolution is sufficient to resolve the peaks from 1, 2, 3 and 4 photoelectrons [380].

The photocathode area is larger than that of the diode.

9.10 MONOLITHIC ACTIVE PIXEL SENSORS

Bump-bonding is a demanding operation, and obtaining high yields is a challenge. It is therefore attractive to integrate at least the first stages of the analogue readout into the pixel sensor. Such a detector is called a Monolithic Active Pixel Sensor (MAPS). Current approaches exploit the wide availability of CMOS technology for this approach. This technology is widely used to make CMOS imagers as used in mobile phone cameras. The same sensors could in principle be used as detectors for charged particles but there are several disadvantages:

- Imaging chips only require a thin sensitive region, due to the short mean free path for visible photons. The signal from minimum-ionising particles in thin sensors is small;

- The demands for speed in imaging chips are usually not high, so that it suffices to rely on diffusion rather than drift. Particle detectors, in particular after irradiation, need to be faster;

- Image sensors do not usually have 100% 'fill-factor', i.e. not all the area of each pixel is sensitive.

For applications at low rates and low radiation damage the slow signals from diffusion can be used. An epitaxial layer is used on a p-doped substrate which has a much higher conductivity. The signal is collected on an n-well with an n^+ terminal. Without a bias, there is only a small depletion region at the edge of the n-well, and most of the signal comes from electron diffusion in the epitaxial layer. The collection n-well can then be connected to CMOS transistors on the same wafer, which can be part of an amplifier or a buffer circuit.

An epitaxial layer is grown on a substrate layer so that it is formed with one or more well-defined orientations with respect to the seed layer. Epitaxy results in improvement of the electrical characteristics of the epitaxial layer in a highly controlled manner.

The signal from the thin epitaxial layer is small, $\mathcal{O}(1000\ e^-)$, but a good S/N can still be achieved because of the very small capacitance of the collection electrode (as the noise is proportional to the capacitance, see eq. (3.11)). As the signal is from diffusion rather than drift, the signal is slow. Typical collection times are $\mathcal{O}(100\ \text{ns})$.

Other transistors, which are part of the active electronics can be contained in the same chip as the sensor, but any n-wells need to be insulated by a p-well, so that the transistor is not affected by parasitically collected charges. By embedding NMOS (PMOS) transistors in p-wells (n-wells) we can use the full functionality of CMOS electronics. Therefore all functions of the readout ASIC can in principle be incorporated into the MAPS chip.

The difference between a DEPFET and a MAPS detector is that in the former case the charge is trapped temporarily in the depletion region and its field is controlling the drain-source current in the FET, whereas in the latter the charge is collected and then conducted to a gate of a transistor.

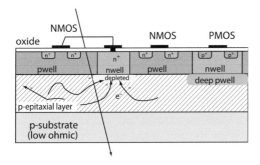

Figure 9.32: Schematics of a MAPS detector [256].

Another advantage of MAPS over hybrid pixel detectors is that they are much thinner as there is only the combined sensor/readout chip compared to the separate sensor and readout ASIC required for hybrid detectors. In addition, MAPS chips can be thinned as the bulk silicon is not used. This allows MAPS to be an order of magnitude thinner than hybrid pixel detectors. This approach has been used very successfully for vertex detectors for heavy ion collisions [210, 300].

However, diffusion-type MAPS cannot be used for high rate applications like the LHC, where we need collection times of $\mathcal{O}(10\ \text{ns})$. In addition the radiation tolerance of MAPS is very poor because of charge trapping and the long collection time. Therefore we need fast charge collection by drift caused by the field created in a depletion region, rather than by diffusion.

These type of MAPS are therefore called 'Depleted MAPS' or DMAPS.

Figure 9.33: Schematics of a DMAPS detector [256].

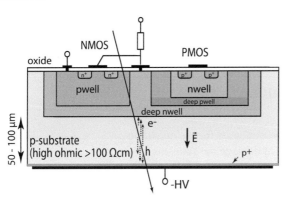

Following eq. (9.4) the depletion depth at a *pn* junction scales as

$$d \propto \sqrt{\rho V_{\text{bias}}} \,,$$

where ρ is the resistivity and V_{bias} the bias voltage. A large depletion depth is thus achieved for high resistivity and bias voltages.

Hence, there are two approaches for DMAPS:

The resistivity ρ is the inverse of the conductivity σ, and the conductivity in semiconductors $\sigma = e(\mu_n n_n + \mu_p n_p)$, with the electron charge e, the mobilities for electrons and holes $\mu_{n,p}$, respectively, and the charge carrier densities $n_{n,p}$.

- *High resistivity (HR) MAPS*. High resistivity silicon is used, which allows a sufficiently large depleted region to be created at low voltage;

- *High voltage (HV) MAPS*. A special high voltage process, HVCMOS, is used to allow a bias voltage of $\mathcal{O}(100 \text{ V})$ to be applied, which generates a significantly deep depletion layer.

The HVCMOS process is also required for the automotive industry and thus has become commercially relevant.

The highest charge collection efficiency is achieved for devices with high resistivity and high bias voltage, and if a high fill factor can be achieved.

In the DMAPS the drift field is generated by a deep *n*-well. The size of this electrode determines the fill factor of the sensor. The advantage of a small collection electrode is that it represents a small capacitance and therefore results in lower noise (eq. (3.11)). However, the drift distances are longer than for the large collection electrode, and therefore it will be more sensitive to charge trapping after radiation damage. A large collection electrode will have higher capacitance and therefore higher noise than a small collection electrode.

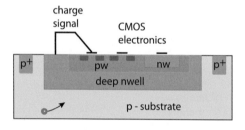 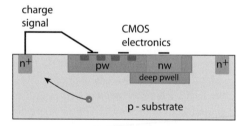

Figure 9.34: Schematics of different DMAPS geometries. Left: large collection electrode. Right: small collection electrode [256].

While there has been very significant progress in the design of radiation tolerant DMAPS, achieving the required performance for use at the LHC is very challenging. This is, however, a very active field, so more progress should be expected in the coming years.

9.11   RADIATION DAMAGE IN SEMICONDUCTOR DETECTORS

Radiation damage can have very serious consequences for the operation of semiconductor detectors. The same type of damage mechanisms affect all semiconductors, but we will again focus on silicon detectors as they are the most widely used detectors in high radiation environments.

In semiconductors the main effect of radiation is due to bulk damage to the crystal lattice. Charged particles interact with the silicon nuclei by Coulomb scattering, and neutrons by the strong (or nuclear) force. If a primary particle has sufficient energy, it can knock out a silicon atom from the lattice. This will result in a 'vacancy' in the lattice where the knocked-out atom came from. The knocked-out atom can become trapped in the lattice, which results in an interstitial state. The vacancy or interstitial state can also interact with impurities present in the lattice to produce more complex defects.

The threshold energy for this process is calculated and the implications for radiation damage by low energy neutrons and electrons are considered in exercise 7.

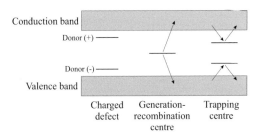

Figure 9.35: Mid-band energy levels.

If the energy of the knocked-out atom is sufficiently high, then this atom can knock out further silicon atoms from the lattice, resulting in large defect clusters. In general, these defects will result in the creation of new energy levels in the band gap between the valence and the conduction band.

The impact of these new energy levels will depend critically on where they are in the gap. As we have seen in section 9.5, those defects near the centre of the band gap will contribute to an increase in leakage current. Energy levels near the conduction or valence band on the other hand will affect the effective doping concentration. Other energy levels in the band gap can result in trapping centres for mobile electrons or holes. This effect reduces the distance free carriers can drift and therefore reduces the measured signal.

Some of the defects can be mobile and therefore there will be a complex temperature dependent time evolution of the radiation effects. Vacancies and interstitial defects can 'annihilate', resulting in 'beneficial' annealing. However, neutral defects can combine to create new charged states and this affects the net effective doping density with serious implications for the long term operation of silicon detectors in high radiation environments. This is called 'reverse annealing'.

The bulk radiation damage is a function of the type and energy of the particle causing it. Detectors at hadron colliders are exposed to a complicated mixture of different particle types, each with different energy spectra. Particles lose energy in the sensor material by ionisation, but also by other, non-ionising, processes. The non-ionising energy loss in a crystal like silicon will be converted into phonons and create crystal defects. The Non-Ionising Energy Loss (NIEL) scaling hypothesis assumes that degradation phenomena in silicon detectors under irradiation are proportional to the NIEL in the material. It relies on the assumption that damage scales with the displacement energy. It does not consider the actual microscopic details of the damage for different particles and/or energy.

This is a good approximation for silicon but is much less reliable for compound semiconductors like GaAs.

Using the NIEL scaling hypothesis, we can scale the expected displacement damage caused by particles of different energy to that of one particular particle at a fixed energy. It is conventional to use 1 MeV neutrons for this normalisation. Therefore all fluences will be expressed as ϕ_{eq}, where this refers implicitly to 1 MeV neutrons.

The unit used is often written as n_{eq}/cm^2, or (1 MeV) n_{eq}/cm^2.

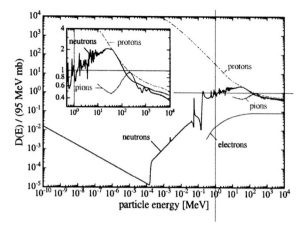

Figure 9.36: Displacement damage function for different particles in silicon as a function of the particle energy [337].

The displacement damage function $D(E)$ relates to the NIEL (in units of $MeVcm^2/g$) as

$$NIEL(E) = \frac{N_A}{m_A} D(E),$$

where N_A is Avogadro's number and m_A the molar mass of the material.

In addition to the effects of bulk damage, radiation will create electron-ion or electron-hole pairs in semiconductors. The magnitude of the ionising dose is given by the energy absorbed per unit mass. As the energy absorbed per unit mass depends on the atoms involved, a meaningful dose figure must specify the material. As the common material in silicon detector systems is silicon it is standard to use Gy(Si). Damage from ionising dose is usually not significant for the silicon sensors but it can be for the electronics or structural materials.

The SI unit for ionising dose is the Gray (1 Gy = 1 J/kg). The older unit of Rad is also still used, 1 Gy = 100 Rad.

The effects of radiation on the electronics have been discussed in section 3.9. Data on radiation hardness of different materials can be found in a series of CERN technical reports [471, 470, 134, 340].

LEAKAGE CURRENT AS A FUNCTION OF IRRADIATION

Radiation damage will create mid-band states and therefore increase the leakage current. The increase in leakage current generated per unit sensor thickness as a function of the fluence ϕ in a volume V can be parameterised by

$$I_{leak} = \alpha \phi V, \tag{9.31}$$

where the linear rise reflects the NIEL scaling hypothesis.

Annealing of the mid-band states will always reduce the leakage current, and higher temperatures accelerate the annealing. Temperatures during and after irradiation therefore need to be carefully monitored to make leakage current measurements comparable.

Also, the dependence of the leakage current on the instantaneous sensor temperatures as described in section 9.5 needs to be taken into account.

The increase in leakage current due to irradiation has two serious implications for the operation of silicon detectors:

- It will result in an increase in the shot noise (see eq. (3.4)), which will degrade the S/N;

- As discussed in section 9.5, if the leakage power exceeds the heat removal capability of the thermal design, the sensor temperature can become unstable and the detector can go into thermal runaway.

DOPING CHANGES

Energy levels in the band gap close to the bottom of the conduction band will tend to have electrons promoted to the conduction band by thermal motion, thus leaving behind fixed positive charges. Similarly, energy levels

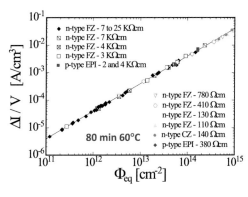

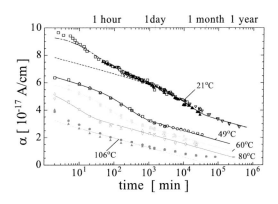

Figure 9.37: Left: Leakage current increase as function of particle fluence for various silicon materials with different resistivities and conduction type. The current was measured after a heat treatment of 80 min at 60 °C and is normalised to the current measured at 20 °C. Right: Current-related damage rate α as a function of cumulated annealing time at different temperatures [378].

close to the top of the valence band will result in fixed negative charges. Defects created by radiation can combine with dopants to remove some of the fixed charges.

The net result of these complex processes is that there will be a net change in the effective charge number density n_{eff}. The dominant effect in n-type silicon (initially with positive space charge) is to create acceptor states that can change the silicon from n-type to p-type due to irradiation. If the radiation levels are further increased, the effective charge density, which is now negative, will increase. This will result in an increase in the voltage required for full depletion (see eq. (9.4)) and if the applied high voltage cannot be increased sufficiently, there will be signal loss. For example, an initial n-type silicon type will invert to become p-type. We observe an initial decrease in the full depletion voltage up to type inversion and then an increase.

If we start with initial p-type silicon, there will be no type inversion, the value of n_{eff} (and therefore the full depletion voltage) will increase monotonically with fluence, in accordance with the NIEL hypothesis.

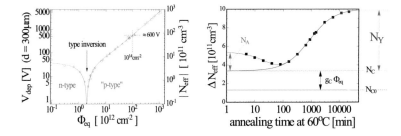

Figure 9.38: Left: Effective doping concentration (depletion voltage) as a function of particle fluence for a standard n-type silicon detector. Data were measured with a short annealing time. The data were extrapolated to zero annealing time and then used to fit a parameterisation of the depletion voltage versus fluence. The points shown are calculated values based on this parameterisation. Right: Evolution of the effective doping concentration as a function of annealing time. The data shown here were taken at room temperature while the annealing took place at 60 °C [378].

The value of n_{eff} depends on the temperature and time as well as fluence, because of interactions amongst mobile defects. Over short periods of time after irradiation there is 'beneficial' annealing and the value of n_{eff} will decrease. Over long periods of time (years) there can be 'reverse' annealing which results in n_{eff} increasing. Both beneficial and reverse annealing happen faster at higher temperature. In general, in order to limit the growth of the full depletion voltage (V_{FD}) for a p^+-in-n detector operating for many years in a high radiation environment (e.g. an LHC silicon detector) it is essential to keep the detector cold once the detector has received a significant fluence.

This complex annealing behaviour can violate NIEL scaling. This has been used to optimise the radiation tolerance of silicon by adding impurities. These studies showed that the value of n_{eff} after irradiation was lower for detectors with higher oxygen concentrations [336].

CHARGE TRAPPING

Moving charge carriers (which generate the induced signal in our detectors) can be captured by trapping levels. If they are not released quickly enough this will result in a loss of signal. If the mean trapping time is τ_{eff}, then the signal will decrease with time exponentially

i.e. in a time shorter than the integration time of the detector readout.

$$Q(t) = Q_0 \exp\left(-\frac{t}{\tau_{\text{eff}}}\right). \tag{9.32}$$

Values of τ_{eff} have been measured separately for electrons and holes as a function of fluence. The variation with the fluence ϕ can be simply parameterised as

$$1/\tau_{\text{eff}} = 1/\tau_{\text{eff}}^0 + \beta\phi. \tag{9.33}$$

For irradiated detectors, the initial value of $1/\tau_{\text{eff}}^0$ is usually negligible.

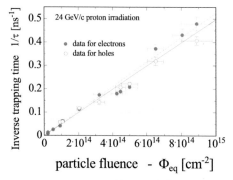

Figure 9.39: Inverse trapping time as function of particle fluence as measured at 0 °C after an annealing of 30 to 60 min at 60 °C. [378].

For very high fluences ($> 5 \times 10^{14}$ $n_{\text{eq}}/\text{cm}^2$) the loss of signal is the limiting factor in the operability of silicon detectors. As electrons have higher mobility than holes in silicon, they will have shorter charge collection times. Therefore electrons are less sensitive to charge trapping than holes. This makes n^+-in-p sensors less affected by charge trapping than p^+-in-n for high fluences ($\sim 10^{15}$ $n_{\text{eq}}/\text{cm}^2$). This is the reason why ATLAS and CMS have decided to use n^+-in-p rather than p^+-in-n silicon detectors for the upgrades for HL-LHC.

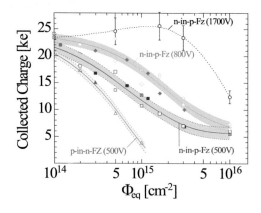

Figure 9.40: Collected charge as function of ϕ_{eq} for 300 μm thick ministrip sensors irradiated with 23 GeV protons, 26 MeV protons, and reactor neutrons for different doping configurations with the indicated bias voltages [378].

PREDICTION OF RADIATION DAMAGE EFFECTS IN SILICON

We have seen that radiation damage effects have a serious impact on the long-time operability of silicon detector systems, with the ultimate loss of S/N. It is therefore essential for the design of silicon detector systems in high intensity environments to be able to predict the radiation damage effects, in particular in response to the temperature history of the system and the profile of fluence over the lifetime of the detector.

Particle fluences at different locations in an experiment can be simulated with advanced Monte Carlo event generators and particle transport codes as implemented in GEANT, and can be translated into 1 MeV neutron equivalent fluences using the NIEL hypothesis. The leakage current can be predicted from radiation damage parameterisations, like the Hamburg/Dortmund model [377, 320] or the Sheffield-Harper model [281].

Such studies have been made for the silicon trackers for ATLAS, CMS and LHCb, and compared to measured leakage currents, together with data on instantaneous luminosity and device temperature as a function of time, to check the validity of the models and to verify the long-term prognosis of performance for the systems in the high-intensity LHC environment. Excellent agreement has been observed.

The prediction models can also be combined with models of the thermal behaviour as described in section 9.6.

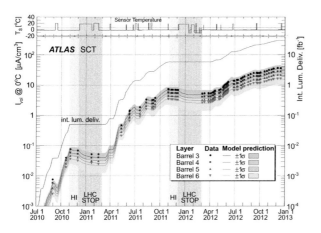

Figure 9.41: Comparison between data (points) and Hamburg/Dortmund model predictions (lines with uncertainties shown by the coloured bands) of the leakage current per unit volume scaled to a sensor temperature of 0 °C of the four layers of the ATLAS barrel SCT. The integrated luminosity and the average sensor temperatures are also shown. The blue shading and label 'HI' indicate periods of heavy-ion running, while extended periods with no beam in the LHC during which the SCT was off are shaded grey [78].

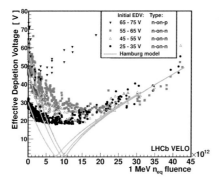

Figure 9.42: The effective depletion voltage against fluence for sensors for the LHCb VELO detector [27]. The depletion voltage from data is compared to the predictions by the Hamburg model. Note type inversion at about $10^{13}\ n_{eq}/cm^2$.

9.12   3D SENSORS

A silicon detector design aimed at improving the radiation tolerance is the
3D sensor. Here the heavily doped implant regions are not on the surface
as in the planar sensor, but penetrate as pillars through the thickness of
the sensor. The spacing of the pillars is smaller than the thickness of the
wafer. In both approaches the silicon sensor can have the same thickness,
and therefore the MIP signal before irradiation will be similar. However,
after doping changes due to radiation damage the planar sensor will require too high a bias voltage to be fully depleted and the charge trapping
will also reduce the signal. In the 3D sensor the required depletion depth is
smaller and the drift distance is shorter.

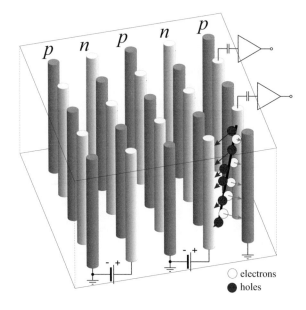

*Figure 9.43: Schematic design of a 3D
sensor.*

The other important advantage of 3D sensors is that they can have 'active edges', unlike planar detectors which need some type of guard ring
structures to avoid leakage current from the sensor edges.

The fabrication of 3D sensors is more complicated than that of planar sensors. The 3D sensors require a process called Deep Reactive Ion
Etching (DRIE) to form the columns [315]. In one approach to DRIE a
short pulse of a suitable plasma is used to etch the silicon isotropically.
The walls of the resulting hole are then passivated with a gas such as C_4F_8.
When a second plasma etch is applied the etching is only in the vertical
direction because of the passivation at the sides. Therefore the two steps
can be repeated multiple times to create the columns that are the full depth
of the silicon. Thermal-diffusion steps are then used to drive in dopants to
form the n^+ and p^+ electrodes.

*See for example [316] for an overview
of the process steps involved.*

As 3D sensors are more radiation tolerant but more expensive than
planar sensors, it makes sense to use them for the innermost layers in a
collider detector. 3D sensors were used for the highest radiation area of
the innermost pixel layer of the current ATLAS tracker [217] and will
be used for the innermost barrel layer for the ATLAS tracker for the
HL-LHC [283].

9.13   OTHER SEMICONDUCTORS

Although silicon has many properties that make it the best semiconductor for tracking detectors, the low atomic number ($Z = 14$) makes it less suitable for high resolution X-ray detectors. Germanium is better suited for this purpose, as it has a higher atomic number ($Z = 32$). Germanium has a much smaller bandgap than silicon, $E_g = 0.67$ eV. This results in one complication in the use of Germanium detectors compared to silicon: even without radiation damage they need to be cooled. Compound semiconductors are also an attractive material for X-ray detectors. CdTe and $Cd_{1-x}Zn_xTe$ (abbreviated as CZT) have high atomic numbers (Cd: $Z = 48$, Zn: $Z = 30$, Te: $Z = 52$) and a large band gap (1.44 and 2 eV, respectively), but they are challenging to produce in the required purity, and the size of these crystals which can be grown is still limited to a few cm^3.

x represents the fraction of Zn in the compound.

The radiation damage effects seen in silicon have prompted research on alternatives for the use in tracking detectors. One alternative semiconductor considered was GaAs (Ga: $Z = 31$, As: $Z = 33$). Initial radiation damage studies using low energy neutrons showed superior radiation resistance. However, further studies with higher energy hadrons showed a reduction of the charge collection efficiency (CCE) due to trapping of electrons and holes [116].

GaAs has a wide band gap of 1.41 eV, and a relatively low electron–hole pair creation energy (4.2 eV).

One benefit of GaAs is the very high mobility for electrons (8500 cm^2/Vs), which makes GaAs attractive for timing applications, and allows for thinner sensors, but it requires n^+-in-n or n^+-in-p geometries [419]. Another advantage of GaAs is that it has a direct band gap, which means that it absorbs and emits light efficiently.

GaAs has the highest efficiency for photovoltaic conversion of light to electricity.

Other semiconductors that are investigated for radiation-hard detectors are SiC or group III-V compounds like GaN [453]. The challenge with these materials is the production of thick enough sensors and fabrication methods are not yet mature for large scale detectors.

Another idea for very high radiation tolerance is to use diamond. Diamond has a bandgap of 5.5 eV, and is therefore considered to be an insulator. The average energy required to create a electron-hole pair is 13.1 eV, a factor of about four larger than for silicon. The resulting signal is also reduced by the very short achieved carrier lifetimes of $\mathcal{O}(100$ ns$)$.

This is not very significant for fast LHC detectors.

As diamond is an insulator, there is no need for a bias voltage to deplete the substrate and as the bandgap is so large, there is no need for cooling even for irradiated detectors. Radiation damage does result in the shortening of the mean free path of charge carriers, but the rate is smaller than in silicon. Crystalline diamond can be made using Chemical Vapour Deposition (CVD), but the production of material with sufficient thickness and uniformity is challenging, and this technology is still very expensive. Diamond detectors have been used at the LHC for very small systems required in the highest radiation regions for beam protection monitoring and luminosity measurements.

For a summary of the performance achieved see [18].

The losses from the very high intensity beams hitting the collimators can potentially result in dangerously high fluxes of secondary particles that can damage the detectors. In order to minimise the risk of damage to the silicon detectors the beam protection system enables the beams to be aborted if the fluxes are too high.

Key lessons from this chapter

- The fundamental idea of the most common silicon detectors is to use a reverse bi-ased *pn* junction to create a high electric field region depleted of free carriers. A fast charged particle passing through such a detector creates a large number of electron-hole pairs which drift in the electric field. The resulting induced current is detected and amplified.

- A depletion depth of about 300 μm results in sufficient signal, so that no internal amplification is required.

- Internal amplification can be achieved by introducing another doping layer. Gains from a few tens to 10^5 and higher can be achieved. Internal amplification can be used to reduce sensor thickness, for improved timing performance and detection of photons (even single photons).

- The standard silicon detector configuration is p^+-in-n. n^+-in-n give higher radiation hardness, at the cost of increased manufacturing complexity. Very good radiation tolerance can also be achieved with n^+-in-p sensors which are easier to manufacture.

- The reverse biased diode junction displays a thermally generated leakage current. The leakage current causes shot noise, and the heat generated by the leakage must be re-moved by a cooling system, as the system can otherwise become thermally unstable.

- Bulk radiation damage results in the creation of energy levels in the forbidden gap.

  - This will result in significant increase in the detector leakage current, therefore it is essential to operate irradiated detectors cold.

  - The defects will also change the effective charge density in silicon and therefore change the required full depletion voltage.

  - The effects of charge trapping will reduce the signal after a very high radiation fluence. This is the ultimate limit on how large a fluence can be tolerated.

EXERCISES

1. *Charge carrier density in pure silicon.*

   Undoped pure silicon has an intrinsic charge carrier density of $n_i \simeq 10^{10}$ cm^{-3} at room temperature (see section 9.11).

   a) Find the number of electron/hole pairs per unit thickness in a detector element with an area of 1 mm^2.

   b) Compare this with the ionisation produced by the passage of a minimum-ionising charged particle and comment on the realisability of a practical detector (density of silicon: $\rho = 2.33$ g/cm^3, energy required to produce an electron/hole pair: $E = 3.6$ eV).

2. *Position resolution of a strip detector.*

   a) For a silicon microstrip detector with strips of 80 μm width and 120 mm length, assuming binary readout, what is the resolution in the strip direction (z) and the orthogonal direction ($r\phi$). Why might the resolution in practice be better than indicated by the simple estimate?

b) Improved resolution in the z direction can be achieved by using two back to back sensors with their strips running at different angles. The angle between the strip directions in the two sensors is called the stereo angle. Show with the aid of a simple sketch how if there are multiple hits on a sensor pair, we can reconstruct the correct hit location, but there will be additional false locations found in a high-rate environment ('ghost hits').

c) How does the z-resolution and the rate of ghost hits depend on the stereo angle? For a collider detector, why are the resolution requirements different for the z and $r\phi$ directions?

3. *Properties of a pn junction.*

a) Perform the integral of the distribution of the number of electron states as a function of energy $n(E)$ to verify the result of eq. (9.1), justifying any approximations you make. The non-relativistic density of states is given by

$$\rho(E) = 4\pi \left(\frac{2m^\star}{h^2} \right)^{3/2} E^{1/2}.$$

b) Derive eq. (9.3).

c) Verify the results in figure 9.2.

4. *Fermi energy in silicon.*

a) Find an expression for the Fermi energy E_f in an *intrinsic* semiconductor, in terms of E_c, E_v and the effective masses of negative and positive charge carriers.

b) Repeat the calculation of part a) but now for a *doped* semiconductor with a volume density of donor dopants of N_d.

5. *Pulse shape for a simple pad detector.*

Consider a simple silicon p^+-in-n pad detector. Consider MIPs crossing the detector orthogonally. The detector is 300 μm thick and the acceptor density is $n_A = 10^{12}$ cm^{-3}.

a) Ignoring the built-in voltage, what is the full depletion voltage?

b) The mobility of electrons and holes are 450 cm^2/(Vs) and 1400 cm^2/(Vs). Assuming electron/hole pairs are created uniformly along the path of the MIP, what is the induced current with time in a pad detector?

c) Calculate the longest charge collection time for electrons and holes.

d) Now assume that the detector is 'over-depleted' by 100 V. Sketch the resulting electric field in the detector and compare it with that of the just-depleted case. Calculate the pulse shapes for electrons and holes for the over-depleted case and compare this to the results of the just-depleted detector.

6. *Time resolution.*

Consider a detector with an intrinsic rise time t_d, read out with an amplifier with a rise time t_a.

a) What is the rise time at the output of the amplifier?

b) The noise output of the amplifier is $\sigma_V = e_n^2 \Delta f$, where e_n^2 is the noise spectral density and Δf is the noise bandwidth. The noise bandwidth $\Delta f = a/t_a$, where a is a constant. Show that the timing jitter

$$\sigma_t = \frac{\sigma_V}{\left(\frac{dV}{dt} \right)} \propto \frac{\sqrt{t_a^2 + t_d^2}}{\sqrt{t_a}}.$$

c) Hence show that the optimal time resolution is obtained when $t_a = t_d$.

7. *Displacement damage to silicon lattice.*

The energy imparted to a silicon atom needs to be at least 25 eV for the silicon atom to be displaced from its location in the lattice. Determine the minimum kinetic energy required for this process for (a) a proton, (b) an electron and (c) a silicon atom.

Comment on the implications of these results on the displacement damage.

8. *Leakage current and shot noise.*

Consider a silicon microstrip detector with strip widths of 80 μm, length 120 mm and thickness 300 μm. Calculate the leakage current in one strip for two cases:

a) For an unirradiated silicon detector with a generation lifetime of $\tau_g = 1$ ms.

b) After an irradiation with a fluence of $\phi_{eq} = 10^{14}$ n_{eq}cm^{-2} (ignoring annealing effects).

c) Estimate the shot noise for these two cases assuming a shaping time of $\tau = 10$ ns and considering the silicon temperature to be $-20\,°$C, $0\,°$C and $+20\,°$C.

Compare these values with the signal expected from a MIP traversing the detector (the average energy required to create an electron-hole pair in silicon is 3.6 eV and the energy loss for MIPs in silicon is 1.664 MeVg^{-1}cm^2 and the density is 2.329 gcm^{-3}). Discuss the significance of these results for the operation of silicon detectors.

9. *Linear model of thermal behaviour of silicon systems.*

a) For the linear model discussed in section 9.6 and based on the condition for thermal stability shown in Figure 9.17, show that for the critical sensor temperature $T_{S,\mathrm{crit}}$ the following relation holds:
$$T_{S,\mathrm{crit}}^2 + (T_A - 2T_0)T_{S,\mathrm{crit}} - T_A T_0 = 0.$$
Hence show that eq. (9.24) applies.

b) Sketch the relation between the sensor temperature and the coolant temperature described by eq. (9.25). What is the slope at the end of the stable region? Invert this relation, and find eq. (9.26), which gives the critical sensor temperature if the leakage power Q_{ref} at a temperature T_{ref} is known.

c) Show using this model that to maintain thermal stability at end of life the local sink temperature for the ATLAS SCT ($R_s = 0.945$ K/W, $R_c = 2.164$ K/W, $Q_h = 6$ W, sensor area 163 cm^2) for an expected sensor leakage power of 2 W at $0\,°$C must be at most $-17\,°$C.

d) How much would the coolant temperature have to change if the leakage power doubles?

10. *Responsivity of a photodiode.*

What is the reflectivity of an air-silicon boundary at a wavelength of 850 nm? If a photodiode uses an anti-reflective coating to reduce the reflected power to be negligible compared to the incident power, what is the minimum thickness of the active i layer to ensure that the responsivity $R > 0.5$ A/W? You may assume that all converted photons contribute to the signal.

The index of refraction for Si at 850 nm is 3.65. The attenuation constant for silicon at 850 nm is 535 cm^{-1}.

11. *Depletion voltage and resistivity.*

We wish to be fully deplete a p-in-n silicon detector with a thickness of 300 μm at a bias voltage of 100 V.

a) Show that the resistivity is related to the carrier density n and mobility μ (assume a linear relationship between drift speed and electric field strength) by $\rho = 1/(ne\mu)$.

b) Neglecting the 'built-in' voltage of the diode calculate the carrier density required to fully deplete the sensor at a bias voltage V. Hence estimate the minimum resistivity of the n-type silicon required.

c) Why would the minimum resistivity for n^+-in-p detectors be larger?

The relative permittivity of silicon is 11.7 and the electron mobility is 1400 cm^2/(Vs).

10 Tracking

Tracking detectors are designed to reconstruct the trajectories of charged particles as they traverse the tracking volume. Usually a magnetic field will be present in this volume. In this field, the trajectories will curve and we can reconstruct the momentum of the charged particles from the curvature of the track. From the direction of the curvature we can determine the sign of the charge of the particle.

The space points used in track reconstruction usually come from a combination of gaseous (chapter 7) and/or semiconductor detectors (chapter 9). Space points can also be measured in a liquid TPC (chapter 8).

After reconstruction of the trajectories of the charged particles, we can reconstruct the 'primary' vertex. In a collider experiment, this will be where the beam particles interacted, in a fixed target experiment, this will be where the beam particle interacted with the target. Particle trajectories not coming from the primary vertex can be associated with secondary and tertiary vertices. These can arise from the decays of long lived particles like hadrons containing b or c quarks, or τ leptons.

A sudden change in direction ('kink') of the charged particle can also indicate a decay of the original charged particle, for example a decay like $\pi \to \mu \nu_\mu$, or an electron emitting bremsstrahlung.

If the energy deposited by the charged particle for each spacepoint is recorded, we can combine these measurements to estimate the energy loss dE/dx and use this for particle identification (see section 12.3).

The reconstructed tracks can be extrapolated into the electromagnetic calorimeter and compared to the location of energy depositions there. This is used in electron identification (see section 12.7). The trajectory of the tracks can also be extrapolated to 'muon' chambers outside the calorimeters. This can be used to improve muon identification and measurement (see Section 12.8).

Fast track reconstruction has been used in some detectors for low level trigger decisions. In general, track reconstruction can be used in high level triggers (see chapter 13).

Tracking detector designs will be very dependent on the particular accelerator at which they are employed. Tracking detectors for low energy collisions will emphasise reducing the material in the detector to minimise multiple scattering. For high energy collisions such as at the LHC it will be necessary to have a large enough tracking volume to measure the momenta of high momentum particles with sufficient precision. For detectors operating with a large number of interactions per bunch crossing ('pile-up') the detector needs to have sufficient granularity to be able to reconstruct trajectories in very dense track environments. This implies that we need detectors with sufficient granularity to ensure that the fraction of elements hit ('occupancy') is sufficiently low.

For tracking detectors using a few high precision layers of silicon detectors, a rule of thumb is that the occupancy should be below 1%. For tracking detectors using many more layers of lower precision gas chambers much higher occupancies can be tolerated.

We will start this chapter with a discussion of the finding and reconstruction of tracks. We will also review the incorporation of precise timing measurements in the track reconstruction ('4D tracking'). An important ingredient for the connection of space-points is the knowledge of the position of the individual detector components, which we refer to as the 'alignment'. We will then discuss the measurement of the momentum of a charged particle from the curvature of its track in a magnetic field and

DOI: 10.1201/9781003287674-10

we will look at common geometries for spectrometer magnets. Finally, we will discuss the reconstruction and measurement of interaction vertices.

10.1   TRACK RECONSTRUCTION

A powerful way to find the kinematic properties of charged particles even in an environment with many particles is the measurement of the position at multiple locations along the trajectory, and the combination of the individual space-points to a track. Depending on the detector used, the tracks are typically reconstructed from either a limited number of points with excellent space point resolution (typical for silicon trackers), or many points with a somewhat poorer resolution (e.g. in a TPC). Independent of the nature of the track, the usual steps in the reconstruction are

For a detailed review of pattern recognition and track reconstruction see [253].

1. Find combinations of space points that are likely to belong to one track ('pattern recognition');

2. Find for each track the track parameters, and calculate the associated covariance matrix to estimate the errors on these parameters. This is usually done by a fit of a track model to the data.

In a homogenous magnetic field the track model will be a helix. In more complex magnetic fields the track model can be found from stepwise numerical integration (for example using the Runge–Kutta–Nyström algorithm [347]).

3. Test the track hypothesis, and identify outliers and wrongly assigned points and remove them from the track fit. This includes detection of kinks in the track, which are indicative of a decay vertex.

The first track finding step can be performed with global or local methods. A popular example of a global method is the Hough transform [253]. A simple demonstration for a Hough transform can be given for the case of hits lying along a straight line $y = cx + d$. The hits are transformed from 'image' to 'parameter' space using $d = -xc + y$. This results in lines in parameter space which overlap at the correct values for c and d in parameter space. Therefore values of (c, d) corresponding to candidate tracks can be found from peaks in the 2D histogram of c versus d. A general discussion of local methods will be given below, and a specific example in the case of ATLAS will be discussed.

For very complex events, e.g. in high pile-up at LHC, there are several complications. One constraint is the limited computer resources, so one approach is to use a limited number of layers to look for 'seed' tracks and then extrapolate the seed tracks to find additional hits on the track. After candidate tracks have been found there will be many cases of hits shared between different tracks. An algorithm is needed to assign these ambiguous hits to the best track. After this process, some tracks will have too few hits and will be rejected. Therefore there will be a trade-off between the track finding efficiency and the number of spurious tracks.

The quality of the track parameters will be affected by measurement errors of the particle position in the different detector layers, errors on the knowledge of the position of the sense elements with respect to each other and a global coordinate system, and disturbances of the particle's path, like for example by multiple scattering, decays and energy loss of the particle. Track reconstruction is complicated in high-rate environments by ambiguities in the assignment of individual hits to specific tracks.

In principle, a simple least square fit of the track model to the measured datapoints gives the optimal estimate of the track parameters (if the measurements are strictly linear functions of the track parameters, and all stochastic disturbances are Gaussian). However, in practice the least squares method requires inversion of large matrices. The most common approach in track reconstruction is therefore to use a Kalman filter [252], which is a two-step recursive method, consisting of a prediction, where the track hypothesis is extrapolated to the next detector layer using the track model and the track parameters determined up to this point, and an update, where the information from this layer is used to update (and ideally improve) the estimate of the track parameters. This sequence is repeated, until all the detector layers are included, or if no suitable hits are found over several layers. After this forward filtering backwards smoothing is performed [253].

A least square fit minimises

$$\chi^2 = \sum_i \left(\frac{x_i^{meas} - x_i^{model}(\vec{p})}{\sigma(x_i)} \right)^2 ,$$

where the x_i^{meas} are the measured positions in 1, 2 or 3 coordinates, $\sigma(x_i)$ are the uncertainties on the measured positions and x_i^{model} are the positions predicted by the track model for the set of parameters $\vec{p}$, and the sum runs over all measured points.

Due to the iterative approach outliers can be rejected at each step, and the algorithm can be used to find the most plausible direction of the track in the case of ambiguities by branching out track hypotheses at ambiguous points and choosing the most plausible combination of hits at the end (determined from the χ^2).

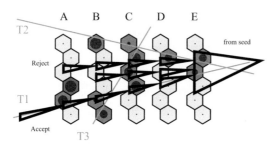

Figure 10.1: Example for the use of a combinatorial Kalman filter for the track reconstruction in the HERA-B outer tracker [357]. The track reconstruction proceeds from the right to the left and the black triangles indicate the track parameters at each step. The reconstruction has been seeded with a first estimate on T1. The detector is a drift chamber with hexagonal drift cells and the hits are indicated by circles with a radius corresponding to the measured drift time.

TRACKING IN LHC INNER DETECTORS

Tracking is very challenging for the general purpose LHC experiments because of the very large pile-up. It is particularly challenging in the core of jets, where the particle density is significantly higher. We will briefly review some of the ideas used for tracking in the ATLAS ID to illustrate how tracking is performed there.

The primary pattern recognition uses an 'inside-out' algorithm. 3D space points from the precision silicon detectors are reconstructed in a global coordinate system. Candidate track segments (seeds) are formed from combination of hits in the pixel detector. The seeds are then used to form 'roads' and these are used with a Kalman filter to successively add hit coordinates to the track. At each stage 'outliers' are rejected using a χ^2 cut on the fitted track segment. This process results in a very large number of ambiguities in the resulting track candidates. The ambiguities are resolved by ordering the tracks according to a measure of quality. This uses the fit χ^2 but it also includes information about whether particular layers were missing hits. Hits that are shared between tracks are generally assigned to the track with a higher score and the quality score is re-evaluated. At the end of an iterative process, track candidates with a low score are discarded. The candidate tracks are then extrapolated into the straw tracker (TRT) and compatible hits there are added to the track.

This is straightforward for the pixel detectors but for the SCT microstrips this requires matching hits on the two stereo sides of a module.

The outlier is rejected, if the χ^2 after inclusion into the track is larger than a threshold value. The threshold depends on the number of degrees of freedom and the probability to falsely reject a genuine hit [252].

This inside-out tracking can fail to find tracks that arise from decays or conversions. Therefore an outside-in tracking stage is also performed to recover this class of tracks. The track finding in the TRT uses Hough transforms in the two 2D views. Global track candidates are simply found by looking for peaks in the 2D histograms in the transformed parameter space. These track candidates are used to seed another Kalman filter to find track segments in the straw chambers of the TRT. These tracks are then extrapolated into the silicon trackers to add any hits on the track that were not already used by the inside-out tracker. This procedure then re-covers the tracks from photon conversions and decays that might have been missed by the inside-out tracking.

The TRT uses longitudinal and radial straw tubes in the barrel and endcaps. Therefore, the barrel (endcap) provides information in the transverse $r - \phi$ $(z - \phi)$ view.

The tracking is particularly challenging in the core of high-p_T jets where the track density is very high. In this environment there are often cases of clusters in the silicon detectors which are assigned to multiple track candidates. The efficient resolution of these ambiguities is the key to maintaining good performance in dense environments. For the case of pixel clusters, a neural network is trained to find merged clusters. The input uses the measured charge in the pixels and the angles of the candidate tracks crossing the cluster.

The tracking efficiency for muons can be estimated from Monte Carlo simulations. The efficiency varies with muon η and p_T. In order to minimise systematic uncertainties associated with inaccuracies in the simulation, data-driven methods are used to correct the simulation. This is done using the 'tag and probe' method, investigated in exercise 6.

It is more difficult to make a data-driven efficiency measurement for pions and therefore this is usually estimated using Monte Carlo simulated samples. An example plot showing the efficiency as a function of the angular separation between the track and the jet axis [79], ΔR is shown in Figure 10.2. We can see that there is a significant decrease in efficiency in the core of a jet which is due to the very high local particle density. The loss of efficiency becomes more significant as the jet-p_T increases because of the increase in multiplicity of jets with p_T and because the jet becomes more collimated. The efficiency 'plateau' at large values of ΔR is significantly below 100%. This is mainly due to hadronic interactions in the material of the inner detector. The efficiency for muons which do not undergo these interactions and are not in the core of jets is very much closer to 100%.

We can use the same type of tag and probe methodology to make many other measurements such as trigger and electron finding efficiencies. There is no low-background source of di-pion events.

$\Delta R = \sqrt{(\Delta\phi)^2 + (\Delta\eta)^2}$, where $\Delta\phi$ is the distance in the azimuthal angle between the track and the jet axis, with an equivalent definition for the pseudorapidity interval $\Delta\eta$.

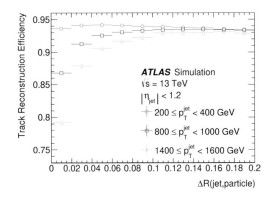

Figure 10.2: The track finding efficiency as a function of angular separation from the jet axis, ΔR for jets in three different momentum ranges [79].

Electrons can undergo large energy loss by bremsstrahlung and the radius of curvature of the electron in the magnetic field will decrease when this happens. The classic Kalman filter approach cannot accommodate this sudden change. In these cases ATLAS uses a modified Kalman filter that

The probability will be significant in any tracking detector which has $\mathcal{O}(1)$ radiation length.

allows for up to 30% energy loss at each material intersection [80]. Finally for tracks loosely matched to electromagnetic clusters, a Gaussian sum filter (GSF) algorithm can be used to obtain optimised estimates of the track parameters.

GSF is a non-linear generalisation of the Kalman filter.

10.2　4D TRACKING

Traditional particle detectors reconstruct the tracks of particles accurately in space but not in time. An ideal detector would provide accurate information in 3 spatial and 1 time coordinates, which would allow full 4D tracking. As we have discussed in chapter 9 the pixel and strip detectors that provide precise spatial resolution have modest time resolution. On the other hand, silicon detectors that have been optimised for very good time resolution (for example LGADs, see section 9.9) use pad detectors giving very modest spatial resolution, and these are going to be introduced into the LHC general purpose experiments.

The LHC tracking detectors had just enough time resolution to assign hits to the correct bunch crossing.

At the general purpose LHC experiments the number of interactions in the same bunch crossing (μ) is very large at peak luminosity. We wish to reconstruct the particles from the 'hard-scatter' that triggered the event and not be affected by the particles from pile-up. The jet reconstruction can result in fake jets from particles from pile-up. If we can resolve the different primary vertices, we can reject these pile-up jets by requiring that a large fraction of the tracks associated to a jet are from the primary vertex. This works well at LHC but would not be effective at HL-LHC. At HL-LHC the maximum pile-up is $\mu = 200$ interactions, and the mean density of vertices along the beam direction will be around 1.4 mm^{-1}, with fluctuations up to a density of 3 mm^{-1} [82]. This figure is too high for the tracking detectors to efficiently separate these vertices.

ATLAS measured a luminosity weighted average value in 2018 of $\mu = 36.1$ [81].

The effects of pile-up can be mitigated by the addition of a timing layer in the forward and backward directions. Tracks from very closely spaced 3D vertices can be efficiently separated if there is timing information. This can be seen from the following simple example: the vertex location (z_0) for collisions between particles at the 'head' of the bunches in both beams will result in the same value of z_0 as for collisions between particles at the tail of the bunches in both beams. However, the time of the interaction will be earlier for the first case.

The resolution in the longitudinal impact parameter is worse in the forward directions.

A sufficient separation of vertices can be achieved with a timing resolution per track of 30 ps, which can be achieved with LGADs (see section 9.9).

A more ambitious use of 4D detectors has been proposed for the LHCb Upgrade II, with the aim of creating a full 4D tracker [399]. Different technologies are being considered for the various systems. 3D silicon pixel detectors are a very attractive option for the detectors closest to the beam pipe as they have the best radiation tolerance and can have thicker active depths and thus give larger signals which improves the timing resolution. The aim is to achieve a few tens of ps resolution per particle. This would allow the same performance to be achieved as for Upgrade I but for a further increase in luminosity by a factor of 7.5.

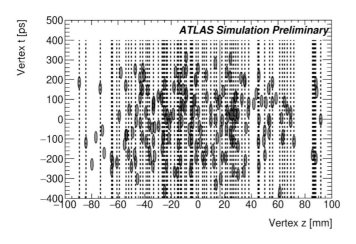

Figure 10.3: Simulation of an event with a hard scatter (at $z \simeq 15$ mm) and 200 pile-up interactions. The plot shows vertex position z versus time. The dashed vertical lines represent the positions of reconstructed vertices and the ellipses represent the positions of 'truth' vertices. The size of the ellipses represent the approximate expected resolution in z and time, 1 mm and 30 ps respectively [82].

10.3   ALIGNMENT

To correctly link the space points recorded in different parts of the tracking system, the position of all detection elements in a global coordinate system must be known to a level that it does not introduce a significant degradation of the spatial resolution beyond the intrinsic element resolution. Several approaches are possible.

For a semiconductor tracker this requirement can be $\mathcal{O}(1\ \mu m)$.

Build accuracy

During assembly, the position of the detection elements can be controlled by precision machining (in reductive or additive manufacturing), or by precision moulds or jigs (for composites or gluing). However, the typically required accuracy can usually only be achieved for items of limited size.

Assembly surveys

After construction, the position of detector elements can be measured, either by an optical or mechanical survey, or using X-rays.

For an example of the use of X-ray tomography in the survey of large wirechambers see [449].

For both these approaches, precision build and/or assembly survey, the limitation is that deformations of the geometry after the build, whether during subsequent stages of the installation or due to mechanical or thermal loads during the operation of the detectors, are not captured. However, for objects of limited size and high internal stiffness assembly-based spatial information can remain accurate, so that the information can be useful to reduce the number of parameters to be determined by other alignment methods.

In a magnetic spectrometer changes in the state of the magnetic field can also be a significant source of deformations.

Hardware alignment

During operation deformations of the mechanical structures supporting the detection elements can be tracked by hardware alignment systems. Such systems are typically optical, and can achieve sub-μm absolute accuracy, although the accuracy will degrade with the size of the system, and be worse for large muon systems. Examples for measurements in the direction of the optical axis are Frequency Scanning Interferometry (FSI) as used in the ATLAS SCT [208], or for measurements perpendicular to the

In a magnetic spectrometer optical alignment beams can be interpreted as infinite momentum tracks.

optical axis the Red Alignment System Nikhef (RASNIK) technology as used in the ATLAS muon system [266]. Standalone optical alignment systems have the disadvantage that they require a step of extrapolation from the position of the optical elements to the detector elements. This can be overcome by using the detector elements themselves for the detection of the optical alignment reference as done for (some set of the modules of) the CMS tracker [457].

For a more detailed discussion of the alignment of the ATLAS muon system, see [23].

Track-based alignment

In track-based alignment (also known as software alignment) a high-quality subset of the data recorded in actual physics events is used to reconstruct the positions of the detector elements offline. The optimal software alignment algorithm uses the global χ^2 fit approach. A χ^2 is calculated as the sum of the residuals for all the hits in the sample of high-quality track fits and the optimal alignment constants are obtained by minimising this χ^2. This technique has been used in the alignment for the inner trackers of the LHC experiments (see for example ATLAS [83], CMS [198]) with great success, demonstrating that the position of the modules in these large systems can be aligned at the μm level, and that software alignment is even capable of reconstructing the shape of individual detector modules.

A high-quality subset comprises typically high-momentum tracks with small errors on the track fit parameters.

If information from a hardware alignment system or tracks from cosmics are available they can be included in the χ^2 fit.

Track-based alignment works best with detector systems that have a high level of internal stiffness and high point position capability. It is working so well in the LHC silicon trackers that the dedicated hardware alignment systems described above are not required to determine sensor positions.

The number of degrees of freedom for a large silicon detector system can be several 10^5, and it is typically beneficial to introduce a hierarchy of sub-structures, which matches the actual hierarchy of mechanical support structures, starting at large units (barrels, endcaps, etc.) with limited degrees of freedom, and becoming progressively more detailed. Experience shows that movements of large units are most frequent, while internal deformations of these units are usually small, so that internal realignment is required infrequently (only every several months).

One of the challenges in the more refined levels of alignment is that the global χ^2 fit leaves certain classes of coherent degrees of freedom (twists, telescoping, etc.) undefined, with more or less arbitrary position parameters. These deformations are called 'weak modes', and they need to be constrained by other means (cosmics, higher-level analysis like mass peaks, etc. and their invariance in different directions).

Probably the most important of these weak modes is curl because this can result in equally good track fits but with very different values of the reconstructed p_T and even result in charge asymmetries. These weak modes can be very well constrained by cosmic rays (mainly muons) in the barrel region that pass through the entire detector. The reconstruction of a common track with one value of p_T very strongly constrains the curl. Another powerful technique to constrain the curl is to measure the energy E and momentum p of electrons. Any curl will bias the distribution of E/p.

Curl refers to correlated rotations in the azimuthal angle of different radial layers in such a way that the particle trajectories still give good helical fits.

Cosmic rays are mainly vertical and thus cross the barrel layers of a collider experiment. In the endcaps, detectors are typically perpendicular to the beam line and therefore the cosmic rays do not cross many endcap layers. For the endcap detector, beam halo tracks can be used instead.

Naively for ultra-relativistic electrons $E/p \simeq 1$. However the measurements are severely affected by interactions like bremsstrahlung in the detector material which biases E/p. Therefore a very accurate knowledge of the detector material is essential for this technique to work.

Other weak modes are described in [235]. Several types of measurements can be used to constrain these modes including the reconstructed masses of decays like $K_s^0 \to \pi^+\pi^-$ and reconstructing beam gas events.

Software alignment is the most powerful in the direction where the tracking layers are the most precise (typically in the detector plane for planar detectors); it is less capable in the direction perpendicular to these layers.

The most important structural design requirement to support track-based alignment is stability over periods that correspond to offline alignment cycles. Typical loads under which that stiffness needs to be achieved are (in order of relevant timescale):

Currently at the LHC these periods range typically from $\mathcal{O}(24\,\text{h})$ for high level structures to a few times per year for small structures.

- Vibrations (external from ground vibrations or other experimental equipment, or internal, for example from coolant flow);

- Thermo-mechanical loads, due to differential coefficients of thermal expansion and local and/or temporary variations of the temperature;

An example for deformation due to differential thermal expansion is the ATLAS IBL [6], which shows significant detector movements as the cooling temperature is adjusted and as the luminosity varies during a fill of the LHC. Dedicated software procedures are required to correct for these changes [83].

- External, 'seismic' shocks, due to drastic changes of the environmental state of the experiment (cooling system shutdowns, planned or unplanned magnet ramps etc.);

- Long-term deformations, like long-term dimensional changes caused by humidity variations, or relaxation processes in polymer components (plastics and glues).

These relaxation processes are sometimes referred to as 'creep'.

Usually, these loads are small or can be made small for the static experiments typical in particle physics, allowing for the aggressive optimisation of support structures to reduce mass.

10.4 MOMENTUM MEASUREMENT IN MAGNETIC FIELDS

In a homogeneous magnetic field B the trajectory of a particle with charge ze becomes a helix with radius R and pitch angle λ. The momentum is then given from the Lorentz force

Often a dimensional version of this equation is used
$$p_t/\text{GeV} = 0.3z(R/\text{m})(B/\text{T}).$$

$$p_\text{t} = p\cos\lambda = zeRB.$$

In the case of three equidistant position measurements, the middle station provides a measurement of the sagitta with respect to the base line set up by the two outer stations. The sagitta of the track in the magnetic field relates to the momentum,

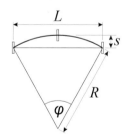

$$s = R - R\cos\frac{\varphi}{2} \simeq R\frac{\varphi^2}{8} \simeq \frac{R}{8}\frac{L^2}{R^2} = \frac{L^2}{8R} = \frac{zeBL^2}{8p_\text{t}}.$$

If the sagitta measurement is used to determine the momentum, then the error on the momentum measurement is related to the error on the sagitta

$$\frac{\sigma_{p_\text{t}}}{p_\text{t}} = \frac{\sigma_R}{R} = \frac{L^2}{8sR}\frac{\sigma_s}{s} = \frac{\sigma_s}{s} = 8R\frac{\sigma_s}{L^2} = \frac{8}{ze}\frac{p_\text{t}}{BL^2}\sigma_s.$$

Figure 10.4: Sagitta geometry.

The sagitta error is closely correlated with the position measurement error in each layer. The exact relation depends on the number and position of the point measurements. For the simple case discussed here with three equidistant stations $s = x_2 - (x_1 + x_3)/2$ and therefore, assuming the same position measurement error σ_x in the three stations,

L is sometimes called the 'lever arm' of the momentum measurement.

For a large number ($n > 10$) of equidistant planar position measurements on a parabolic track [264]
$$\sigma_s = \frac{1}{8}\sqrt{\frac{720}{n+4}}\sigma_x.$$

$$\sigma_s = \sqrt{\sigma_x^2 + 2\frac{\sigma_x^2}{4}} = \sqrt{\frac{3}{2}}\sigma_x.$$

For more general expressions, including other geometries, see [420].

It should be obvious that for a good momentum measurement a large
tracking volume, a large magnetic field and good space point resolution
are required.

At low momenta, the effects of multiple scattering must also be considered. If we stay with the simple system of three thin stations for the
sagitta measurement with no scattering material between, then for the
three stations

*This estimation is only a justification
for the functional form of the result. The
challenge of all calculations of multiple
scattering is the integration of the effect
for thick slabs of material. An analytical
calculation is difficult and the practical
alternative is to obtain the multiple
scattering distributions using Monte
Carlo.*

$$\Delta x_1 = 0, \quad \Delta x_2 = \frac{L'}{2}\Delta\theta_1, \quad \Delta x_3 = L'\Delta\theta_1 + \frac{L'}{2}\Delta\theta_2,$$

where $L' = L/\sin\Theta$, where $\Theta = \pi/2 - \lambda$ is the angle between the track
and the magnetic field, and the $\Delta\theta_{1,2}$ are the deflection angles in the first
and second station, respectively. Therefore

$$\Delta s = -\frac{L'}{4}\Delta\theta_2 \text{ and } \sigma_s = \frac{L'}{4}\sigma_\theta \simeq \frac{L'}{4}\theta_0,$$

where θ_0 is given by eq. (2.53).

We simplify eq. (2.53) further (which, as discussed, already ignores
the large-angle scattering tails), by dropping the logarithmic part to get

$$\left(\frac{\sigma_{p_t}}{p_t}\right)_{MS} = \left(\frac{\sigma_s}{s}\right)_{MS} \simeq \frac{\frac{L'}{4}\frac{13.6\,MeV}{\beta cp}z\sqrt{\frac{L'}{X_0}}}{\frac{zeBL^2}{8p\sin\Theta}} \propto \frac{1}{\beta B\sqrt{LX_0}\sin\Theta}.$$

The combined transverse momentum resolution can thus be described by

*For very low momenta energy loss
fluctuations in detector material in front
of the spectrometer can also be relevant.*

$$\frac{\sigma_{p_t}}{p_t} = C_1\frac{p_t\sigma_x}{BL^2} \oplus C_2\frac{1}{\beta B\sqrt{LX_0}\sqrt{\sin\Theta}}. \tag{10.1}$$

*'$\oplus$' denotes a quadratic addition of
errors.*

Hence, multiple scattering will be the dominant contribution at small momenta, while the position measurement error will dominate at high momenta and therefore the momentum resolution will increase linearly for
high momenta. A good resolution requires large B and L, and good space
point position resolution (and detector alignment) and small amounts of
scattering material.

*Note that a large L usually implies a
large magnet, and the stored energy
of the magnet is proportional to the
volume.*

For the resolution on total momentum, the error on the pitch angle λ
needs to be included [483],

$$\frac{\sigma_p}{p} = \frac{\sigma_{p_t}}{p_t} \oplus \sigma_\lambda \tan\lambda. \tag{10.2}$$

The fractional momentum resolution of the ATLAS inner detector [84] as
a function of p_T shows a constant value at low values of p_T as expected
from multiple scattering. At large values of p_T the resolution is dominated
by the error on the sagitta measurement and therefore the fractional error
increases with p.

10.5   SPECTROMETER MAGNETS

In addition to the need for a high magnetic field and a large size to allow
for a long lever arm, other considerations are relevant for the design of the
spectrometer magnet: the field should be homogeneous, to simplify the

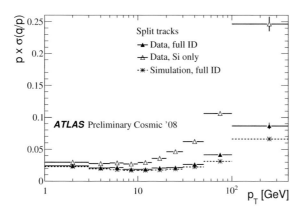

track fitting and reduce non-uniformity of the response of the tracking detectors (for example due to changes of the Lorentz angle). The field lines have to loop back outside of the tracking volume. One way to do this in a controlled fashion is by providing a return yoke made from iron. The return yoke for central solenoid fields in collider experiments can be instrumented with detectors for the tracking of muons, as the iron acts as a muon filter, stopping all other particles (except neutrinos).

In practice, the magnetic field will have significant non-uniformity. For example, a solenoid gives a very uniform field in the centre of the magnet, but a weaker and non-axial field near the edges of the solenoid.

These are usually gaseous muon detectors.

In particular in collider experiments the coil must fit into the available space, and material in magnet components crossed by tracks must be minimised to limit multiple scattering. There is a fundamental design choice in collider experiments whether to place the electromagnetic calorimeter (see chapter 11) within the coil to minimise the material in front of it with the penalty of constraining its depth, or outside. In the latter case, the deterioration of the energy resolution can be reduced by the use of a pre-shower detector in front of the calorimeter. To reduce the size of conductors and supply cables spectrometer magnets in collider experiments are usually superconducting. Finally, cost is a major factor limiting the size and strength of spectrometer magnets.

For a detailed discussion of superconducting detector magnets see [368].

Three magnet geometries are in general use:

- *Dipole:* Large-volume dipoles are primarily used in fixed target experiments, where they provide a large unimpeded opening for the entry and exit of the close-to-parallel tracks in these experiments. The coils and return yoke can be placed outside of the detector acceptance.

  In collider experiments dipole magnets have the disadvantage that the bending occurs only in one direction and their use there is only historic (for example in UA1 [75]), with the exception of forward-oriented experiments like LHCb [54].

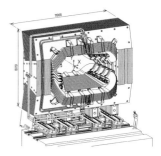

Figure 10.6: Dipole magnet of the LHCb experiment [332]. Note that in the drawing the rear aperture (upstream) is smaller than the front to minimise the field volume in accordance with the acceptance of the experiment.

- *Solenoid:* Solenoids are today the standard geometry for the central tracking systems in collider experiments. They provide high field uniformity and invariance of the magnetic field under rotations by the azimuth angle ϕ. The drawbacks are that this geometry only provides a direct measurement of the transverse momentum p_T, and low momentum tracks can be trapped in the field and cause high occupancies in the tracking detectors.

The momentum resolution is thus poorer for tracks in the forward direction. However, this is also the direction where additional material increases multiple scattering.

The superconducting magnet with the largest stored energy currently is the CMS solenoid with a stored energy of 2.6 GJ [285]. It has a free bore diameter of 6 m and a maximum magnetic field of 4.0 T that is achieved with a current of 19.5 kA. The return flux of the solenoid is through the iron in the hadronic calorimeter. The large bore diameter allows for the inner tracker together with the electromagnetic calorimeter to be contained inside the solenoid, thus minimising the 'dead' material in front of the electromagnetic calorimeter.

In CMS operation the magnetic field is 3.8 T (Superconducting magnets are usually operated slightly below the maximum current, to minimise the risk of a quench, a sudden loss of superconductivity).

The solenoid represents passive material between the electromagnetic and hadronic calorimeter. The thickness is $3.9\,X_0$.

An example of a smaller solenoid is that used by ATLAS [508], which has a magnetic field of 2.0 T using a current of 8.4 kA. The solenoid has an inner diameter of 2.3 m and a length of 5.3 m. The stored energy is 44 MJ. The inner tracker is inside the solenoid, but the electromagnetic calorimeter is outside. Therefore, minimising the thickness of the magnet (in units of radiation lengths), resulting in a somewhat lower field, has been critical in order not to degrade the resolution for electrons and photons. The impact of the passive material of the solenoid on the electromagnetic calorimeter is reduced by the use of a pre-shower detector.

The solenoid represented a thickness of $0.66\,X_0$ at normal incidence [85].

- *Toroid:* Toroidal magnet geometries again are in principle symmetric in the azimuth angle, and thus are attractive for collider experiments, and they have the benefit that they provide a direct measurement of the momentum. However, particles have to cross the inner part of the coil, and thus the most common application for toroid magnets is for muon spectrometers. Because of the size of such systems, it is usually not practical to realise the field with one continuous coil, but the toroid field is approximated by several discrete coils, which introduces non-uniformities of the field.

Another feature of a toroidal field is that in the centre there is no field, which has been exploited in the CLAS fixed target experiment at CEBAF to allow for the use of dynamically polarised targets [365].

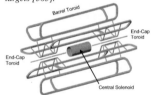

  The largest toroid magnet system today is part of the ATLAS experiment, with an air-core toroid consisting of 8 coils in the barrel, and two endcaps with 8 coils each, surrounding the central solenoid.

Figure 10.7: ATLAS magnet systems [508].

For all practical magnets the field will show significant non-uniformities over the detector volume. Therefore it is essential to accurately measure the magnetic field over the relevant volume. This can be done using Hall and NMR probes [39].

10.6   VERTEX RECONSTRUCTION AND MEASUREMENT

Vertices can be classified in different categories:

- A *primary vertex* is the point of collision of the particles in the two beams in collider experiments, or the position of the target particle in a fixed target experiment. In high-intensity environments more than one primary collision can take place close in time (e.g. within the same bunch-crossing at a collider). Primary vertices are constrained by the size of the beams and, when applicable, the target.

Even though the collisions involve quantum mechanical states with non-localised wavefunctions, on the scale of the detector resolution the concept of a well-localised particle collision is meaningful.

- A *secondary decay vertex* is the point where an unstable particle decays into daughter particles. These can be short-lived particles like B or D mesons, or τ, in which case the secondary vertex will usually

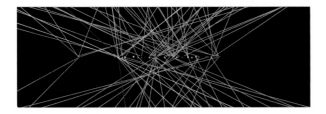

Figure 10.8: $H \to ee\mu\mu$ candidate event at a collision energy of 13 TeV [100]. The Higgs boson candidate is reconstructed in a beam crossing (beams from left and right) with 25 additionally reconstructed primary vertices from minimum bias interactions. Tracks above 1 GeV are shown. Red lines show the path of the two muons, green lines show the paths of the two electrons, and the yellow tracks are the remaining charged particles from the Higgs boson candidate vertex. Grey tracks originate from other vertices.

be well before the first tracking station, or long-lived particles like K or Λ, which decay typically within the instrumented volume of the experiment.

- A *secondary interaction vertex* is the point of the interaction of a particle created in the primary collision with material in the detector. Examples for such interactions are bremsstrahlung, pair production, and inelastic hadronic interactions. These vertices need to be understood, as they otherwise can be a source of mis-measurements. The identification of conversion vertices is important for reducing the backgrounds to prompt electrons. Photons converting in the tracker material can be identified from the tracks of the e^+e^-. The identification of bremsstrahlung is used for an improved determination of electron energies.

This is particularly important in the ATLAS and CMS detectors for which the total radiation length of the tracking detectors is $\mathcal{O}(1X_0)$.

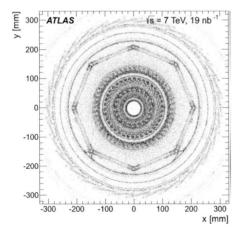

Figure 10.9: Position of secondary vertices in the x-y plane for the inner parts of ATLAS [1]. $x < 140$ mm is the pixel system, 140 mm $< x < 250$ mm are support structures and outside of these is the first layer of the SCT. Vertices inside the beam pipe are masked out.

Vertex reconstruction makes use of reconstructed tracks. It typically consists of vertex finding, in which compatible tracks are assigned to a vertex candidate, and vertex fitting, where the position of the vertex is found, usually by a least square fit, or again by a Kalman filter. Additional constraints, either due to the decay kinematics or the beam size can be included (typically as Lagrange multipliers).

More robust techniques de-weight or ignore tracks with lower compatibility.

For short-lived particles the full reconstruction of the decay vertex is difficult and can result in loss of efficiency. Therefore, a parameter that is often used to determine secondary vertices is the impact parameter of a track, which is the distance of the primary vertex to the point on the parameterised track that is closest to the primary vertex (sometimes called the perigee). In a cylindrical tracking detector geometry the impact parameter can be split into a transverse impact parameter δ_t, and a longitudinal impact parameter δ_z. We assume that the resolution due to multiple scattering is uncorrelated with the intrinsic detector resolution. It can be shown (for example in [228]) that the error on the impact parameters can be parameterised as

See exercise 5.

$$\sigma_{\delta_t} = C_1 \oplus \frac{C_2}{p_t\sqrt{\sin\Theta}},$$

and

$$\sigma_{\delta_z} = C_3 \oplus \frac{C_4}{p_t\sqrt{\sin^3\Theta}},$$

where C_1 and C_3 parameterise the resolution due to measurement errors in the detector and C_2 and C_4 parameterise multiple scattering. In complex detector systems the track is also extrapolated to match it with signals in other sub-detectors, most importantly the calorimeters, but also possible particle identification detectors or other tracking systems (in particular for muons).

Key lessons from this chapter

- Track fitting combines several space points to obtain overall kinematic properties of the track from a primary particle. In a magnetic field this includes the momentum, which can be determined from the curvature.

- Track reconstruction involves pattern recognition, track fitting and rejection of signals that do not belong to the track.

- The momentum resolution perpendicular to the magnetic field can be parameterised as

$$\frac{\sigma_{p_T}}{p_T} = A p_T \oplus B,$$

  where A depends on the space point resolution of the detector and B on the multiple Coulomb scattering (i.e. material budget).

- Secondary vertices can be reconstructed from tracks. Decay vertices of short-lived particles can be characterised by the impact parameter. The impact parameter resolution can be parameterised as

$$\sigma_\delta = A' \oplus \frac{B'}{p_T}.$$

- Good knowledge of the positions of all the detector elements is required. For the case of large tracking detectors in collider experiments, the data can be determined from software alignment, provided that the detector is sufficiently stable.

EXERCISES

1. *Determination of sign of charge for charged particle.*

   Assume a tracker with a large number ($N > 10$) of space points with a point resolution σ_x in a uniform magnetic field.

   a) Neglecting multiple scattering, find the maximum momentum for which the sign of the particle can be determined with a significance of 3 σ. (Hint: Find the relation between $\sigma_{R^{-1}}/R^{-1}$ and σ_s/s and then use the expression for σ_s for large n, to find $\sigma_{R^{-1}}$.)

   b) What is this momentum for a tracker with $N = 12$, a lever arm of $L = 1.2$ m, and a point resolution of $\sigma_x = 10$ μm, in a magnetic field of 2 T? What is this momentum for a tracker with a similar lever arm, but for $N = 50$, $\sigma_x = 100$ μm, and $B = 1$ T? Which detector technologies could be used to achieve these specifications?

2. *Momentum measurement in a dipole spectrometer.*

   A dipole spectrometer consists of a dipole magnet between two tracking detectors of length D. The magnetic field is assumed to be uniform over a length L and zero outside. The momentum of an incoming particle can be inferred from the deflection angle θ between the straight particle track before and after the magnet. Before the magnet the angle of the track to the axis of the experiment, which is perpendicular to the magnetic field, is α, and after the magnet it is $\beta = \alpha + \theta$. You can assume that all angles are small.

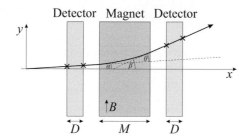

Detector Magnet Detector

a) Show that, if the tracking detectors consist of two tracking layers, separated by the distance D, and each with a point measurement error of σ_y perpendicular to the magnetic field, the error on the slope m of the track measured by one tracking detector is given by

$$\sigma_m = \sqrt{2}\,\frac{\sigma_y}{D}.$$

b) Use this to find the error on the measurement of α, β and hence on θ.

c) Show that $p = 0.3BR$, where p is the momentum transverse to the magnetic filed is units of GeV/c, B is the magnetic field in T, and R the radius of curvature of the track in the magnetic field in m. Find $dp/d\theta$, and then, using your result from b), the momentum resolution σ_p/p. For a spectrometer with $\sigma_y = 100\ \mu$m, $D = 1$ m, $B = 1.5$ T and $L = 2$ m, what is the momentum resolution for muons with a momentum of either 10 or 100 GeV/c?

d) The best estimate for the slope of a straight track measured with N spacepoints with coordinates (x_i, y_i) can be obtained from a least-square fit and is given by

$$m = \frac{N\sum(x_i y_i) - \sum x_i \sum y_i}{N\sum x_i^2 - (\sum x_i)^2},$$

where the sums run from 1 to N. Show that the uncertainty on the slope for equally spaced measurements from 0 to D, which all have an point measurement accuracy of σ_y is given by

$$\sigma_m = \frac{\sigma_y}{D}\sqrt{\frac{12(N-1)}{N(N+1)}}.$$

e) Hence show how the resolution of the dipole spectrometer improves when additional detection layers are inserted into the upstream and downstream detectors (while maintaining the overall depth). What performance drawback that is not covered in these equation will this introduce?

3. *Impact parameter measurement.*

The figure shows a B meson produced at primary vertex V_1 decaying at vertex V_2. The impact parameter δ for a lepton from the decay is defined in the figure.

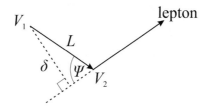

a) Explain the meaning of L and Ψ, and derive the relation

$$\delta = \frac{c\tau_B \beta \sin \theta^*}{1 + \beta \cos \theta^*},$$

where c is the magnitude of the velocity of the B meson in the laboratory frame, τ_B is the proper lifetime of the B meson and θ^* is the angle between the direction of the lepton and the B meson in the latter's rest frame.

b) For a typical B meson lifetime of 1.6×10^{-12} s, $\theta^* = \pi/2$ and assuming that the B meson takes 80% of the b-quark energy, calculate δ. What quantities must be measured to determine δ and what detector components are used for this in a typical collider experiment?

4. *Impact parameter resolution.*

A simple vertexing system at a collider experiment consists of two concentric cylindrical layers with radii of $R_1 = 50$ mm and $R_2 = 70$ mm around the collision point, respectively. The point measurement resolution of this system in the r-ϕ direction is $\sigma_{r\phi} = 10$ μm.

a) Show that, neglecting multiple scattering, the uncertainty of the impact parameter (the closest distance of the reconstructed track to the beamline in the plane perpendicular to the beam) is given by

$$\sigma_\delta = \frac{\sqrt{R_1^2 + R_2^2}}{R_2 - R_1} \sigma_{r\phi},$$

and evaluate this for the parameters given. How does σ_δ change if (i) R_1 is increased by 10 mm, (ii) R_1 and R_2 are both increased by 10 mm? What limitations are their to decreasing R_1, or increasing R_2?

b) Assume that each layer has a thickness of 1% X_0. How does multiple scattering affect the impact parameter resolution? For what momentum would the uncertainty on the impact parameter introduced by multiple scattering equal the uncertainty caused by the point measurement error? What other source of multiple scattering would need to be considered in a real experiment?

5. *Multiple scattering and impact parameter resolution.* Consider a cylindrical collider detector with a radius of the first layer, R. The first layer has a thickness of s in units of radiation length. Assume that the trajectory is accurately reconstructed after the first layer. Consider a particle produced at the centre of the detector with a momentum p and polar angle θ. Let α be the multiple scattering angle at the first layer.

a) How does the uncertainty in α, $\sigma(\alpha)$ scale with p and θ?

b) The uncertainty on the transverse impact parameter is $\sigma(d_0) = \sigma(\alpha)L$, where L is the distance from the vertex to the detection point in the first silicon layer. How does $\sigma(d_0)$ scale with s, R and θ?

6. *Tag-and-probe method to determine efficiencies from data.*

Consider a sample of muons from single muons triggers which is reconstructed in both the tracking and muon detectors. We can then search for an additional muon in the event using muons found using only the tracking detector and for for each pair we calculate the invariant mass. Cuts are then made on $m_{\mu\mu}$ to obtain a sample of $Z \to \mu^+\mu^-$ events. We can then use the triggered muon as the 'tag' and the other muon found in the tracking detector as the 'probe' to measure the efficiency of identifying a muon in the muon detector (the procedure can be reversed to find the efficiency of identifying a muon in the tracking detector). Let the number of probes be N_{probes} and let the number that pass the muon identification be N_{pass}.

a) Assume that all muon pairs come from $Z \rightarrow \mu^+\mu^-$ decays. What is the average muon identification efficiency? What is the statistical uncertainty?

b) How can this method be extended to measure the efficiency as a function of muon p_T or η?

c) How can this method be made to work for a real sample of muon pairs which originate from an unknown mixture of $Z \rightarrow \mu^+\mu^-$ and background events?

d) In a sample of $Z \rightarrow \mu^+\mu^-$ events, in which no charge selection has been made, there are N_{SS} same-sign and N_{OS} opposite sign events. Ignore background contributions, what is the probability to wrongly measure the charge sign of a muon? Why is the probability of wrongly measuring the charge of electrons generally much larger than that for muons?

11 Calorimetry

In the previous chapters we have seen how the quasi-continuous small energy loss of charged particles enables the detection of these particles in tracking detectors, typically made of thin detector layers. A different approach is taken in calorimeters, where the particle ultimately is fully absorbed. If the particle has a high energy it is lost in a cascade of secondary interactions until all particles in the cascade have come to a stop. Despite the name, calorimeters in particle physics usually do not measure the heat generated in the detector material, as this is usually too small to be detected, except at extremely low temperatures, and the typical signal is either ionisation charge or scintillation light.

We will start this chapter by looking at the energy measurement of γ rays from nuclear decays, as this does not involve the more complex effects in electromagnetic cascades that are instigated by high-energy incoming particles. We will see that the resolution of this measurement is defined by the counting statistics of secondary quanta in the detector. One way to improve this measurement therefore is to rely on quanta that require little energy for creation and are thus produced in abundance. Thus detection of lattice vibrations (phonons) allows for very good energy resolution at low energies, but they can only be detected at ultra-cold temperatures of a few mK.

For incoming particles with high energies, the secondaries created by the interaction with the detector matter have enough energy to create further particles with sufficient energy to create a cascade of particles with progressively lower energy. For incoming particles with higher energy the number of secondary particles in the cascade will be larger and therefore the relative energy resolution improves with energy. We will first discuss cascades caused by electromagnetic processes, and consider some examples of homogeneous and sampling electromagnetic calorimeters. We will then discuss the more complex cascades induced by hadrons and the implications of this complexity for hadronic calorimeters. We will finally look at examples of hadronic calorimeters and the different approaches to optimising the resolution.

As the processes causing the cascades apply for neutral as well as charged particles, calorimeters are usually the only type of detector that can detect stable neutral particles, and measure their kinematic properties (i.e. their energy and direction).

Historically, in early nuclear physics, indeed the heat created in radioactive decays was measured (see for example [393]). However, this was done for the radiation from sources comprising a large number of particles. The name was then applied to detectors that detected neutral particles through a cascade of interactions as described later in this chapter. The refinement for particle physics is then that the particle's energy is converted into a proportional signal. This is the meaning of the term in particle physics today.

We will see that the statistical fluctuations are actually often smaller than expected from pure Poisson fluctuations, due to correlations in the creation of the secondary quanta.

One special case of a neutral particle that can be reconstructed in a tracking detector is a photon that undergoes pair conversion in the material of the tracking detector, which allows reconstruction of the tracks of the e^+e^- pair and hence the measurement of the γ energy and direction.

11.1   GAMMA SPECTROSCOPY

Photons from nuclear transitions have a well-defined energy between a few keV and about 10 MeV. As discussed in section 2.1, these photons will interact with the sensor material in a detector, depending on their energy, dominantly by photo-electric effect, Compton scattering or pair production. Each of these processes will create free electrons (and positrons) that will deposit the energy they accrued from the original interaction of the gamma ray, by ionisation and excitation of the sensor material.

DOI: 10.1201/9781003287674-11

The active detector volumes for the detectors for low energy γs are usually too small to guarantee full containment so that, depending on the interaction mechanism, not necessarily all the energy of the original photon will be deposited in the detector, and the measured spectrum contains several characteristic features:

For high-energy calorimeters as discussed later, these features will not be visible.

Due to the finite energy resolution of the specific detector some of these features may not be visible in the measured spectrum.

- *Photopeaks* are full energy peaks, where the photon transfers all its energy to an electron in the detector material by the photoelectric effect. The range of the electron is typically smaller than the size of the detector, and thus all the energy of the incoming γ is collected.

In solids the range of a 1 MeV electron is typically less than 1 mm.

- The *Compton edge* is the upper endpoint of a continuum of deposited energies from electrons released by Compton scattering. The Compton edge energy is given by eq. (2.11). While the electron from the Compton scattering has a short range and usually deposits all its energy in the detector, the scattered photon often does not interact again in the detector, and this results in the continuum of recorded energies.

In nuclear spectroscopy sometimes an 'anti-Compton shield' or 'Compton-suppression shield' is used, which is a thick scintillator surrounding the gamma detector, which vetoes events in which a photon gets scattered out of the detector into the scintillator.

- If the photon interacts by pair-production the resulting electron-positron pair will share the energy available after their creation, $E_\gamma - 2m_e c^2$. Both will lose their energies in the detector material by ionisation and excitation until they come to a stop. The positron will then annihilate with an atomic electron, creating a pair of photons, each with an energy of $m_e c^2 = 511$ keV. The probability for these photons to interact in the detector is limited, causing *single-escape* and *double-escape peaks* that are 511 keV and 1022 keV below the photopeak, respectively.

In addition, there can be features that originate in the material surrounding the sensor, typically from photons generated in the surrounding material:

Again, in addition to the geometry and type of material around the sensor, the height and width of these features, and whether they are visible at all, will depend on the size, efficiency and resolution of the detector.

- There can be a symmetric *reverse Compton edge*, which is caused by photons that have been backscattered into the sensor from surrounding material, and then deposited their remaining energy by the photoelectric effect.

- When the photons undergo a photoelectric effect in surrounding materials, the emitted electron leaves a vacancy that gets filled by another electron with the emission of an X-ray with an energy characteristic for the surrounding material. This can be captured by the sensor, resulting in *X-ray peaks* with energies of a few to a few ten keV.

- The photon can also interact by pair-production in the surrounding material, in which case a *511 keV photon* from the positron annihilation can be backscattered and detected in the sensor.

The finite size and the material affect the efficiency of the detector, which also depends on the photon energy.

For germanium, for example, the attenuation length varies from about 30 μm at 10 keV to about 7 mm at 10 MeV.

Common gamma spectrometers are either based on scintillation in crystals like NaI(Tl), BGO, or LaBr$_3$:Ce, read out by photomultiplier tubes (PMTs) or silicon photomultipliers (SiPMs), or they can be based on high-Z semiconductors like Ge, CdTe, or CZT. Scintillation detectors

See section 5.3.

See sections 4.4 and 4.6.

See section 9.13.

can easily be produced in large size, but have a somewhat poorer energy resolution because of the large energy required to produce a scintillation photon. Germanium can also be produced in large, pure crystals, but has a small band gap, which requires cooling by Liquid Nitrogen (LN$_2$), whereas the type III-V semiconductors have a larger band gap, but are difficult to produce in larger size.

The best resolution for scintillators is achieved for LaBr$_3$:Ce.

A review of the use of Ge detectors for gamma spectroscopy can be found in [229].

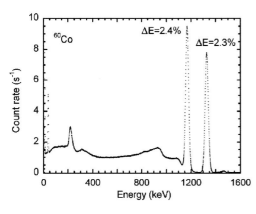 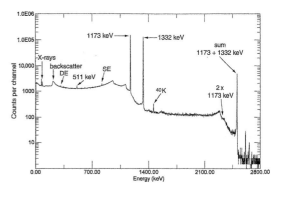

Figure 11.1: *Gamma spectra from ^{60}Co. Left: Spectrum from a LaBr$_3$:Ce scintillator [412], Right: Spectrum from a Ge detector [312]. 'SE' and 'DE' are the single and double escape peaks, respectively, for the 1332 keV peak. '^{40}K' is a background peak from naturally occurring radioactive ^{40}K, 'backscatter' corresponds to the reverse Compton edge. Data above 1400 keV are from double-event pile-up due to a finite integration time of the sensor readout.*

The energy recorded in the detector material is the sum of many individual energy transfers. However, usually only a fraction of the energy will end up as detectable signal, with the rest ultimately ending up in heat. As the energy transfer is quantised by the finite energy required to create an electron-ion or electron-hole pair or a scintillation photon, there will be a finite, integer number of energy transfers, the exact number of which will vary from event to event. A simple estimate of the energy resolution can therefore be obtained from Poisson counting statistics.

The best energy resolution is achieved with cooled Ge detectors, 0.1% for photons of $\mathcal{O}(1\,\mathrm{MeV})$. This is smaller than the resolution expected if the pulseheight would be Poisson-distributed. The improved resolution is due to the fact that the individual electron-hole pair creations are not uncorrelated. In practice, the reduction can be parameterised using the Fano factor F (see section 2.6), as

It takes about 2.9 eV to create an electron-hole pair in Ge. So, for a 1 MeV photon, $N \simeq 10^6/2.9$ and from Poisson statistics $\sigma_N/N = 1/\sqrt{N} \simeq 1.7 \times 10^{-3}$. For the energy resolution $\sigma_E/E = \sigma_N/N$, not including any other sources of experimental errors. For these see for example [394].

$$\frac{\sigma_E}{E} = \sqrt{F}\left(\frac{\sigma_E}{E}\right)_{\mathrm{Poisson}} = \sqrt{F}\,\frac{\sigma_N}{N} = \sqrt{\frac{F\Delta_E}{E}}, \qquad (11.1)$$

with $0 \leq F \leq 1$, and Δ_E the energy required to create a secondary quantum. The value of the Fano factor can be predicted from Monte Carlo simulations. For semiconductors it is found to be about 0.1.

In scintillators, the energy used to create the signal, scintillation light, is a small fraction of the energy deposited by the photon, hence the Fano factor is close to 1.

Fluctuations in the internal amplification process in a PMT, as is often used to detect the scintillation photons, will degrade the resolution even further.

11.2 SUB-KELVIN CRYOGENIC DETECTORS

As noted above, in the common detectors collecting ionisation charge or scintillation light a significant fraction of the energy of the incoming particle will be lost to heat, and the fluctuations in the partition of the energy drive their energy resolution.

At a microscopic level the heat manifests itself as lattice vibrations, or phonons. At room temperature phonons from particle interactions in the detector material cannot be detected above the background of thermally created phonons. However, at very low temperatures (typically well below 1 K) this background is sufficiently depressed to allow for their observation. In addition, the low energies required for the creation of these quasi-particles result in low detection thresholds, and the operation at low temperatures reduces electronics noise.

A signature of the heat deposited in the detector is an increase of the temperature due to the heat capacity of the material. At low temperatures the heat capacity of an insulator can be described by the Debye model,

$$c \simeq \frac{12\pi^4 R}{5} \left(\frac{T}{T_D} \right)^3, \qquad (11.2)$$

with the Debye temperature T_D and $12\pi^4 R/5 \simeq 1944 \ \mathrm{JK^{-1}mol^{-1}}$. The small heat capacity at very low temperatures makes the heat from particle interactions detectable. For example, an energy deposition of 10 keV increases the temperature of 1 kg of silicon at 10 mK by 6 μK.

The typical arrangement to measure heat is a bolometer, consisting of a block of detector material with finite heat capacity C that is connected through a thermal impedance R_{th} with a heat sink at constant temperature. A deposition of the energy Q results in a temperature spike with $\Delta T = Q/C$, which decays with a time constant given by $\tau = R_{th}C$. The lifetime of thermal phonons in the detector is given by this time. The time required for the phonons to thermalise and the time required to restore the temperature limits the counting rate of thermal calorimeters to a few Hz.

Fluctuations in the total number of phonons in the absorber have a variance C/k_B, which yields a fundamental limit of the energy resolution of $\langle E^2 \rangle = k_B T^2 C$, which is independent of the energy and proportional to the heat capacity of the absorber [361]. An absorbed heat P will raise the temperature by $\Delta T = PR_{th}$. Fluctuations in the power flowing to the sink have a spectral density $\mathrm{d}p^2/\mathrm{d}f = 4k_B T^2/R_{th}$. Larger thermal impedance leads to lower noise power, but also to longer response time and reduced heat removal capability.

The most common phonon sensors to measure the temperature increase are resistive thermometers, for which the resistance changes as a function of temperature. Examples are semiconducting thermistors or superconducting transition edge sensors (TES).

A thermistor is a heavily doped semiconductor slightly below the metal-insulator transition. Good doping homogeneity and reproducibility is required, and can be achieved with Neutron Transmutation Doped (NTD) devices. Germanium with NTD (Ge-NTD) is a commonly used sensor material. The typical resistance at the working point is between 1 and 100 MΩ.

For a more detailed summary of cryogenic detectors, see [505].

Such detectors are also called 'non-equilibrium' detectors, as the signal is acquired before the deposited energy is transferred to other, undetectable, channels (e.g. heat).

Phonons can be interpreted as quasi-particles with energies down to meV (thermal phonons). Other low-energy quasi-particle states that can be used for particle detection are for example superconducting Cooper-pairs.

In a metal there is a contribution from the electron heat capacity that is proportional to T, which dominates at temperatures below 1 K.

Table 11.1: Debye temperature in common cryogenic detector materials.

Material	T_D [K]
Silicon	645
Germanium	374
NaI	165
CaWO$_4$	355
LiF	735
Al$_2$O$_3$	1042

The classical bolometer is used for the measurement of continuous (infrared) radiation. In this section, we are considering energy depositions by discrete particle passages. It would therefore be more correct to call this a 'calorimeter', but the designation 'bolometer' has stuck.

Strictly speaking, temperature is an equilibrium concept and is not applicable until complete thermalisation of the phonons. However, the principles outlined in this section still apply in the near-equilibrium case.

These expressions constitute the limits of resolution for detectors that have reached thermal equilibrium. If there is insufficient time to allow for complete thermalisation, then eq. (11.1) applies.

The resistance of a semiconductor thermistor at a temperature T_0 can be described by $R(T) = R_0 \exp(\sqrt{T_0/T})$, where R_0 is the resistance at T_0.

NTDs can be produced by thermal neutron irradiation of Ge sensors to produce dopants like Ga, As and Se. This method can result in very uniform doping [360].

A TES consists of a superconducting layer that is operated at a temperature close to the transition between normal and superconductivity. This transition occurs over a short temperature range, which makes the device highly sensitive to temperature changes. The temperature range of the transition is small and thus the sensor has to be operated at a very specific temperature. A simple readout scheme uses the change in resistance to steer the current in an inductance, causing a very small change in the magnetic flux that can be measured by a superconducting quantum interference device (SQUID).

Table 11.2: Transition temperatures for common superconductors in TES.

Material	T_c [K]
Tungsten	0.012
Iridium	0.14
Titanium	0.39
Molybdenum	0.92
Aluminium	1.14

The readout uses a feedback circuit which actively adjusts the temperature using Joule heating.

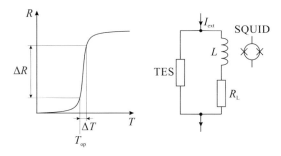

Figure 11.2: Principle of a TES (after [410]). Left: Resistance as a function of temperature in a superconductor close to the transition temperature. Right: Simple readout circuit. R_L is a load resistor with fixed value.

The sensitivity of a temperature sensor can be described by the logarithmic temperature sensitivity $\alpha \equiv d\log R / d\log T = (T/R)\,dR/dT$. For semiconductor thermistors α is typically around 5 or 6, and for TESs it is two orders of magnitude higher. In a TES the energy required to break up a Cooper pair in the superconductor is of the order of meV, and the Fano factor typically about 0.2, and thus sub-K calorimeters with TES readout can achieve the best calorimetric low-energy resolutions, down to eV for a signal of several keV.

This resolution is often difficult to achieve in practice, due to imperfections of the device material. In particle physics, sub-K detectors have been employed for precision measurements of the neutrino mass, and the search for neutrinoless double beta decay or weakly interacting massive particles. However, scaling up the detector technology to the large active detector volumes required for these searches is challenging, and today multi-ton noble liquid detectors are a strong contender in this field (see section 8.1).

Bolometric detectors are in wide use in astrophysics, for example in detectors to study the cosmic microwave background.

11.3 ELECTROMAGNETIC CASCADES

In particle physics photon energies are often significantly higher than in nuclear decays, and the electrons (and positrons) created in pair-production dominant at these photon energies will have an energy above the critical energy, above which radiative energy loss will dominate over ionisation and excitation energy loss. The result is a cascade with an increasing number of electrons, positrons and photons, with decreasing energy as the shower develops. At the beginning, the shower will be dominated by the sequence of pair production and bremsstrahlung, but as the shower progresses, the secondaries will become softer and the probability for the creation of lower energy photons increases. For photon energies below a few MeV pair production stops, and the number of positrons ceases to increase, while more free electrons continue to be produced in Compton and photoelectric scattering events. At the same time, once the

See section 2.1.

The bremsstrahlung photon energy spectrum is shown in Figure 2.34.

Typically, about three quarters of the charged particles in the electromagnetic cascade are electrons, most of them with low energies (<1 MeV).

energies of electrons and positrons drop below the critical energy, creation of new photons by bremsstrahlung becomes insignificant. At this point the shower will cease to grow, although the particles in the shower will continue to lose their energy to ionisation and excitation, until all the energy is absorbed.

The signal in the calorimeter is generated by the ionisation or excitation energy loss of the charged particles in the shower. The overwhelming majority of the charged particles depositing this signal are soft (for materials with large Z on average 40% of the energy is deposited by particles with an energy below 1 MeV [497]), and their range in the calorimeter material is short.

A minor role is played by photonuclear interactions [117]. The cross-section for such interactions peaks at a few 10 MeV due to the 'giant dipole resonance', but is still typically only about 10^{-2} of the cross-section for pair production. A consequence of these reactions is the presence of neutrons from nuclear de-excitation in the cascade (see also Figure 11.13).

Similar electromagnetic cascades will be instigated by high-energy electrons or positrons hitting a sufficiently deep block of matter. Muons will only cause electromagnetic showers at much higher energies, because of the lower rate of bremsstrahlung due to their larger mass. Similarly, charged hadrons will not directly initiate electromagnetic showers by bremsstrahlung, but they will will start a hadronic cascade (which will include an electromagnetic component) as discussed in section 11.9.

Ionisation in this section is understood to include Cherenkov radiation in the material, if the conditions for its emission are met.

In solids the range of a 1 MeV electron is less than 1 mm.

The giant dipole resonance (GDR) is a high-frequency collective excitation of atomic nuclei, typically a coherent movement of all neutrons against all protons. It can decay by nuclear fission, or emission of neutrons or gamma rays.

At extremely high energies ($>10^{20}$ eV), the cross-section for e^+e^- pair-production decreases due to the Landau-Pomeranchuk-Migdal (LPM) effect [60, 311], whilst the cross-section for photonuclear interactions increases, so that cascades at these energies are predominantly hadronic.

11.4 ELECTROMAGNETIC SHOWER DIMENSIONS

As discussed in section 2.1, the dominating interactions for high-energy particles in the initial stages of the cascade, pair production and bremsstrahlung, do have a common characteristic length scale, the radiation length X_0. It thus makes sense to express the longitudinal development of the shower in this unit.

Because the number of charged secondaries that are generated in high-energy electromagnetic interactions per X_0 is small ($\sim$1 to 2), a significant number of X_0 are required to contain the shower. For example, in lead 25 X_0 are required to contain 99% of a shower of 100 GeV. An incoming high-energy photon will on average undergo a pair-production event after 9/7 X_0, which will divide its energy between the electron and the positron. Hence, for a photon of twice the energy, the depth of the shower will increase linearly by such a distance. Thus, the depth of a high-energy shower scales logarithmically with the incoming energy. To achieve shower containment in an experiment the size of the calorimeter needs to scale logarithmically with the scale of the energy of the experiment. Thus calorimeters are powerful detectors in particular for high energy experiments.

The scaling of the shower depth with the radiation length is not perfect. In materials with higher Z the critical energy is smaller. Hence, the increase in the number of shower particles due to pair production extends to lower energies, and thus to later stages in the shower development. In these materials the shower maximum will occur later, and the shower tail is longer.

A simple model for the initial stages of an electromagnetic cascade is discussed in exercise 1.

For comparison, to maintain the momentum resolution in a spectrometer the size of the spectrometer needs to scale with L^2, where L is the lever arm of the spectrometer (see section 10.4).

While 25 X_0 are needed for 99% containment in lead, in iron 21 X_0 and in aluminium 18 X_0 are sufficient.

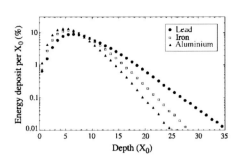

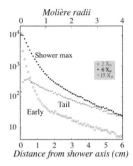

Figure 11.3: Electromagnetic shower dimensions calculated using Monte Carlo simulation [496]. Left: Longitudinal profiles of 10 GeV e^- showers developing in aluminium ($Z = 13$), iron ($Z = 26$) and lead ($Z = 82$). Right: Radial distributions of the energy deposited by 10 GeV electron showers in copper at various depths. Monte Carlo techniques to simulate electromagnetic showers will be discussed in section 11.12.

Laterally, in the early stages of the shower the expansion of the shower is driven by multiple scattering of the charged high-energy secondaries. The scale length for this is called the Molière radius, which can be found from integration of eq. (2.53),

$$R_M \simeq \frac{21.2\,\mathrm{MeV}}{E_c} X_0. \tag{11.3}$$

Note that the radius of containment is thus independent of the energy of the incoming particle. The Molière radius is also only weakly dependent on Z, as both radiation length and critical energy scale approximately with Z^{-1}.

Several versions of this equation using approximate expressions for E_c and X_0 are in use. Typically, 90% of the deposited energy of a shower is contained in a cylinder with a radius of one R_M.

In the later stages of the shower development the low-energy photons, which have a significant probability to be Compton-scattered by larger angles but still have a significant range, cause a broadening of the tail of the shower.

11.5   ENERGY RESOLUTION OF HOMOGENEOUS ELECTROMAGNETIC CALORIMETERS

In homogeneous calorimeters the whole volume is actively contributing to the signal from the detector. Typically, homogeneous calorimeters are made from transparent crystals that scintillate or produce Cherenkov radiation, or from liquid noble gases where the ionisation electrons are collected in a homogeneous electric field. The readout elements for homogeneous calorimeters (PMTs or APDs for photon readout, or electrodes and electronic readout for ionisation chambers) are usually outside of the acceptance or behind the calorimeter.

Thin electrodes for ionisation chamber readout can sometimes be inside the active volume, resulting in a quasi-homogeneous geometry (see for example [244]).

Similarly as in section 11.1, in high-energy calorimeters the signal, which is proportional to the energy of the incoming particle, is the sum of a large number N of more or less independent small energy transfers (to create ionisation charges or scintillation photons, etc.). Using similar arguments, we again expect that stochastic fluctuations of the measured energy can be described by

$$\frac{\sigma_E}{E} \propto \frac{\Delta N}{N} = \frac{\sqrt{N}}{N} = \frac{1}{\sqrt{N}} \propto \frac{1}{\sqrt{E}},$$

For the exact proportionality factor again statistical constraints will need to be considered in the form of a Fano factor, while additional stochastic contributions, for example from the readout, will degrade the resolution.

and indeed a stochastic term $\propto E^{-1/2}$ is usually a major contribution to the energy resolution of a calorimeter. Other contributions to the stochastic term come from particles in the cascade escaping detection, because they traverse sections without instrumentation or because the shower is not completely contained. Typically, in homogeneous calorimeters a stochastic term of a few $\%/\sqrt{E/\text{GeV}}$ is achieved.

More generally, the energy resolution of a calorimeter can be parameterised as

$$\frac{\sigma_E}{E} = \frac{A}{\sqrt{E}} \oplus \frac{B}{E} \oplus C. \tag{11.4}$$

The first term contains the stochastic contribution, and is often (but not always) the most relevant. The second term, which corresponds to an energy-independent error typically is caused by readout noise. The final term is associated with inhomogeneities in materials, calibration errors and some forms of radiation damage.

An example for a large homogeneous electromagnetic calorimeter system is the CMS calorimeter [196], which uses $PbWO_4$ scintillating crystals with APD readout. The properties of $PbWO_4$ as a scintillator have been discussed in section 5.3. This material was chosen for the homogeneous electromagnetic calorimeter in CMS because of its short radiation length, its high light yield (LY), the high speed, and good uniformity and radiation tolerance.

The design energy resolution in the barrel part of this detector using the parameterisation in eq. (11.4) is $A/\sqrt{E} = 2.7\%/\sqrt{E/\text{GeV}}$, which is dominated by photostatistics due to fluctuations in the amplification in the APD ($2.3\%/\sqrt{E/\text{GeV}}$), $B = 210$ MeV at high luminosity operation, which is dominated by electronics noise (150 MeV), and $C = 0.55\%$, which is dominated by calibration effects (0.4%) [199]. The performance was studied with prototypes in testbeams, where stochastic, noise and constant terms of $A/\sqrt{E} = 2.8\%/\sqrt{E/\text{GeV}}$, $B = 127$ MeV and $C = 0.3\%$ were observed [22]. However, in LHC operation there are additional effects which will degrade the resolution; 'pile-up' (multiple pp collisions in the same bunch crossing) and radiation damage. In addition it is more difficult to achieve a small value of the constant term in a large calorimeter than in a small calorimeter used in a testbeam.

The main effect of radiation damage in $PbWO_4$ is from ionising radiation [273]. When $PbWO_4$ crystals are exposed to ionising radiation pre-existing point-structure defects may act as traps for electrons or holes. These defects act as colour centres that absorb light strongly over some range of wavelengths, which reduces the attenuation length for the scintillation light. The light absorption coefficient grows linearly with the dose rate. The attenuation results in an increase of the constant term, which becomes the dominant contribution to the energy resolution (10% after irradiation by 1.3×10^{14} p/cm^2) [19].

The effect of neutron damage was studied using nuclear reactors, but the observed damage was attributed to γs, not neutrons. Fast hadrons generate locally extended defects in the crystal. This results in a characteristic variation of absorption with wavelength, $\mu \propto \lambda^{-4}$. There was no observed effect on the production of scintillation light.

As the radiation damage only effects the light transmission, a very strong correlation between measured light yield and absorption was observed [273]. Therefore the effects of radiation damage can be monitored and corrected for by in-situ measurements of the absorption length. This

It is common to express the coefficients A, B and C in this expression in percent, assuming that the energy is given in GeV.

We keep the designation 'stochastic' for the term proportional to $1/\sqrt{E}$ even though the source for the other terms can also be stochastic.

For APDs see section 9.9.

The LY needs to be sufficiently high so that the stochastic term in the energy resolution is small enough to obtain very good resolution for high energy photons and electrons ($E > \sim 10$ GeV). The LY for $PbWO_4$ is more than 100 photons/MeV.

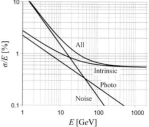

Figure 11.4: Design energy resolution for the CMS electromagnetic calorimeter [199].

The effect of radiation damage on the attenuation length of $PbWO_4$ crystals can be mitigated by specified impurities such as La, Y, and Nb [273].

This is the expected fluence for the CMS endcap calorimeter after an integrated luminosity of 3000 fb^{-1}. This corresponds to the expectations for HL-LHC and is an order of magnitude larger than that expected for LHC phase 1. Therefore for the phase 2 upgrade this calorimeter will be replaced by a lead/silicon calorimeter [200].

was achieved using lasers with a wavelength of 420 nm (the peak of the scintillation spectrum) to inject light into each crystal [196]. The resulting signal was measured using the front-end electronics, and the laser power using *pn* silicon photodiodes. The stability of this system was also controlled by using a laser at $\lambda = 796$ nm (no significant change in attenuation is expected at this wavelength).

Homogeneous electromagnetic calorimeters have also been made using lead glass. For this type of calorimeter, the signal comes from the Cherenkov light generated in the glass by the charged secondaries in the shower, which can be detected by PMTs or other sensitive photon detectors. The light yield from Cherenkov radiation is much lower than that from scintillation in inorganic scintillators. Hence, the electron energy resolution is typically inferior to that of homogeneous crystal calorimeters that are relying on scintillation. However, lead glass has been chosen for homogeneous calorimeters for its low cost and ease of handling and manufacture compared to crystal scintillator materials.

Ordinary glass has a refractive index of $n \sim 1.5$. The addition of lead results in an index of refraction of up to ~ 1.7.

We have seen in 2.8 that Cherenkov radiation constitutes only a small fraction of the charged particle energy loss.

A large lead glass calorimeter was employed in the OPAL experiment, where the energy resolution for testbeam prototypes of the barrel calorimeter without material in front was $\sigma_E/E = 6.3\%/\sqrt{E/\text{GeV}} + 0.2\%$ between 5 and 100 GeV [392]. A more recent example is the lead glass component of the PrimEx calorimeter, for which an energy resolution of

$$\sigma_E/E = \frac{2.3\%}{E/\text{GeV}} \oplus \frac{5.4\%}{\sqrt{E/\text{GeV}}}$$

was measured in the energy range from 1 to 5 GeV [322].

In addition to the energy measurements calorimeters sometimes also need to provide information about the position of the shower (for track shower matching, or to reconstruct the 4-momentum of a neutral particle). In general, the position resolution has a similar energy dependence as the energy resolution, and in particular is often dominated by a $1/\sqrt{E}$ term caused by stochastic fluctuations. The scale of the position resolution is given by the Molière radius R_M or the cell size of the calorimeter, whichever is larger.

Hence the cell size in electromagnetic calorimeter is often chosen to be similar to R_M.

Finally, for some calorimeters also the time of the shower signal can be important, if it is needed to correlate the shower with signals in other parts of the experiment. The time resolution in that case can usually also be described by a similar functional form as in eq. (11.4).

It should be noted, that even though resolution, in particular for energy, is used as a benchmark number for calorimeters, other considerations also influence the choice of design, like the energy range of the particles to be detected, space, the amount of material in front of the calorimeter, complexity of readout, ease of construction and/operation, and cost. Resolution numbers are often obtained in dedicated test beams of prototypes that are often performed under highly idealised conditions, and the performance of the calorimeter in the actual experiment might differ significantly. It is also common practice to quote for the resolution the standard deviation of a Gaussian fit to the measured energy distributions, thus ignoring any non-Gaussian tails.

The idealised conditions in a testbeam often allow to make geometrical selections, or to use calibration procedures that might not be available in the real experimental conditions.

11.6   SAMPLING CALORIMETERS

To minimise the depth of the material required to contain a shower, thick layers of a dense solid (usually a metal) are desirable as absorber. However, in general these dense absorbers do not provide a measurable signal as dense materials are usually opaque and conductive or impure insulators. Hence, a common approach is to intersperse layers of dense absorber with less dense active detector layers. This is called a 'sampling calorimeter'. Because of the size of the required instrumentation, the typical readout technologies are scintillators in planes or fibres, ionisation gaps filled with liquid noble gases, or more recently, semiconductors. Semiconductors in the past have been considered prohibitively expensive for the use in calorimeters, but as will be discussed in section 11.13 there are benefits from the fine segmentation such readout can provide, which leads to their consideration in modern detector systems. Wire chambers are less useful, because of the small amount of active material, resulting in significant Landau fluctuations, and additional fluctuations from the internal avalanche amplification process.

This section deals with electromagnetic sampling calorimeters. For the use of sampling in hadronic calorimeters see section 11.8.

Typically, the absorber layers will have a high Z, whereas the detector layers contain lower Z elements.

Although the obvious choice is to arrange these in layers, other geometries also work, as long as size of the sampling is smaller than the shower size.

The price to pay for the compactness are additional fluctuations of the energy deposited in the detector layers, which increase the stochastic term in the energy resolution (these are called 'sampling fluctuations'). As discussed, in an electromagnetic calorimeter most of the energy is deposited by low energy charged secondaries. Because of the low energy these have a range of less than 1 mm and their directions are random. They are typically produced by low-energy photons and, in particular in high-Z absorbers, these photons will also have a very limited range due to the strong dependence of the absorption cross-section on Z. Consequently, most of the detected signal in the sampling calorimeter will be generated within a thin layer at the absorber/detector interfaces in the detector gap of the sampling calorimeter. The resolution will still be dominated by Poissonian fluctuations of the number of deposited energy quanta, but the number of these quanta is now limited by this geometrical acceptance reduction.

The energy response of a sampling calorimeter is given by the measured signal in the active layers, and not by the energy deposited over a sampling layer. The two are generally not equivalent.

See section 2.1.

Several fit functions to predict the energy resolution of electromagnetic sampling calorimeters have been proposed, however, none of these are completely satisfactory[1]. Qualitatively, the following points emerge:

---

[1] A parameterisation that is common in the literature is $\sigma_E/E = \sqrt{\Delta E/E}$, where ΔE is the energy loss in one absorber/detector layer (often the equation is written for E in GeV and ΔE in MeV with a corresponding scale factor of 3.2%). This equation is flawed, because it relies on Rossi's 'approximation B' [432] (see exercise 1), without taking into account the energy scale of energy deposition in the electromagnetic cascade and the range of particles responsible for it, and it does not reproduce experimental observations on the dependence of the resolution on the detector layer thickness (for a detailed discussion see [497]).

A modification of this expression has been given in [53], which takes into account the reduction of the total path length of all particles in the shower due to a finite cut-off for the minimum kinetic energy E_{cut} of an electron (positron) that can be detected in the calorimeter, and the effect of different angles of the charged secondary tracks in the detector layer,

$$\frac{\sigma_E}{E} = 3.2\% \sqrt{\frac{\Delta E/\text{MeV}}{F(E_c, E_{cut})\langle\cos\theta\rangle}} \frac{1}{\sqrt{E/\text{GeV}}} =$$

$$= 3.2\% \sqrt{\frac{E_c/\text{MeV}}{F(E_c, E_{cut})\langle\cos\theta\rangle}} \sqrt{\frac{t}{E/\text{GeV}}}, \quad (11.5)$$

where $F(E_c, E_{cut})$ is the correction factor for the total path length, and $\langle\cos\theta\rangle$ the average of the cosines of the angle of the charged secondary tracks. Expressions for both factors are

- The energy resolution of electromagnetic sampling calorimeters is usually dominated by a stochastic term $\sigma_E/E = A/\sqrt{E}$. The proportionality constant A is larger than for homogeneous electromagnetic calorimeters, typically between $5\%/\sqrt{E/\text{GeV}}$ and $20\%/\sqrt{E/\text{GeV}}$.

- For a larger sampling fraction, if the detector thickness is kept constant and the same absorber thickness is maintained to achieve the same level of containment, the energy resolution improves, as this increases the number of interfaces between absorber and detector.

The sampling fraction is the ratio of the energy loss in the detector layer and the energy loss in the combined absorber/detector layer.

- For the same reason, a higher spatial sampling frequency improves the resolution.

A higher spatial sampling frequency can be achieved if the sampling is divided into finer units, while maintaining the same total material for absorber and detector, respectively.

- It also improves for absorbers with a higher Z value, as this increases the number of particles in the shower.

A semi-empirical parameterisation that aims to describe the reduction of the counting statistics due to the geometrical arguments has been proposed by Livan et al. [341]

It might seem counterintuitive that the resolution improves with a thinner detector layer, but the key constraint here is that the sampling fraction is maintained, so that this only increases the spatial frequency of sampling layers.

$$\frac{\sigma_E}{E} = 2.7\% \sqrt{\frac{d_{\text{det}}/\text{mm}}{f_{\text{samp}}}} \frac{1}{\sqrt{E/\text{GeV}}}, \tag{11.6}$$

where d_{det} is the thickness of the detector layer (fibre diameter for scintillating fibre calorimeters) and f_{samp} the sampling fraction. The numerical front factor has been derived from a fit to the performance published for several electromagnetic sampling calorimeters. For these data the result of this parameterisation is within 10% of the actual value.

For practical purposes this parameterisation uses the sampling fraction for a MIP,

$$f_{\text{samp}} = \frac{d_{\text{det}} \left(\frac{dE}{dx}\right)^{\text{det}}_{\text{mip}}}{d_{\text{det}} \left(\frac{dE}{dx}\right)^{\text{det}}_{\text{mip}} + d_{\text{abs}} \left(\frac{dE}{dx}\right)^{\text{abs}}_{\text{mip}}}.$$

Despite the correlation with data eq. (11.6) also appears incomplete, as it does not contain a dependence of the resolution on Z, and the interchangability of fibre diameter and plate thickness appears fortuitous.

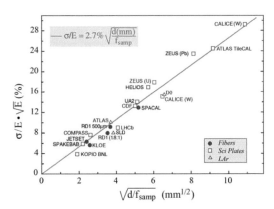

Figure 11.5: Stochastic term of the energy resolution for different sampling calorimeters as a function of the ratio $d_{\text{det}}/f_{\text{samp}}$ [342]. The energy in the figure is given in GeV. The line shows the parameterisation given in eq. (11.6).

Eq. (11.6) can be used for sampling fractions between 1 and 10%, and a sampling layer thickness between 0.1 and 1 X_0. For thicker detector layers the assumption that the dominant contribution to the signal comes from secondaries produced in the absorber is not valid any more and the statistics of particles generated in the detector layer itself becomes relevant. For very thin detector layers, and this applies to calorimeters read out with

---

given in [53]. The last equality uses another result from 'approximation B', $\Delta E/E = t E_{\text{c}}/E$, where t is the thickness of a combined absorber/detector layer in radiation lengths. However, the modifications do not fully address the flaws in the original expression. Nevertheless, this equation is widely quoted, with various omissions and approximations.

typical gaseous or semiconductor detectors, the energy resolution does degrade less strongly as a function of the sampling fraction than predicted by this equation [497].

EXAMPLES OF ELECTROMAGNETIC SAMPLING CALORIMETERS

In this section we will discuss three different sampling calorimeter examples covering a wide range of sampling fractions and different detection mechanisms.

An example of a sampling calorimeter with a very large sampling fraction and scintillating fibre readout is the KLOE calorimeter [20]. There the requirement was for good energy resolution for photons from $K_{L,S} \rightarrow \pi^0\pi^0$ decays. The energies of photons generated in these decays are low ($20 < E < 510$ MeV), so that one requirement was to minimise material in front of the active volume. Full solid angle coverage was required, calling for a compact calorimeter. This motivated the choice of a sampling calorimeter with lead absorber and scintillating fibre readout. The fibres were 1 mm diameter with a spacing of 1.35 mm, resulting in a very high sampling fraction (lead/fibre/epoxy 42/48/10 by volume). The energy resolution therefore was excellent, with a stochastic term of $5.7\%/\sqrt{E/\text{GeV}}$ and a negligible constant term. This calorimeter also achieved a good time resolution (54 ps$/\sqrt{E/\text{GeV}}$).

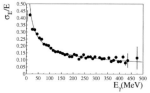

Figure 11.6: Energy resolution of the KLOE electromagnetic sampling calorimeter as a function of the energy [20].

The ATLAS electromagnetic calorimeter consists of lead absorber plates in a a zig-zag ('accordion') shape with gaps filled with Liquid Argon (LAr) as the active medium read out in an ionisation chamber geometry [109].

Liquid argon was an attractive option for the active layer of an electromagnetic calorimeter at LHC because it is inherently radiation-tolerant. A challenge associated with this approach is the long drift time. In ATLAS the width of the gap for the LAr is typically 2 mm, which results in a maximum drift time of 0.45 μs [108], which is much longer than the time between bunch crossings (25 ns). Therefore, readout with a slow amplifier would result in too much noise from 'pile-up' background (energy deposited by particles in other bunch crossings). This is addressed by the use of a fast pre-amplifier followed by an $RC - CR^2$ filter to provide a bipolar shaper (the RC acts as a differentiator and the CR^2 as an integrator). The electronics noise decreases with the rise time but the pile-up noise increases with it (this is the same effect as discussed for semiconductor detectors in chapter 9). The rise time selected is longer than the time between bunch crossings as further digital filtering is performed by the receiver electronics. The overall filtering allows the assignment of an energy deposition to a particular bunch crossing. The bi-polar pulse shaping implies that there is no net DC effect on the baseline current, independent of the level of pile-up background. A drawback of this shaping is that it reduces the size of the integrated charge signal and therefore requires very low noise electronics.

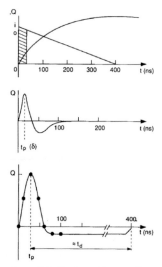

Figure 11.7: Current and integrated charge from electron drift in the ATLAS EM calorimeter (top). The effect of pulse shaping with a bipolar shaper is shown for a δ-function pulse (middle) and for the triangular current pulse expected from a uniform ionisation charge in the gap (bottom) [108].

In a 'classical' lead/liquid argon calorimeter, in which the absorbers are plates perpendicular to the trajectories of the incoming particles, the signals have to be read out by long cables. This results in a large capacitance which would increase the noise (see chapter 3), and a large

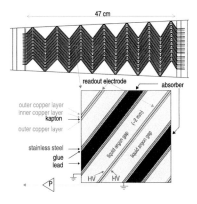

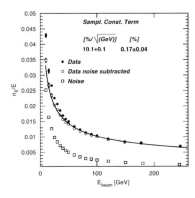

Figure 11.8: The ATLAS accordion electromagnetic calorimeter. Left: Geometry [86]. Right: Energy resolution as a function of energy measured using a prototype in a test beam [31]. The curve is a fit to the expected scaling with energy, $\sigma_E/E = A/\sqrt{E} + B$.

inductance that would slow the response. These two issues were resolved by the use of an 'accordion' geometry. In this geometry the signals are read out by copper/polyimide flexible tapes which have much lower capacitance and inductance than cables. The pre-amplifiers for the electromagnetic calorimeters are placed outside the cryostat behind the active volume.

If the pre-amplifiers are in the cold volume the noise would be slightly lower, but this results in more radiation damage and makes maintenance more challenging. In the ATLAS hadronic endcap calorimeter (HEC) calorimeter a different design choice has been made, due to the larger cell size that results in a larger capacitance and hence more noise. The design goal of being able to see MIP signals from muons required the use of cold pre-amplifiers for the HEC.

Another benefit of the absorber geometry is that it achieves a projective readout (organised in towers, which are oriented in the direction of the incoming particle) while minimising non-uniformities across the width of the readout cell. This geometry allows for a calorimeter without cracks between neighbouring cells.

The stochastic term of the energy resolution measured with prototypes in beam tests was $10.1\%/\sqrt{E/\text{GeV}}$ and the linearity was within $\pm 0.1\%$ in the energy range $10\,\text{GeV} < E < 180\,\text{GeV}$ [31]. The constant term was $(0.17 \pm 0.04)\%$, but the constant term determined from an in-situ analysis in the experiment was larger (see Figure 11.11).

The real figure of merit for a calorimeter should be the full system performance, not the resolution achieved in a relatively small prototype in a test beam.

Sampling calorimeters with a very small and adjustable sampling frequency have been studied by the SICAPO collaboration using sandwich calorimeters with silicon detector readout. For 7 mm thick ($2\,X_0$) thick tungsten absorber plates the visible energy was measured to be $E_{\text{vis}}/\text{MeV} = (5.558 \pm 0.004)E/\text{GeV} + (1.3 \pm 1.5) \times 10^{-3}$ for an incoming electron with an energy in the range of 5 to 50 GeV [112]. For a calorimeter made up of $24\,X_0$ (12 cm total length), which was sufficient to ensure good longitudinal containment, a stochastic term of $24.9\%/\sqrt{E/\text{GeV}}$ was observed. The collaboration did also study other absorber materials and the dependency on the thickness of the absorber τ (where τ is defined as the sampling fraction measured in units of X_0 for the absorber) and found scaling of the resolution with $\sqrt{\tau}$.

The small error on the slope and the small constant term indicate good linearity of the response.

This energy resolution is comparatively poor due to the small energy fraction sampled in the silicon. However, the strength of the silicon readout is the very fine achievable segmentation, which can be used for particle flow analysis (see section 11.13).

11.7   CALIBRATION OF ELECTROMAGNETIC CALORIMETERS

The signals from calorimeters are processed by suitable electronic circuits to provide amplification and filtering and finally digitised by ADCs. A calibration procedure is therefore required to convert back the raw ADC data to energy.

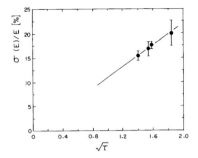

Figure 11.9: Energy resolution as a function of $\sqrt{\tau}$ for a lead/silicon calorimeter for a fixed energy of $E = 4$ GeV [157].

The first step in this procedure is to measure the ADC counts with no signal, and these values are subtracted from the raw ADC values. A very simplified calibration scheme is then given by

These are called 'pedestals'.

$$E_i = F_{\text{el}}(ADC_i) \times F(I_i), \qquad (11.7)$$

where E_i is the reconstructed energy in cell i with an ADC value of ADC_i, $F_{\text{el}}(ADC_i)$ converts the ADC value to a measured current, and $F(I_i)$ converts the current to energy. The function $F_{\text{el}}(ADC_i)$ is determined using electronic calibration circuits in which known signal currents are injected into the amplifier and the measured currents are recorded. The conversion from measured current to reconstructed energy $F(I_i)$ requires a test beam study in which the response is measured for electrons of known energy.

Here we have implicitly assumed the use of ionisation chambers to measure currents. A similar scheme for an active region that was measuring charge would replace current with charge. This will apply in particular for calorimeters using scintillators, which ultimately record charge in the photodetectors. For an example of a more detailed scheme, see [87].

In this very simplified scheme the energy of an electron would simply be given by summing the value of E_i for the cells assigned to the electron. If the electromagnetic calorimeter is divided into longitudinal samples, the performance can be improved by modifying eq. (11.7). The current for an electron cluster is first calculated in each layer. The reconstructed energy is given by

For a large detector, it is impractical to use this procedure for all calorimeter cells. The energy scale needs to be transferred from the cells calibrated in a test beam to the rest of the calorimeter. This can be achieved by various techniques, including radioactive sources and laser pulses.

$$E = \sum_{\text{layer}} w_{\text{layer}} E_{\text{layer}}. \qquad (11.8)$$

The weights w_{layer} are determined empirically using data from a test beam by minimising

$$\chi^2 = \sum_i \left(\frac{E_i - E_{\text{beam}}}{\sigma_{E_i}} \right)^2, \qquad (11.9)$$

where the sum runs over the events in the sample, E_i is the estimated electron energy for each event using eq. (11.8), E_{beam} the beam energy, and σ_{E_i} is the estimated uncertainty in the reconstructed energy.

In a test beam the energy can usually be measured for reference by an upstream magnetic spectrometer with high accuracy.

In-situ calibration of electromagnetic calorimeters in collider experiments can be performed using the large sample of $Z \to e^+e^-$ events. We will consider the calibration of the ATLAS electromagnetic calorimeter as an example [87].

We will first assume that the energy scale is the same for e^- and e^+.

The discussion given here is a very simplified version of the full procedure described in [87].

The scale of the measured electron energy is affected by upstream material in the Inner Detector (ID) and services. The interactions in the upstream material will result in energy being absorbed before the calorimeter and, more importantly, some of the lower energy electrons created in the upstream material will be deflected by the magnetic field so that the energy is no longer contained in the electron cluster reconstructed in the electromagnetic calorimeter. The material in the active volume of the ID can be studied using data from, for example, γ conversions.

The probability of a gamma conversion depends on the material thickness in units of radiation lengths. Therefore a map of the reconstructed gamma conversion vertices acts like an 'x-ray' of the detector material.

The next step is to use the energy sharing between the first two longitudinal samplings to refine these estimates. After the upstream material in the simulation is tuned to agree with this data, a data-driven estimate of the electron energy scale is performed using $Z \rightarrow e^+ e^-$ events. The correction factors α_i in a given slice of pseudorapidity are defined by

$$E^{\text{data}} = E^{\text{MC}}(1 + \alpha_i), \qquad (11.10)$$

where E^{data} is the measured energy and E^{MC} is the corresponding energy from the Monte Carlo simulation.

The correction factors α_i in bins i of η are determined from $Z \rightarrow e^+ e^-$ events, using

$$m_{ij}^{\text{data}} = (1 + \alpha_{ij}) \, m_{ij}^{\text{MC}}, \qquad (11.11)$$

$$\alpha_{ij} \approx (\alpha_i + \alpha_j)/2, \qquad (11.12)$$

where m_{ij}^{data} (m_{ij}^{MC}) is the Z mass peak in data (Monte Carlo) for which the electrons are in the pseudorapidity bins i and j. The values of the α_i are then determined by a numerical χ^2 minimisation [87].

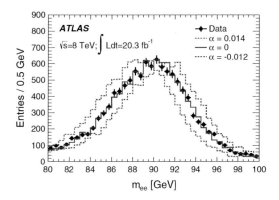

The energy sharing is also sensitive to the services between the ID and the LAr calorimeter, which are less well known than other types of material (for example the magnet). The effect of more material in this region will be to start the showers earlier and therefore increase the relative energy deposited in the first longitudinal sampling compared to the second.

This procedure generates correction factors to be applied to the simulated data. These correction factors are evaluated in relatively coarse regions of phase space (typically η and p_T). Therefore the Monte Carlo simulations are essential because they allow for variations in the detector response over smaller regions of phase space.

Histograms of real data and simulated data are created in the (i, j) bins of η. For each bin a χ^2 value is computed comparing the measured and simulated samples. The overall χ^2 is defined as the sum of the χ^2 for each bin. This χ^2 is minimised by varying the α_i values.

Figure 11.10: Reconstructed invariant $e^+ e^-$ mass in the ATLAS electromagnetic calorimeter for different values for the scale factor α for one electron with $1.63 < \eta < 1.74$ and the other electron with $2.3 < \eta < 2.4$ bins [87]. The variation of α in the plot illustrates the sensitivity of the reconstructed mass peak to this parameter.

To first order the energy scale for photons should be the same as for electrons but the sensitivity to upstream material is different. After correcting for these effects the energy scale for photons is validated using samples of $Z \rightarrow e^+ e^- \gamma$ and $Z \rightarrow \mu^+ \mu^- \gamma$ [87].

A similar methodology is used for the in-situ measurement of the electron energy resolution [88]. This uses a comparison of the measured widths of the distributions of reconstructed masses from a sample of $Z \rightarrow e^+ e^-$ in real and simulated (Monte Carlo) data. The correction factor c_i in a given slice of pseudorapidity is defined by

$$\left(\frac{\sigma(E)}{E} \right)^{\text{data}} = \left(\frac{\sigma(E)}{E} \right)^{\text{MC}} \oplus c_i, \qquad (11.13)$$

where E^{data} is the measured energy and E^{MC} is the corresponding energy from the Monte Carlo simulation. The c_i values are also determined from $Z \rightarrow e^+ e^-$ events, using the difference in quadrature between the mass

Additional validation of the electron energy scale is obtained from samples of $J/\Psi \rightarrow e^+ e^-$.

resolution measured in data and Monte Carlo

$$\left(\frac{\sigma(m)}{m}\right)^{\text{data}} = \left(\frac{\sigma(m)}{m}\right)^{\text{MC}} \oplus c_{ij}, \qquad (11.14)$$

$$c_{ij} \approx (c_i + c_j)/2, \qquad (11.15)$$

where $\sigma_{ij}^{\text{data}}$ (σ_{ij}^{MC}) is the Z resolution in data (Monte Carlo) for the pseudorapidity bin (i, j). The values of c_i are then determined by a numerical χ^2 minimisation using the measured and simulated m_{ee} distributions [88]. In practice simultaneous fits are performed to obtain α_{ij} and c_{ij} values.

In order to make precise and unbiased physics measurements we need the resolution in the Monte Carlo simulation to be the same as in data. Therefore the values of c_i are used to smear the Monte Carlo values. This is done by generating random numbers (g_i) from a Gaussian distribution with $\mu = 0$ and $\sigma = 1$ for each electron and adding an energy $\Delta E = g_i c_i E$.

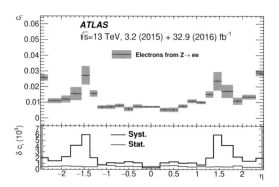

Figure 11.11: Fitted values of the energy resolution correction term c_i as a function of pseudorapidity (top panel) in the ATLAS electromagnetic calorimeter. The statistical and systematic uncertainties are shown separately in the bottom panel [88]. The resolution is worse and the uncertainty is larger in the transition region between the barrel and endcap calorimeters because of the material from the services exiting the detector.

11.8 HADRONIC CALORIMETERS

In the previous sections of this chapter we have seen that electromagnetic calorimeters can make precise measurements of the energy and directions of $e^{\pm}$ and photons by detecting the cascade of secondary particles produced by electromagnetic interactions. Hadronic calorimetry works on a similar principle that hadrons produce secondary particles in interactions with nuclei, which in turn can undergo further interactions. The energy of the incident hadron can be estimated by detecting the products of this avalanche. However, there are several factors that make hadronic calorimetry more challenging than electromagnetic calorimetry and typically result in poorer resolution:

For a more detailed, yet still educational discussion of hadronic calorimeters see [497].

- The interaction length for nuclear interactions in dense media is much longer than the radiation length (see section 2.12). Therefore hadronic calorimeters will need to be much larger or denser in order to achieve sufficient shower containment. This factor alone makes construction of a homogeneous hadronic calorimeter impractical, and all hadronic calorimeters are sampling calorimeters.

- π^0s, which are copiously produced in hadronic interactions, decay into photons, which instigate electromagnetic cascades within the hadronic shower. The underlying physics of the electromagnetic cascade and the way particles in the electromagnetic shower deposit the energy is different than for hadronic cascades. In general this leads to a different calorimeter response for the electromagnetic shower component compared to the hadronic cascade.

- The most significant issue is that a large fraction of the incident energy in a collision of a hadron with a nucleus goes into effectively invisible modes (see section 2.12). The fluctuations of this energy fraction severely limits the achievable resolution.

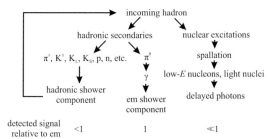

Figure 11.12: Components of a hadronic cascade.

11.9 HADRONIC SHOWER DEVELOPMENT

The secondary particles from hadrons interacting with nuclei will contain $\pi^\pm$ and π^0. The $\pi^\pm$ will induce further hadronic interactions. The π^0 will decay promptly, $\pi^0 \to \gamma\gamma$, with the photons initiating electromagnetic cascades.

We will see that this mixture of purely hadronic and electromagnetic shower components will in general degrade the resolution of a hadronic calorimeter.

A high energy hadron will induce many 'generations' of hadronic interactions, each producing the same average fraction of π^0 and $\pi^\pm$, until the energy decreases below the threshold for π^0 production, and the energy of the hadrons will be deposited in the medium as ionisation. In a first approximation, we assume that on average 1/3 of the secondary hadrons are π^0 and 2/3 $\pi^\pm$.

In this picture, after one generation of hadronic interactions, the electromagnetic fraction of the energy deposited is 1/3. Assuming this process is repeated, then after the second generation the electromagnetic energy fraction is $1/3 + 2/3 \times 1/3$. This process then clearly results in an increase in the electromagnetic fraction as the number of generations increase. Therefore in this naïve model we expect the hadronic energy fraction to vary with the number of generations of hadronic interactions n as $f_{\rm h} = (2/3)^n$ [274]. The number of generations n increases with energy, therefore the hadronic energy fraction in a shower will decrease with increasing energy.

This assumes that all the secondary particles will be $\pi^\pm$ and π^0.

The electromagnetic energy fraction $f_{em} = 1 - f_h$ increases with increasing energy.

This argument must break down when the energy of the hadrons are below threshold for pion production. A more detailed analysis predicts a relationship of the form

$$f_{\rm h} = 1 - f_{\rm em} = (E/E_0)^{k-1}, \qquad (11.16)$$

where E_0 is a constant related to the energy threshold for pion production and k is a constant with a value around 0.85 [255]. This more accurate calculation still shows that the hadronic energy fraction in a shower will decrease with increasing energy. In the high energy limit the shower becomes purely electromagnetic.

The value of E_0 depends on which nucleus is used, but is of the order of 1 GeV.

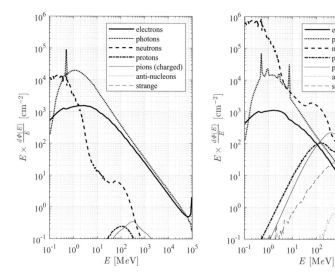

Figure 11.13: Particle spectra produced in lead in electromagnetic cascades initiated by 100 GeV electrons (left), and in hadronic cascades initiated by 100 GeV protons (right), calculated from a FLUKA simulation. The y-axis shows the energy times the flux for a given energy $E\,\mathrm{d}\Phi(E)/\mathrm{d}E$ (which equals the flux per unit lethargy, $\Phi(u)$. Lethargy is defined as decrease in neutron energy, $du = dE/E$, to preserve the visual feeling on how many particles each bin counts for, given that the E axis is logarithmic. The integral of each curve gives the relative fluence for a given particle (Data courtesy of A. Ferrari).

11.10   HADRONIC SHOWER DIMENSIONS

As for electromagnetic calorimeters, the depth required to give good containment of the showers grows logarithmically with energy. For particles in the hundreds of GeV range a total absorber thickness of at least $\sim 10\,\lambda_{\text{int}}$ is needed.

The longitudinal shower profiles for pions of different energies are shown in Figure 11.14. The profiles for proton-induced showers are slightly different because of the larger pN inelastic cross-section. Shower profiles in different material show approximate scaling in λ_{int}. The development of the shower after the maximum is dominated by the neutron component [237].

As it is common that the hadron calorimeter is used as a muon filter (see section 12.8), an additional advantage of making the hadronic calorimeter thicker is that it will reduce the background of pions generating hits in the muon chambers behind the hadron calorimeter ('punch-through').

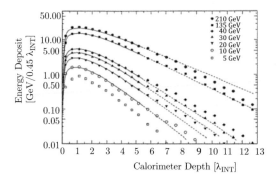

Figure 11.14: Hadronic shower profiles for different pion energies for the WA78 calorimeter (U/scintillator up to 5.4 λ_{int}, Fe/scintillator after that) [186].

The widths of hadronic showers are also much larger than for electromagnetic showers. As for electromagnetic showers, the width grows with shower depth. We can understand the approximate lateral size at the end of the shower by considering the threshold for pion production. The threshold energy for pion production $\pi^{\pm} p \rightarrow \pi^{\pm} \pi^{+} n$ is 312 MeV (see exercise 4). As discussed in section 2.12, the transverse momentum in hadronic interactions is ~ 400 MeV/c. The hadrons produced by high energy incident hadrons at the start of the shower will tend to be emitted at small angles to the incident hadron (see section 2.12). The pions (or other secondary hadrons) produced at the end of the shower will tend to have momenta comparable to the threshold for pion production. Therefore these hadrons will have similar values for the longitudinal and transverse momentum and therefore be produced with an approximately isotropic angular distribution. As these pions will travel for a distance of about one λ_{int}, the width of the hadronic shower will grow by about this amount. This is guiding the choice of transverse granularity in a hadronic calorimeter as there will be no gain in having a granularity very much finer than one λ_{int}.

A useful rule of thumb [330] is that 95% of the hadronic shower will be contained within a radius of 1 λ_{int}.

11.11 ENERGY MEASUREMENT IN HADRONIC CALORIMETERS

As discussed, hadronic interactions inevitably result in 'invisible' energy. In addition, the low energy protons and light nuclei from spallation will be very heavily ionising and therefore they will have a very short range in a dense absorber. Hence, most of their energy will be deposited in the passive rather than the active layers of a sandwich calorimeter. Therefore, in general we expect that the calorimeter response to electrons will be greater than that of hadrons at the same energy.

In order to understand the implications for the calorimeter energy resolution it is convenient to define the ratio of response of electrons to pions, and of electrons to hadrons that do not generate any photons or electrons in their interactions (i.e. they produce $\pi^{\pm}$ but not π^0). We define the corresponding ratios symbolically as

$$e/h \equiv \frac{E(e)}{E(h)} \quad \text{and} \quad e/\pi \equiv \frac{E(e)}{E(\pi^{\pm})}, \tag{11.17}$$

As will be discussed below, these ratios are a property of the calorimeter, given by its geometry, materials and readout.

where $E(e)$ is the detected signal (energy) from an electron, and π and h indicate the same property for a real $\pi^{\pm}$, and for a hadron creating only a non-electromagnetic (purely hadronic) response, all for the same incoming particle energy.

Most of the visible energy in a high energy hadronic shower will result from low energy hadrons towards the end of the cascade where there are more particles. Let f_{em} be the fraction of electromagnetic energy released in an interaction of a $\pi^{\pm}$ with a nucleus. We can then write down an expression involving real $\pi^{\pm}$ as

$$e/\pi = \frac{E(e)}{f_{\text{em}} E(e) + (1 - f_{\text{em}}) E(h)}, \tag{11.18}$$

which we can rewrite in a more useful form as

$$e/\pi = \frac{e/h}{1 - f_{\text{em}}(1 - e/h)}. \tag{11.19}$$

The first issue for high-resolution hadronic calorimetry is that there are large event-by-event fluctuations in f_{em} and therefore, if the value of e/h is significantly different from 1, this will result in fluctuations in e/π. This in general degrades the resolution of the energy measurement of a hadron and the arguments which led to the resolution of an electromagnetic calorimeter varying as $\sigma_E/E = A/\sqrt{E}$ will no longer be valid and the energy resolution will generally improve more slowly with energy.

f_{em} can be measured for each event in dual readout calorimeters, as discussed below.

This is often accommodated by adding a (large) constant term to the $A/\sqrt{E}$ term in the parameterisation of the resolution.

The second issue is that f_{em} increases with energy, as shown in eq. (11.16). This will mean that the response is non-linear if the value of e/h is significantly different from 1. If we were only interested in measuring single hadrons, then the non-linearity could in principle be corrected by a calibration factor. However, we usually need to measure a 'jet' of hadrons containing a mixture of hadrons with unknown energy, so the non-linearity cannot be simply corrected.

However, if we can ensure that our calorimeter satisfies $e/h = 1$ then from eq. (11.19) $e/\pi = 1$ and therefore we are completely insensitive to the value and fluctuations in f_{em}. In sampling calorimeters we can tailor the response to the different shower components, which allows to achieve this goal. This goes by the general name of 'compensation', and we will review different approaches in the following.

For homogeneous calorimeters e/h would always be >1. Hence, homogeneous calorimeters, even if they would be dimensionally feasible, would always be uncompensated.

LOWERING RESPONSE TO ELECTRONS

One approach to the problem is to lower the average response to electrons to achieve the desired condition $e/h = 1$. To consider this approach quantitatively it is useful to define in a similar fashion as before symbolically the ratio of the response of electrons and MIPs, e/MIP.

To a good approximation we can use muons to measure the calorimeter response to MIPs.

As discussed in section 11.3, it is the low-energy interactions towards the end of the electromagnetic shower development that have a significant effect on the signal recorded in the calorimeter, and the energy recorded will be different for the neutral and the charged component in the shower. For low energy electrons generated in the passive layers of a sampling calorimeter the stopping power due to ionisation/excitation energy loss is high and they will have a very short range. However, electrons generated near the edge of the passive absorber can reach the active layers and make a large contribution to the 'visible' energy deposited. For photons the cross-section for Compton scattering scales with Z (the atomic number) and the cross-section for the photoelectric effect scales more rapidly with Z. Therefore, if the atomic number of the passive layers ($Z_{passive}$) is equal to that of the active layers (Z_{active}) we expect $e/MIP = 1$. However, in practice we need $Z_{passive} > Z_{active}$ to make a compact calorimeter, and in this case $e/MIP < 1$.

The scaling for photoelectric effect is in the range Z^3 and Z^5.

In typical calorimeters $e/MIP \simeq 0.6$.

These insights provide a possible way to further reduce the value of e/MIP: if we add a thin lower-Z absorber behind the higher-Z absorber, low energy electrons produced in the high-Z absorber cannot reach the active layer, as they are stopped in the low-Z absorber, but in the low-Z absorber fewer electrons are produced due to the lower interaction cross-section for photons. The net result will be to lower the value of e/MIP. This effect was studied systematically by the SICAPO collaboration who used silicon as the active layers, high-Z absorbers (uranium or tungsten) and added thin layers of low-Z material (G10) downstream of the high-Z absorbers [330].

G10 is a high-pressure fiberglass laminate that is a base material for the fabrication of printed circuit boards.

RAISING THE RESPONSE TO HADRONS

The fraction of energy lost to nuclear breakup and the fraction of energy given to the neutrons are both approximately constant with energy. Therefore, if we can efficiently detect the neutrons in the active layers, there is the possibility to compensate for the lost energy and therefore achieve the compensation condition, $e/h = 1$.

Most of the neutron energy comes from low-energy evaporation neutrons with energies ~ 1 MeV. We can transfer some of this neutron energy to nuclei by elastic scattering. The average energy transferred to the struck nucleus with mass number A, by a neutron of energy E_n is

See exercise 5.

$$\langle \Delta E \rangle = \frac{2A}{(A+1)^2} E_n. \tag{11.20}$$

Therefore the most efficient case corresponds to $A = 1$, i.e. hydrogen. In practice the most convenient scheme to introduce hydrogen is to use plastic scintillator for the active layer.

Organic scintillators use plastic base material such as polystyrene, $(C_8H_8)_n$, which have a high hydrogen content.

The path length of low energy neutrons can be $\mathcal{O}(10$ mm) in dense media. Therefore these neutrons can cross several passive and active layers. From eq. (11.20) we can see that the average energy loss for neutrons in collisions in the high-Z (and therefore high-A) absorber will be very small, whereas collisions with hydrogen in the plastic scintillator will result in 50% of the energy being transferred on average. The resulting low energy protons and struck nucleus will have a very short range and therefore deposit all their energy locally. The signal resulting from the neutron elastic scatters in the plastic scintillator will be detected efficiently even if we have thin active layers, but the response to electrons will be smaller because of the reduction of overall active material. Therefore we can increase the hadron response and hence decrease e/h by increasing the ratio of thicknesses of the passive absorber (t_p) to active layer (t_a).

The increase in the signal might be limited by saturation effects. See for example for scintillators the discussion in section 5.1.

Consequently, the energy measurement of hadrons (and jets) will improve for smaller sampling fractions, an observation that might appear counter-intuitive. However, this reduction helps as it boosts the hadronic relative to the electromagnetic response. The energy resolution for electromagnetic showers on the other hand degrades for smaller sampling fraction, as discussed in section 11.6.

A large compensating calorimeter was built by the ZEUS collaboration for the HERA ep collider, using uranium wrapped in stainless steel for the passive layers and scintillator for the active layers [58, 511]. Compensation was achieved by a combination of decreasing the electron response by the use of a combination of high-Z and low-Z absorber (3.2 mm thick uranium plates wrapped in 0.5 mm thick stainless steel) as discussed in the previous section, and by having a large value for t_p/t_a to increase the hadron response.

The (depleted) uranium was radioactive and the stainless steel was also required for radio-protection reasons.

In order to achieve $e/h \simeq 1$ a ratio $t_p/t_a \simeq 4$ was required. The thickness of the scintillator plates could not be less than 2.5 mm, therefore very thick absorber plates had to be used. This had the drawback of increasing the sampling fluctuations for electromagnetic showers.

Very good hadronic energy resolution was achieved in a testbeam experiment [511], but the electromagnetic energy resolution was relatively poor compared to other calorimeters

In a scintillator, light is captured by total internal reflection and transferred to the edge of the plate. In practice, there will be surface imperfections and there will be less than total internal reflections. The number of reflections increase as the plate thickness decreases. Therefore too thin scintillator plates result in larger absorption for the scintillation light.

$$\sigma_E/E = 35\%/\sqrt{E/\text{GeV}} \oplus 2\% \text{ (hadronic)},$$
$$\sigma_E/E = 18\%/\sqrt{E/\text{GeV}} \oplus 1\% \text{ (electromagnetic)}. \tag{11.21}$$

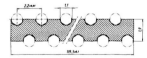

To be able to achieve the optimal ratio of passive to active absorber without compromising the resolution for electrons and photons, a different geometry has been proposed: the 'spaghetti' calorimeter or SPACAL [15]. In this approach lead was used as the passive absorber and 1 mm diameter scintillating fibres were placed in holes running through the calorimeter in

Figure 11.15: Cross-section of a lead absorber plate for the hadronic section of the H1 SPACAL [63]. In this detector $t_p/t_a = 3.4 : 1$, and the fibre diameter was 1 mm.

For a discussion of scintillating fibres see section 5.2.

the direction of the incident particles. The fibres were grouped into bundles and connected to photomultiplier tubes for the readout. A very good resolution for hadrons was obtained,

$$\frac{\sigma_E}{E} = \frac{27.7\%}{\sqrt{E/\text{GeV}}} \oplus 2.5\%. \qquad (11.22)$$

This type of calorimeter was proposed for use at the LHC but this technology was not adopted because it could not compete with alternatives for the electromagnetic energy resolution. In addition, it did not allow for projective geometry which is very important for electrons, photons, τs and QCD jets.

A summary of the resolution achieved for hadronic calorimeters using uranium/scintillators [495] is shown in Figure 11.16. As expected, we obtain the optimal intrinsic resolution for a hadronic calorimeter if the compensation condition ($e/h = 1$) is satisfied.

It was also significantly more expensive than alternative technologies.

The lack of projective geometry in the design proposed at the time would also have been incompatible with the particle flow analysis discussed later in this chapter.

The original motivation for the use of depleted uranium absorbers has been the expectation that neutron-induced fission would boost the hadronic signal (see for example [238]), and thus lead to a compensated calorimeter. While this indeed contributes, today the benefits of this absorber are more seen in its large atomic number Z.

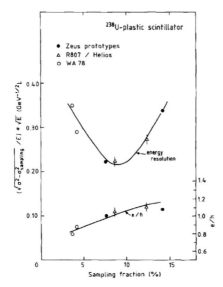

Figure 11.16: Intrinsic resolution and e/h ratio of different uranium/plastic scintillator hadronic calorimeters as a function of sampling fraction [495]. The intrinsic resolution is defined as $\sigma_{\text{intrinsic}}^2 = \sigma_{\text{measured}}^2 - \sigma_{\text{sampling}}^2$, where σ_{measured} is the directly experimentally measured resolution and σ_{sampling} is the expected contribution to the resolution from sampling fluctuations (estimated using eq. (11.1)).

SOFTWARE COMPENSATION

If a hadronic calorimeter has sufficient granularity it is possible to improve the resolution of a non-compensating calorimeter in software by trying to apply different weights to the electromagnetic and hadronic shower components. A first demonstration of this technique was made using a simple non-compensating iron/scintillator sandwich calorimeter for the CDHS neutrino experiment [13]. The idea behind the very simple software compensation algorithm was to de-weight large local energy depositions as they were probably due to electromagnetic energy.

For this type of calorimeter we expect $e/h > 1$ for the reasons discussed earlier.

The weighted cell energies were calculated from the unweighted energies E_i using

$$E_i' = (1 - CE_i) E_i, \qquad (11.23)$$

where C is an empirical constant that optimises the resolution. A significant improvement in resolution is achieved by this simple algorithm. Note

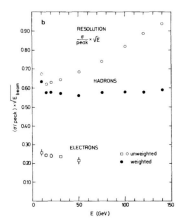

Figure 11.17: Resolution of the CDHS iron/scintillator hadronic calorimeter with and without software weighting [13].

also that the resolution after weighting scales as $1/\sqrt{E}$, unlike before the weighting. This compensation relied on very fine longitudinal sampling which would not be practical for a hadron calorimeter for a collider detector.

Much more sophisticated software compensation algorithms have been developed by the LHC experiments which have more (less) granular calorimeters in the transverse (longitudinal) direction. As an example, the measured jet energy resolution for the ATLAS detector [89] is shown in Figure 11.18. The noise term is a major factor at low value of p_T^{jet} but is very small at high values. The resolution at the highest momenta show a large constant term because the software compensation is imperfect.

The resolution is measured as a function of p_T because the QCD calculations are more accurate for p_T than E and also some experimental systematic uncertainties depend on p_T, not E.

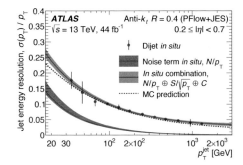

Figure 11.18: Jet energy resolution measured by ATLAS as a function of the jet transverse momentum p_T^{jet} [89]. The algorithm used to cluster the primary objects into jets is described in [174]. $R = \sqrt{(\delta\eta)^2 + (\Delta\phi)^2}$, where $\Delta\eta$ ($\Delta\phi$) is the separation in η (ϕ) between a cluster and the jet axis. Monte Carlo simulations are used to correct the jet energy to that of the particle-level energy of the jet ('JES'). 'PFlow' refers to the use of tracker measurements for charged particles in the jet reconstruction as opposed to calorimeter measurements (see section 11.13).

The resolution is quite good but not as good as could be achieved by a compensating calorimeter (compare to for example [511]). This is another illustration of the fact that detector designs involve compromises which depend on the physics priorities. For ATLAS and CMS, priority was given to the electromagnetic calorimeter resolution, at the cost of poorer hadron reconstruction in the overall calorimeter systems.

The high resolution electromagnetic calorimeters were essential for the discovery and study of the Higgs boson in the $H \rightarrow \gamma\gamma$ and $H \rightarrow ZZ^*$, $Z \rightarrow e^+e^-$ decay modes.

DUAL READOUT CALORIMETERS

The final approach to trying to improve the hadronic resolution for noncompensating calorimeters is to provide, in addition to the measurement of the energy deposited by the hadronic shower, a measurement of the fraction f_{em} of the energy deposited by the electromagnetic component for each event. A way to achieve this is by measuring the light signal from Cherenkov radiation in the calorimeter. This is predominantly sensitive to the electromagnetic shower component because the electrons and positrons in this component are relativistic down to energies of about 200 keV. Most of the non-electromagnetic energy deposition in hadron showers, on the other hand, is from non-relativistic protons generated in nuclear reactions. While these will not generate Cherenkov radiation, they will produce signals by ionisation, for example in liquid noble gas ionisation chamber detectors, or by scintillation. By combining these measurements with the Cherenkov light signal generated in the shower, the electromagnetic shower fraction can be determined for each event, and considered in the reconstruction.

After an independent calibration of the two signals for electromagnetic events (i.e. with electrons), the reconstructed energy for hadrons is given by

$$S = E\left[f_{em} + (h/e)_s(1 - f_{em})\right] \quad \text{(for scintillation),}$$
$$C = E\left[f_{em} + (h/e)_c(1 - f_{em})\right] \quad \text{(for Cherenkov),}$$

where the $(h/e)_{s,c}$ are the ratios of the hadronic response over the response for electrons for the two types of readout. f_{em} can be eliminated from this system of equations, and the hadronic energy is obtained from

$$E = \frac{S - \chi C}{1 - \chi}, \tag{11.24}$$

with $\chi = [1 - (h/e)_s]/[1 - (h/e)_c]$.

This approach avoids several constraints usually imposed by compensation. The absorber material does not have to have a very high Z, copper is sufficient. There is no constraint on the sampling fraction, it can be made large enough to allow for a good energy resolution for electromagnetic showers. Finally, there is no need to detect neutrons, with the associated long time constants. However, it is more challenging to achieve a projective geometry with this approach.

The DREAM calorimeter [327] was a prototype of a sampling calorimeter to demonstrate this concept, with a copper absorber and active elements comprising doped (for scintillation) and undoped plastic or quartz fibres (for detection of Cherenkov photons) that are read out by different PMTs. The volume ratio was Cu/doped fibre/undoped fibre/air 69.3/9.4/12.6/8.7 and the sampling fraction for MIPs was 2.1%. The depth of the prototype was $10\,\lambda_{int}$ [327].

In this prototype χ was about 0.29. The energy resolution was limited by lateral shower leakage due to the limited size of the prototype. However, a clear improvement in hadronic energy resolution after combining the independent signals was observed, with a very small constant term [327]. Furthermore, after combination the signal scales for different particle types do agree, also with the calibration obtained from electrons.

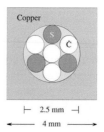

Figure 11.19: Cross-section of the basic cell of the DREAM calorimeter [32]. 7 fibres (3 doped and 4 undoped) are inside a hollow square copper rod.

Dual calorimeter readout has also been investigated for crystal calorimeters, where the different time scales of the prompt Cherenkov and the delayed scintillation signal can be exploited. The main challenge here is the strong attenuation of the short-wavelength Cherenkov photons in the crystal, which leads to a non-linear response of the calorimeter for this signal [33].

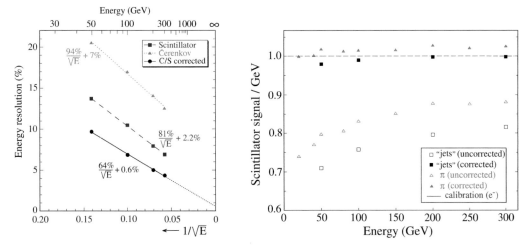

Figure 11.20: DREAM calorimeter prototype [327]. Left: Energy resolution for multi-track events ('jets'). Right: Energy scale for multi-track events and single pions, compared to electrons. 'Corrected' in both cases means results derived with both readouts combined.

11.12   SIMULATION OF CALORIMETERS

As we have seen throughout this chapter, the development of electromagnetic and hadronic showers is governed by complex processes, and the accuracy of analytical predictions of the shower development and the quality of the energy measurement are limited. Therefore, such predictions require the use of Monte Carlo techniques to 'simulate' real showers. The basic idea is to use the Monte Carlo technique to track particles as they propagate through matter and create secondary particles, until the energies of the particles fall below some defined threshold at which point the particles are stopped and their energy deposited locally. After generating the trajectories of all secondaries in the calorimeter, the total energy deposited in small volumes of the active layers are stored. In a separate pass the deposited energy is used to calculate the response of the detector. This can include the response of the full electronics chain used in the readout and therefore can output data in an identical format to that of real data. This allows the same reconstruction software to run on real and simulated data, thus allowing for calculation of all aspects of the detector response (e.g. resolution or efficiency for a given type of particle).

The Monte Carlo method is explored for a very simple example in exercise 7.

One difficulty with this approach is to accurately model the full detector geometry, due to the large number of small segments required in the simulation. This is very challenging for complex collider detectors, where the geometry of the tracking detectors needs to be accurately modelled as well as that of the actual calorimeters. To keep the number of volume elements practical some simplifications are usually required.

Apart from the general difficulty of modelling complex geometries, minimum energies to track produced particles have to be carefully selected. The best accuracy will require the lowest energy thresholds but this will come at a cost in CPU time. Therefore, any application will have to assess this trade-off to select the optimum thresholds.

Simulations of electromagnetic showers are in principle simpler than hadronic showers as the underlying physics processes are less complex.

Rather than just set global thresholds, the thresholds can be adjusted depending on the location of an interaction. This can be used to improve the modelling of the effects of δ rays without too large a cost in CPU time.

The most comprehensive and widely used software for these type of Monte Carlo simulations is GEANT4 [29].

An example for the study of the accuracy of a GEANT4 simulation for electron-induced showers is the comparison of measured and simulated data for a test module of the ATLAS electromagnetic accordion calorimeter [31]. This calorimeter was longitudinally segmented into a 1.7 X_0 thick pre-shower detector followed by a front, middle and back compartment with thicknesses of 4.6 X_0, 17.6 X_0, and 5.0 X_0, respectively. A measurement sensitive to the longitudinal shower development is the fractional energy deposited in the first section of the calorimeter. The transverse width of the shower was studied by measuring the distribution of the fractional energy deposited in the η-strips of the first section of the calorimeter. The Monte Carlo results for the longitudinal and lateral shower shapes are in good agreement with the test beam data.

GEANT4 is widely used in particle physics, but it is had many applications in other domains. For example, it is used in nuclear medicine, where it is used in planning radiotherapy to optimise the delivered doses.
For purely electromagnetic processes, the older EGS4 code can be used. A more modern version of the EGS4 code is maintained by NRC [299]. A systematic comparison between EGS4 and GEANT4 can be found in [384]. Other codes for simulating hadronic interactions such as CALOR and FLUKA can now be used within GEANT4.

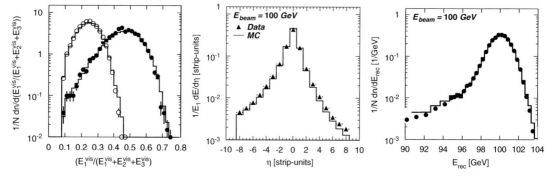

Figure 11.21: Comparison of measured and simulated (GEANT4) shower shapes for the first section of the ATLAS accordion calorimeter [31]. Points are data and lines are the result from Monte Carlo simulations. The shaded area corresponds to the uncertainty in the amount of material before the calorimeter. Left: Reconstructed energy fraction in the first compartment for 10 GeV (open circles) and 100 GeV (full circles) electrons. Middle: Reconstructed energy fraction in the different η (pseudorapidity) slices for 100 GeV electrons. The width of the η slices is 0.025. Right: Reconstructed electron energy for 100 GeV electrons.

A particularly important test of the accuracy of the Monte Carlo simulation is the distribution of measured energy. The data show a Gaussian peak and a 'tail' on the low side. In a test beam the tail is due to electron bremsstrahlung in upstream material resulting in lower energy electrons being deflected out of the beam by the magnets in the beamline, The simulation is in agreement with the data both for the core and the tail of the distribution.

In a collider detector, the tracking detector will be $\mathcal{O}(1\,X_0)$ thick, so there will be a high probability for electrons to emit bremsstrahlung photons before hitting the calorimeter. This can lead to lower energy electrons being deflected more by the magnetic field and not being included in the reconstructed electron cluster.

It is much more difficult to achieve the same accuracy for the more complex geometry of a collider detector. Therefore as discussed in section 11.7 in-situ calibration procedures are required to achieve the optimal performance.

The simulation of hadronic showers is more complex. As discussed in section 2.12 there is no fundamental theory for hadronic reactions in the appropriate phase space region. In addition we need to consider energies ranging from thermal neutrons to TeV. There is no solution that will be suitable for all tasks and many different approaches are required to cover this range.

Even higher energy hadrons need to be considered for simulating detectors for very high energy cosmic rays.

When sufficient data is available, the optimal strategy is to use this to create a data-driven model. An important example of this approach is the use of very extensive data for low energy neutrons. This method provides accurate simulation of neutron transport and isotope production.

When there is some relevant data but the coverage is insufficient for a purely data-driven approach, parameterised models are used. These use fits to experimental data to interpolate between data points and if necessary to extrapolate. An example of this approach is the interpolation of the measured pion-nucleus cross-sections by the GEISHA code [249].

The GEISHA code is included in the GEANT4 package.

Figure 11.22: Pion-nucleus cross-sections as a function of pion momentum. The circles are the measured data and the curves are the model parameterisations. The upper panel is for the total cross-section and the lower panel is for the inelastic cross-section [249].

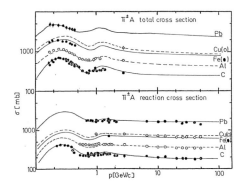

When there is insufficient data for this approach, we rely on theory models. A very wide range of models are available for different energy regions and processes. At high energy, one approach is based on the parton 'string' model. At low energies, models of intra-nuclear cascades are required. The choice of which set of models to use will depend on the application and the available CPU resources.

In the Lund string model the strong gluon self-interactions form a colour flux tube. As the tube is stretched it creates new hadrons.

As an example of the performance of a GEANT4 simulation for high energy pions, Figure 11.23 shows the ratio of measured energy resolution from test beam data of the ATLAS Hadron Endcap Calorimeter (HEC) to that from simulations [308]. There is an approximate agreement between the data and the simulation. As for electromagnetic showers, the best accuracy requires in-situ calibrations during operation as discussed in section 11.14.

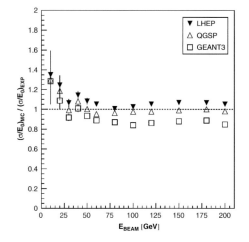

Figure 11.23: Ratio between the simulated and measured energy resolution for the ATLAS HEC as a function of pion energy [308]. LHEP and QGSP refer to different physics models selected in GEANT4. GEANT3 refers to an older version of the code.

11.13   RESPONSE TO HADRONIC JETS

Our discussion of calorimeters has focused on the response to electrons and single hadrons. However, in high energy physics we often need to measure QCD 'jets' rather than single hadrons. In the theory of QCD we can use perturbation theory to predict the cross-sections of events at the quark and gluon level. However, these quarks and gluons must be confined inside hadrons and they therefore hadronise into jets of hadrons. These hadrons will tend to have low values of p_T about the axis of the original quark or gluon and will therefore appear like a narrow jet of particles in a detector.

In order to reconstruct the kinematic properties of the primary parton, we need a jet algorithm that will combine the appropriate energy depositions in different cells of the calorimeter into a cluster of energy belonging to the jet.

We need this if we want to compare data to QCD predictions. We also need this in order to use jets to reconstruct masses of primary objects, e.g. for the Higgs decay $H \to b\bar{b}$ we need to use the measurements of the kinematic properties of the b jets to reconstruct the mass of the Higgs.

The conceptually simplest approach is the cone algorithm. The highest p_T cell in the calorimeter is used as a seed and all cells within a cone of some fixed radius $R = \sqrt{(\Delta\eta)^2 + (\Delta\phi)^2}$ about the seed cell are clustered. The process can be repeated for new seed cells until there are no unused cells above a certain threshold in p_T. There are theoretical problems with this naive algorithm and more sophisticated algorithms are used [174].

$\Delta\eta$ is the separation between a cell and the jet axis in the longitudinal direction with an equivalent definition for $\Delta\phi$ in the azimuthal direction.

JET MEASUREMENTS USING PARTICLE FLOW

An alternative approach to improving the resolution of hadronic jets for detectors not using compensating calorimeters is to use a 'particle flow' algorithm [451]. This method relies on the observation that typically in a jet about 60% of jet energy is carried by charged hadrons, 30% by photons (mostly from $\pi^0 \to \gamma\gamma$), and only about 10% by neutral hadrons (mostly n, K_L^0). If a good charged particle momentum measurement is available (typically from the tracker in a multi-detector experiment), and if the individual particle contributions can be resolved, then the charged hadron component of the jet can be measured more accurately in that way. At the same time, energy resolution for the measurement of photon energies will be better measured in the electromagnetic calorimeter. Only the contribution from neutral hadrons must be obtained from the calorimeters.

The simple idea has many complications in practice; the energy deposited by the charged particles in the calorimeters can overlap with the electromagnetic and neutral hadronic energy and therefore needs to be subtracted to avoid double-counting. Ultimately, the resolution of this method is dominated by 'confusion' in the matching and assignment of energy depositions to different shower components.

This is particularly challenging in the core of a jet where there is a high particle density.

The particle flow approach was studied for the proposed experiments at a future linear e^+e^- collider. Monte Carlo simulation studies showed a jet energy resolution of [473]

$$\frac{\sigma_E}{E} = \left[\frac{21}{\sqrt{E/\text{GeV}}} \oplus 0.7 \oplus 0.004E/\text{GeV} \oplus 2.1 \left(\frac{E/\text{GeV}}{100} \right)^{0.3} \right] \%.$$

$$(11.25)$$

The first term represents the intrinsic calorimeter resolution, and the following terms represent imperfect tracking, leakage and confusion, respectively.

At the LHC the particle flow approach is very challenging because of the large 'pile-up' background arising from interactions in the same beam crossing as the triggered event. The particle flow approach is less sensitive to pile-up than purely calorimetric measurements as the momenta of particles coming from vertices other than the 'primary' can be discarded. However, at high pile-up the vertices are not always sufficiently separated so there can be confusion. Therefore, at low transverse momentum the resolution deteriorates with increasing number of pile-up interactions.

Pile-up contributes a momentum-independent 'noise' source. Therefore the effect on the fractional resolution is larger at low p_T.

The particle flow approach has been used to improve the resolution for hadronic jets at lower energies. At very high energies, the momentum resolution of the charged track measurements deteriorates, which decreases the achievable gain with this approach. An example of the jet energy resolution that can be achieved with this approach in the CMS experiment [201] is shown in Figure 11.24.

The improvement using particle flow was particularly important for CMS compared to ATLAS because of the higher precision tracking detector (due to the large magnetic field) and the limited performance of the hadronic calorimeter.

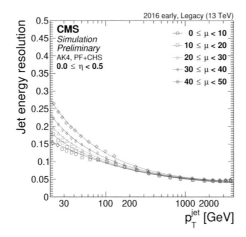

Figure 11.24: Jet energy resolution determined from simulated events by CMS using the particle flow algorithm [201]. μ is the mean number of pile-up interactions.

11.14 CALIBRATION OF HADRONIC CALORIMETERS

As for electromagnetic calorimeters, we require an electronic calibration system to convert ADC counts to charge (typically in fC). We then need a calibration from charge to deposited energy. Small test calorimeters can be used to study the response to single hadrons and determine this conversion constant. The calorimeter in a typical particle physics experiment is divided in an electromagnetic and a hadronic section. Typically a large fraction of the energy of a hadron will be deposited in the electromagnetic calorimeter before reaching the hadronic calorimeter. Therefore these testbeam calibrations will have to include a section of the electromagnetic calorimeter before the hadronic calorimeter as in the final detector.

Hadrons don't 'know' that the calorimeter is divided into electromagnetic and hadronic sections!

As for large electromagnetic calorimeters, it is impractical to calibrate all cells of a very large hadron calorimeter in test beams. The calibration from the part measured in the test beam has to be transferred to the rest of the calorimeter by e.g. using radioactive sources.

The calibration schemes for LHC hadron calorimeters need to allow for many effects, including pile-up, the hadron spectrum within a jet of different flavours and the non-compensating calorimeters used. Finally, if a particle flow analysis (PFA) is used, then this requires an additional calibration step. These procedures need to be developed on Monte Carlo samples in which the 'truth' jet energy can be compared to the 'measured' response.

The Monte Carlo simulation itself will be tuned to agree with the test beam measurements.

The uncertainties arising from the use of Monte Carlo simulations can be reduced by using in-situ calibrations to determine the average response for hadronic jets. Several different samples are used in these studies providing different information. These include

- *Z+jet (with $Z \rightarrow e^+e^-$ or $Z \rightarrow \mu^+\mu^-$).* The Z decay products can be measured with much greater precision and accuracy than hadronic jets in the tracking systems. Therefore they can be used as a proxy for the 'truth'. The response in transverse momentum p_T is defined as $R(p_T) = p_T(\text{jet})/p_T(Z)$. Then, $R(p_T)$ can be measured as a function of p_T in some range of p_T and pseudorapidity η. Similar studies can be performed with γ+jet events, which have a larger cross-section and the analysis can therefore be extended to higher values of p_T before the statistical uncertainties become too large.

There are more subtle second order effects we have ignored, so the analysis is repeated in a Monte Carlo sample to obtain $R^{MC}(p_T)$ and one then looks at the ratio $R(p_T)/R^{MC}(p_T)$ to determine data driven correction factors.

- *Di-jet balance.* There will be insufficient Z+jet events with the jets in the forward (high $|\eta|$) regions of the calorimeter. If the central region is calibrated, then the calibration can be extended to the forward regions by using di-jet balance, with one jet in the central region and one in the forward region. In di-jet events, the p_T of the two jets should be equal to ensure momentum conservation. In this procedure the central jet serves as the proxy for the 'truth' jet and hence determines $R(p_T)$.

- *Multi-jet balance.* There will also be insufficient Z+jet events with very high p_T for the Z+jet calibration to work in this range. Multi-jet events have large cross-sections and we can use events in which one high-p_T jet is recoiling against two lower-p_T jets. If the lower-p_T jets have been calibrated, then they serve as proxy for the 'truth' and so can be used to extend the determination of $R(p_T)$ to higher values of jet-p_T.

Using these in-situ techniques [201] in the LHC general purpose experiments the jet energy scale is known to a precision of 1%.

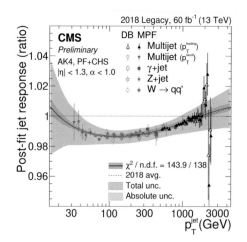

Figure 11.25: Jet energy response measured by CMS [201]. The ratio of the response from data to that from simulation is shown for Z+jet and γ+jet samples as a function of jet transverse momentum.

Finally, $W \to q\bar{q}$ samples can be identified using $t\bar{t} \to b\bar{b}W^+W^-$ events with one W decaying leptonically and one decaying hadronically, $W \to q\bar{q}$. The distribution of the reconstructed $W \to q\bar{q}$ mass using these samples can be compared for data and simulation. The good agreement between data and simulation is an important cross-check for the jet energy calibration.

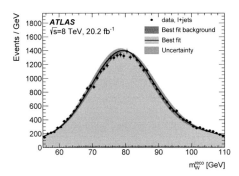

Figure 11.26: Reconstructed W mass from $W \to q\bar{q}$ in $t\bar{t}$ events in the ATLAS experiment. Good agreement is observed between data and simulation [90].

In a similar approach to that used for the electromagnetic calorimeter, we can also use an in-situ analysis to determine the difference in the jet energy resolution between data and Monte Carlo. This difference can then be used to modify the simulated jet energies so that the jet energy resolution is compatible with the resolution observed in the real data. One approach is to use clean di-jet events (i.e. with no third jet) to estimate the resolution. The asymmetry of the two jet energies is defined as [89]

$$\mathscr{A} \equiv \frac{p_T^{\text{probe}} - p_T^{\text{ref}}}{\left(p_T^{\text{probe}} + p_T^{\text{ref}}\right)/2},$$

where p_T^{ref} is the p_T in a well-measured region, and p_T^{probe} is the p_T in the region being studied.

The RMS value of the asymmetry is measured in a sample of events.

This asymmetry is sensitive to the jet energy resolution as well as many physics effects such as additional radiation, so it does not provide a direct measurement of the jet energy resolution. However, these physics effects can be parameterised from the Monte Carlo simulations of the primary particles ("truth" particles). The measured asymmetry, $\sigma_{\mathscr{A}}^{\text{probe}}$, can then be described as a convolution of the detector asymmetry, $\sigma_{\mathscr{A}}^{\text{probe,det}}$, and an asymmetry from the physics effects, $\sigma_{\mathscr{A}}^{\text{truth}}$. Therefore the detector level asymmetry can be determined by subtracting the truth asymmetry in quadrature

The asymmetry will be affected because radiation results in extra jets and the two leading jets will not necessarily have equal values of p_T.

$$\sigma_{\text{det}} = \sigma_{\mathscr{A}}^{\text{probe,det}} \ominus \sigma_{\mathscr{A}}^{\text{truth}}.$$

This deconvolution procedure can therefore be used to determine the intrinsic jet energy resolution [89].

The symbol '$\ominus$' represents a subtraction in quadrature (in a similar way to '$\oplus$' that represents addition in quadrature).

Key lessons from this chapter

- In particle physics calorimeters incoming particles deposit their energy. The signal generated in the calorimeter is a function of this energy.

- Calorimeters are the only detectors capable of efficiently detecting neutral particles.

- The energy is absorbed in the calorimeter in detection quanta (ionisation charges, scintillation photons, phonons, etc.). The resolution of the energy measurement is driven by the counting statistics of these quanta. In the simplest case the fluctuations can be described by Poisson statistics, but correlations due to overall energy conservation can reduce fluctuations by the Fano factor, whereas other instrumental effects can increase the error on the energy measurement.

- For Poisson-distributed fluctuations, the energy resolution will scale with the number of quanta N as $\sqrt{N}$, and thus the relative energy resolution $\sigma_E/E \propto E^{-1/2}$. The energy resolution will be better for larger N, i.e. for smaller energy per quantum.

- Thus the best energy resolution can be achieved if the quanta are phonons or Cooper pairs. However, these can only be detected at sub-K temperatures. The next best energy resolution can be achieved with semiconductors, and scintillation, requiring the most energy per quantum, gives the worst intrinsic energy resolution.

- In high energy calorimeters, the energy deposition of the incoming particle will not be local, but gradual in the form of a cascade (shower) of secondary particles. Two main types of showers can be distinguished by the dominating interactions responsible for the shower development: electromagnetic and hadronic showers.

- These cascades have characteristic scale lengths describing their dimensions. In electromagnetic calorimeters the longitudinal scale length is the radiation length, whereas the transverse dimension is driven by multiple scattering and therefore can be described by the Molìere radius. In hadronic calorimeters the longitudinal and transverse dimensions can be described using the hadronic interaction length.

- While for low Z the radiation length and the interaction length are comparable, the interaction length is much larger than the radiation length for the materials used in typical particle physics calorimeters.

- To maintain containment of the shower in the calorimeter, the size of the calorimeter must scale with $\log E$.

- As the energy degrades in the development of the shower, the interactions of the particles in the shower with the detector material change. Typically, most of the energy is deposited in the latter stages of the shower development. Fluctuations in the early shower development can have a large effect on the shower dimensions and energy reconstruction of individual showers.

- A calorimeter that is made of one uniform material used to absorb the particle and to create the detection signal is called a 'homogeneous calorimeter'. To reduce the size of a calorimeter often dense layers of absorber are interleaved with less dense active layers of detector that create the signal. Such a configuration is called a 'sandwich calorimeter'. In the latter case the energy resolution is degraded by additional sampling fluctuations of the energy between energy deposited in the passive absorber and active detector layers.

- Hadronic showers are more complex, and in addition to an electromagnetic shower component, part of the energy goes into nuclear reactions. This energy fraction usually is lost for detection. Thus, the energy recorded in the calorimeter will typically be different for different particles, and because of the fluctuations in the energy fraction in the different components the energy resolution will be significantly degraded, while the energy response will be non-linear. Hadronic calorimeters that achieve a similar response for different types of incoming particles can alleviate these effects. These are called 'compensated calorimeters'.

- Calibration of energy scale and linearity are a necessity for all calorimeters. The calibration requires electronic systems to convert raw ADC counts to charge or current. The conversion from charge or current to deposited energy has to be determined in a test beam using a prototype or slice of the full calorimeter. Different systems are required to transfer the energy scale to all cells in a large calorimeter. In collider experiments the knowledge of the energy scale, linearity and resolution can be improved with analysis of in-situ data.

EXERCISES

1. *Rossi's approximation B.*

   A simple model of the initial stages of an electromagnetic cascade in matter assumes that photons travel one radiation length X_0 before undergoing pair production and electrons (positrons) also travel one X_0 before undergoing bremsstrahlung. The model also assumes that the energy is shared equally between the two final state particles. In this model further multiplication stops once the energy of the charged particles becomes less than E_c, and the charged particles lose energy to ionisation at a constant rate dE/dx. (This model is known as "Rossi's approximation B" [432].)

   a) Use this model to show that the shower maximum for an electromagnetic cascade from an incident electron of energy E will occur at $\ln(E/E_c)/\ln 2$. Estimate the range of charged and neutral secondaries after multiplication stops. Compare your estimates for an electron of 10 GeV with the results from the detailed simulations in Figure 11.3.

   b) Show that the number of particles in a shower in this model is E/E_c.

   c) What difference is there between the energy loss as a function of distance for electrons/positrons and photons (which is not modelled properly by the assumptions above? What consequence will this have for the energy depositions in showers started by electrons/positrons compared to showers started by photons?

d) What other aspects of the development of the electromagnetic cascade are not modelled by this simple model? Explain qualitatively the difference in the shower shape beyond the shower maximum in the simple model and the detailed simulation.

2. *Pointing accuracy and reconstruction of resonances.*

Consider the decay of the Higgs boson $H^0 \to \gamma\gamma$. The invariant mass of the two photons is M. Show that the resolution in the measurement of M (ΔM) is given by

$$\frac{\Delta M^2}{M^2} = \frac{\Delta E_1^2}{E_1^2} \oplus \frac{\Delta E_2^2}{E_2^2} \oplus \left(\frac{\sin\alpha(\Delta\alpha)}{1 - \cos\alpha} \right)^2$$

where $E_1 (E_2)$ is the energy of the first (second) photon and α is the space angle between the directions of the two photons. The uncertainties in the measurements of E_1, E_2 and α are $\Delta E_1, \Delta E_2, \Delta\alpha$ respectively. You may assume that $\frac{\Delta E^2}{E^2} \approx \left(\frac{\Delta E}{E} \right)^2$.

Consider such decays for which the polar angle of both photons is $\theta = 45°$ and the difference in azimuthal angle between the two photons is $\Delta\phi = 180°$. If the energy resolution of an electromagnetic calorimeter is 1% for photons estimate the angular resolution for photons that would give the same contribution to the invariant mass resolution as that from the energy resolution.

Explain one method that could be used in an LHC detector to make such a precise angle measurement for photons.

3. *Electromagnetic calorimeter resolution.*

An empirical parameterisation for the energy resolution of an electromagnetic sampling calorimeter is given by [221]

$$\frac{\sigma_E}{E} = \frac{\sigma_0}{\sqrt{E}} \left(\frac{d_{abs}}{X_0^{abs}} \right)^{\alpha} \left(\frac{d_{det}}{X_0^{det}} \right)^{-\beta}.$$

σ_0 scales with $\sqrt{Z}$, $0.6 < \alpha < 0.7$ and $0.15 < \beta < 0.3$. Consider a sampling electromagnetic calorimeter consisting of absorber plates of thickness 1 X_0. Using this equation, how does the resolution scale with the value of Z for the absorber? What other factors are relevant for the choice of absorber material for an electromagnetic calorimeter?

4. *Pion production threshold.*

Calculate the threshold energy in the lab frame for $\pi^\pm$ production in the reaction $\pi^\pm p \to \pi^\pm \pi^+ n$.

5. *Neutron scattering.*

Show that a neutron with energy E scattering off a nucleus with atomic number A at an angle $\theta^\star$ in the CMS will transfer an energy to the nucleus (you may assume that the neutron is non-relativistic)

$$\Delta E = \frac{2A(1 - \cos\theta^\star)}{(A-1)^2} E.$$

Hence verify eq. (11.20).

6. *Dual readout calorimeter.*

Consider a dual readout calorimeter. For each readout independently, the calorimeter is non-compensating, i.e. $e/h \neq 1$. The two systems are calibrated with electrons. Stating any assumptions that you make, show that the energy of a hadronic shower can be determined independently of the fraction of electromagnetic energy f_{em} and comment on the significance of your result.

7. *A small Monte Carlo.*

a) To simulate random events from a Probability Distribution Function, $f(x)$, we first define the cumulative distribution as $F(x) = \int_{-\infty}^{+\infty} f(x')dx'$. If the cumulative distribution is normalised to unity, then if we select random numbers y_i from a uniform distribution between 0 and 1, we can obtain random numbers distributed according to the PDF $f(x)$ by inverting the cumulative distribution, $x_i = F^{-1}(y_i)$. Use this method to generate random numbers from an exponential distribution $f(x) = \exp(-x/5)$ for $0 < x < \infty$. Plot a histogram of the generated numbers. Check that the mean of the simulated numbers is consistent with an analytic calculation.

b) Consider particles (A) of mass 5 GeV/c^2 produced with a momentum $p = 5$ GeV/c. The lifetime of the meson is $\tau = 1$ ps. The particles decays isotropically in the particle rest frame to two particles (B) of mass 1 GeV/c^2. These B particles have the same lifetime as A. Write a Monte Carlo code to simulate the decay chain and use this to plot the distribution of distances from the location of the decays of the B particles from the location of the production of the A particles.

12 Particle identification

In this chapter we will look at how we can use the detector techniques we have studied in previous chapters to identify different types of particles. Knowing the type of particle can unlock a powerful unambiguous constraint for the kinematics of a particle, as this defines its mass. Furthermore, it defines the flavour, which can be used as a tag to identify the underlying physics process in an event.

We will start this chapter with discussing four types of dedicated detector techniques that can be used on their own to distinguish between different types of particles: Time-of-flight (TOF), dE/dx, Cherenkov detectors and transition radiation detectors (TRDs). In the second half of the chapter we will then look at techniques to identify specific types of particles, which will usually require the use of the full power of a multi-detector particle physics experiment.

12.1 PRINCIPLES OF PARTICLE IDENTIFICATION DETECTORS

An unambiguous determination of the type of a particle is obtained if its mass can be measured. The mass is not directly accessible, but can be determined, if the momentum is known, from a measurement of the velocity,

The two detectable particles with masses that are too close to be easily directly resolved are the pion and the muon. Here the fact that the muon is a long-lived heavy lepton can be used for identification (see section 12.8).

$$\frac{mc}{p} = \frac{1}{\beta\gamma} = \sqrt{\frac{1-\beta^2}{\beta^2}} = \sqrt{\frac{1}{\gamma^2-1}}. \tag{12.1}$$

Several techniques do exist (measurement of time-of-flight, dE/dx, Cherenkov and transition radiation) and, as we will discuss, each of them performs best in a different regime of momentum of the incoming particle. The figure of merit is the separation power, which describes the discrimination power between two particle hypotheses, and is given in units of the detector resolution σ by

The resolution for the two measurements are not necessarily the same. If they are not, a suitable average can be used.

$$n_{1,2} = \frac{R_1 - R_2}{\sigma},$$

where the $R_{1,2}$ are the average responses of the detector for the two types of particles.

12.2 MEASUREMENT OF TIME-OF-FLIGHT

From the time-of-flight (TOF) t of a charged particle measured between two thin detector layers separated by a distance L the velocity can be determined from $\beta = L/(tc)$, which can then be used in eq. (12.1). The resolution of the mass measurement is then given by (see exercise 1)

Alternatively, the start signal for the time-of-flight measurement can come from other arrival time constraints, for example from the bunch times of the beam producing the charged particle, if they are known with sufficient precision.

$$\left(\frac{\sigma_m}{m}\right)^2 = \left(\frac{\sigma_p}{p}\right)^2 + \gamma^4\left[\left(\frac{\sigma_L}{L}\right)^2 + \left(\frac{\sigma_t}{t}\right)^2\right]. \tag{12.2}$$

DOI: 10.1201/9781003287674-12

Usually, in particle physics we are dealing with relativistic particles, so $\gamma \gg 1$ and thus it is the determination of the velocity that is the dominating contribution. For a typical separation of a few metres $\sigma_L/L \simeq 10^{-3}$ is quite achievable, so that the time resolution becomes the limiting factor for the resolution.

For $L = 3$ m, $\sigma_t/t = 10^{-2}$ requires $\sigma_t = 100$ ps.

The TOF separation power between two particles of type 1 and 2 for the same momentum p is

Using (for $p \gg mc$)
$$E = \sqrt{p^2c^2 - m^2c^4}$$
$$= pc\sqrt{1 - m^2c^2/p^2}$$
$$\simeq pc(1 - m^2c^2/2p^2).$$

$$n_{1,2} = \frac{t_1 - t_2}{\sigma_t} = \frac{L}{c\sigma_t}\left(\frac{1}{\beta_1} - \frac{1}{\beta_2}\right) = \frac{L}{pc^2\sigma_t}(E_1 - E_2) \simeq \frac{Lc}{2p^2\sigma_t}(m_1^2 - m_2^2). \tag{12.3}$$

Hence, the length of the TOF system has to grow as p^2 to maintain the same separation power if the momentum increases. For $L = 1$ m and $\sigma_t = 100$ ps, 3σ π/K separation is achieved up to a momentum of 967 MeV/c.

The most widely used TOF systems rely on plastic scintillators read out with PMTs, as they provide good timing resolution at low cost. In such a system, the timing resolution is the result of several contributions

See section 5.1.

Position resolution requirements in TOF systems are usually moderate.

$$\sigma_t = \sqrt{\frac{\sigma_{\text{ph}}^2}{N_{\text{eff}}} + \sigma_{\text{el}}^2}, \tag{12.4}$$

where σ_{ph} comprises errors for the detection of individual photons (photon emission delay, path length variations, transit time variations in the PMT, etc.), reduced by the effective number of detected photons, N_{eff}, and σ_{el} contains contributions from the electronics (noise, clock jitter, time walk, calibration etc.).

In large TOF systems with long scintillators photon path length variations can be reduced by using the average time of two PMTs recording the light output at the two ends of the scintillator.

The effective number of detected photons, N_{eff}, is smaller than the number of photons created in the scintillator due to acceptance, attenuation and detection efficiencies. An empirical estimate is that approximately 2×10^{-3} photoelectrons will be produced by a primary photon. For a minimum-ionising particle crossing 1 cm of plastic scintillator this gives $N_{\text{eff}} \sim 40$.

Assuming $(dE/dx)_{\text{mip}} = 2$ MeV/cm², $\rho = 1$ g/cm³, and 25 eV required to produce a photon.

Alternative photon detectors can be MCPs (see section 4.4), or LGADs or SiPMs (see section 4.6). Gaseous detectors can also be used as timing detectors, if the gas layers are thin, and the timing resolution can be further improved by stacking several layers, like in multi-gap timing RPCs (see section 7.2). In these systems the timing resolution is given by equivalent expressions as in eq. (12.4).

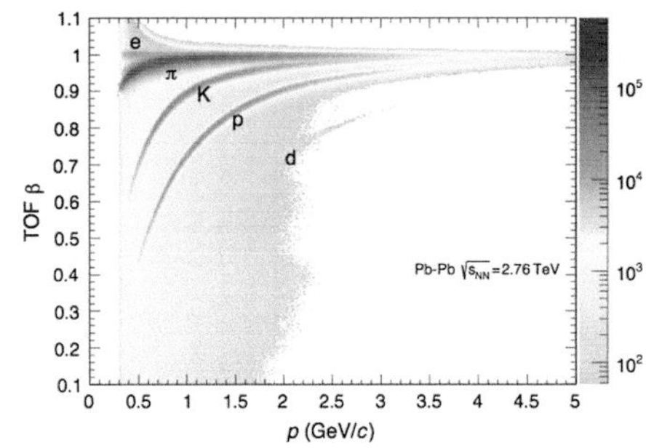

Figure 12.1: Separation in the ALICE TOF system (multigap RPCs with a system time resolution of around 80 ps) [280] (original data from [37]).

12.3　MEASUREMENT OF dE/dx

In section 2.3 we have seen that the ionisation energy loss of charged particles can be parameterised as a function of $\beta\gamma$, independently of the type of particle. If the amount of energy deposited in a detection layer can be measured, then this can be used for a velocity measurement, which can be again combined with a measurement of the momentum in a spectrometer to determine the particle mass.

Due to the shape of the energy loss function, separation is relatively easy for momenta smaller than for minimum-ionising particles, but then becomes very challenging for MIPs. It becomes possible again for momenta in the range of the relativistic rise until it deteriorates again for the Fermi plateau.

Only in gaseous detectors the relativistic rise is large enough (typically 50% to 60%) to allow for separation above the MIP momentum. In solids the rise is limited by the density effect to about 10% (see section 2.3).

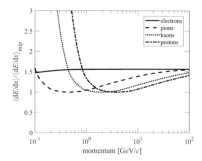
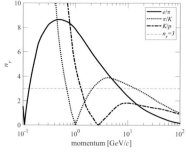

Figure 12.2: Left: Average energy loss as a function of momentum for different particles in Ar/CH$_4$ 93/7 (using the energy loss function shown in Figure 2.16). Right: Resolution power for $\sigma_E/E \simeq 5\%$ for selected discrimination channels.

The challenge for the measurement of the deposited energy by ionisation is the straggling of the energy loss. In particular for gaseous detectors the measurement of the energy loss is severely hampered by the fluctuations of the energy transferred in individual ionisation clusters, and the significant probability for very high energy transfers (δ-electrons) leads to pronounced tails in the straggling distributions.

See section 2.5.

In practice the energy measurement in a gaseous detector is divided into a measurement of the ionisation charge deposited in a number of cells of the gaseous detector, with the charge in each cell collected by one readout channel. Averaging these measurements can reduce the error on the energy measurement. A common approach is to exclude a fixed fraction (typically about 30%) of the energy measurements with the highest energy depositions, to get a result which is less affected by the high energy deposition fluctuations ('truncated mean'). For a fixed length detector there is an optimum number of energy measurements, as the increased straggling when the detection volume is divided into smaller cells negates the advantage of the increased statistics of individual measurements.

The charge collection and amplification process required in gaseous detectors will introduce additional stochastic fluctuations and possible non-linearities for the conversion of energy into an electronic signal.

Sometimes also the lowest 10% of the energy measurements are discarded, to make the distribution even more symmetric.

A likelihood fit approach should in principle be mathematically more powerful than the truncated mean, as it allows to keep all data points. This is particularly relevant if only a small number of samples are taken, for example in silicon detector systems. However, this method is significantly more elaborate, and there is no evidence that the results are significantly better.

Another way to reduce straggling is to increase the pressure of the gas, but this comes at the cost of reducing the height of the relativistic rise due to the increased density effect.

An alternative approach is not to rely on the measurement of total charge deposited and registered in the detector, but to count the number of ionisation clusters, which eliminates the direct effects of cluster size and gas amplification fluctuations [491]. The challenge here is the high

A compromise is a single parameter likelihood, where the shape of the sampled energy distribution is assumed to be universal. However, this assumption does not apply for thin sampling layers [48].

bandwidth (order 100 MHz) required to resolve individual ionisation clusters [183].

In practice, with the truncated mean approach a resolution of $\sigma_E/E \simeq 5\%$ is typically achieved.

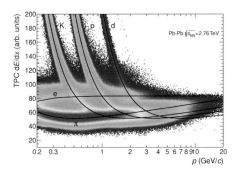

Figure 12.3: Specific energy loss (dE/dx) in the ALICE TPC versus particle momentum in Pb–Pb collisions at $\sqrt{s_{NN}} = 2.76$ TeV. The lines show the parameterisations of the expected mean energy loss [41].

12.4   CHERENKOV DETECTORS

A more powerful particle identification technique at higher momentum relies on the measurement of the photons emitted by the Cherenkov effect in a transparent radiator.

For a detailed review of Cherenkov detectors see [386].

As discussed in section 2.8, the number of Cherenkov photons emitted per unit of track length of the incoming particle is not large, and the number of photons detected will be even smaller, due to attenuation, acceptance and detection efficiencies. Maximising the number of detected photons is thus a central goal of the design of Cherenkov detectors, in particular if the emission angle θ_{Ch} is to be determined and a large number of measurements is needed to reduce the errors associated with the difficult identification of the direction of the photons.

Cherenkov detectors consist of at least two parts, a radiator, in which a sufficient number of photons is emitted, and a photon detector, where they are detected. In many cases optical elements (mirrors or other optical interfaces) are placed in between, which allow to move the detectors outside of the active volume of the detector and/or focus photons from one track into the same region of the photon detector.

Due to the absence of a continuous energy deposition along the path of a photon, its direction can only be deduced from the point of emission and the point of detection of the photon. As detection destroys the photon, the point of creation cannot be directly measured and can only be inferred from other constraints.

RADIATORS

Radiators need to be transparent for photons with wavelengths for which the photo-detector is sensitive (see Figure 7.30). If the Cherenkov angle is measured they also should have a small dispersion in the energy range of detected photons. For observation of Cherenkov radiation a radiator material must be chosen so that the Cherenkov threshold condition $\beta n > 1$ is met for the velocities of particles in the experiment. The Cherenkov angle then increases with β, until it reaches a maximum angle of $\theta_{max} = \arccos(n^{-1})$.

Photon emission from scintillation or fluorescence should be small.

We have seen in section 2.8 that the density of photons emitted per energy and per unit of track length is proportional to $\sin^2 \theta_{Ch}$. Hence, the number of photons increases for larger βn. For relativistic particles this means that this number is larger for optically denser materials, typically solids or liquids. At the same time the Cherenkov angle will be larger for such materials, so that Cherenkov detectors made of such materials will typically be more compact than for materials with smaller index of refraction, where greater depth of the radiator and distance for the separation of the photons is required.

The maximum number of emitted photons is obtained for the maximum Cherenkov angle and can be found from eq. (2.47), assuming no dispersion (n constant),

$$\left(\frac{dN_\gamma}{dx}\right)_\infty = \frac{z^2\alpha}{\hbar c}(E_{max} - E_{min})\left(1 - \frac{1}{n^2}\right).$$

For small angles $\sin\theta_{Ch} \simeq \theta_{Ch}$, and the turn-on is given by

The Cherenkov angles are small in radiators with low indices of refraction (gases).

$$\frac{dN_\gamma/dx}{(dN_\gamma/dx)_\infty} \simeq \frac{\theta_{Ch}^2}{\theta_{max}^2} \simeq 1 - \frac{\gamma_{th}^2}{\gamma^2},$$

where $\gamma_{th}^2 = (1 - 1/n^2)^{-1/2}$, and we have assumed that $\gamma \gg 1$.

Widely used solid radiator materials are transparent crystals like quartz glass (fused silica, SiO_2), LiF or CaF_2. For liquid and gaseous radiators fluorocarbons are a common choice due to their low dispersion (C_6F_{14} is a liquid at STP, whereas CF_4, C_2F_6 or C_3F_8 are gaseous). Large Cherenkov detectors in neutrino physics use water as a radiator. For the region of n between gases and solids/liquids, aerogel can be used.

One drawback of these is that they are potent greenhouse gases.

Aerogel is a synthetic porous ultralight material, most often made of Silica (SiO_2).

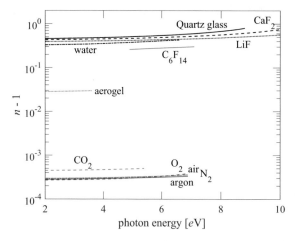

Figure 12.4: Index of refraction for some radiator materials (data from [480, 250]). The index of refraction for aerogel can be tuned between 1.008 and 1.1, depending on the density of the material.

In layers of optically dense materials with a downstream surface perpendicular to the incoming track and backed by a gas (air), the Cherenkov angle is typically so large that the radiation is internally reflected on the surface of the radiator. One way to avoid this is to give the radiator a 'sawtooth' shape [72]. Alternatively, multiple internal reflections in a planar radiator can be used to guide the photons to a photon detector outside the acceptance of the detector (this principle is called 'detection of internally reflected Cherenkov' radiation – DIRC) [16]. To allow for the multiple reflections a high-n material with low photon attenuation and highly polished surfaces has to be used.

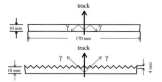

Figure 12.5: Principle of a 'sawtooth' radiator for the CLEO III RICH [72]. Top: planar radiator, bottom: 'sawtooth' radiator.

The extended photon path due to the multiple reflections can help to correct for the chromatic dispersion. In a dispersive medium the phase and

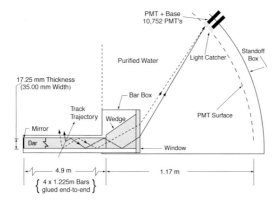

Figure 12.6: Schematic of the fused silica radiator bar and imaging region of the BaBar DIRC detector [16].

group velocities are different. This can be described by two different types of index of refraction, one for the phase velocity, n_φ, and one for the group velocity, n_g. The former defines the Cherenkov angle, whereas the latter determines the propagation velocity of the photon in the medium. They are related by

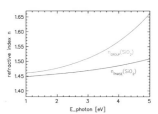

Figure 12.7: Dispersion curves in quartz glass for the phase velocity (n_φ) and for the group velocity (n_g) [280].

$$n_g = \frac{c}{c_g} = c\frac{dk}{d\omega} = \frac{d(ck)}{d\omega} = \frac{d(\omega n_\varphi)}{d\omega} = n_\varphi + E\frac{dn_\varphi}{dE},$$

where E is the energy of the photon. Thus a concurrent measurement of the Cherenkov angle and the time-of-flight of the photon can be used to determine the wavelength of the photon. Examples for DIRC-type detectors using timing information to improve particle identifications are iTOP at Belle II [245], and TORCH, proposed for the LHCb detector [135].

The time resolution required for this measurement is a few tens of ps.

PHOTON DETECTION

In principle, all detector technologies sensitive to optical (few eV) photons discussed in previous sections of this book can be used as photon detectors for Cherenkov detectors, in particular PMTs (section 4.4), silicon detectors like APDs (section 4.6), or gaseous detectors with TEA or TMAE, or with CsI photocathodes (section 7.6). The trend for imaging Cherenkov detectors requires improved segmentation. Early imaging detectors relied on gaseous detectors based on the time-projection principle [57, 17, 9], followed by gaseous detectors with cathode pad readout, enabled by integrated readout electronics [72, 225]. In all these systems the photon detector was in the path of the incoming charged particle and material in the photon detector therefore needed to be kept low. To contain the detector gas, windows transparent in the relevant wavelengths are required.

In fixed-target experiments and DIRC-type detectors the detection elements can usually be placed outside the detector acceptance, so that material is less of a concern. The inclusion of photon time-of-flight measurements introduces new demands on the time resolution of the photon detectors, and the detectors considered for next-generation RICH detectors are MCPs (section 4.4) and SiPMs (section 4.6).

These can be made of quartz glass for TMAE- and CsI-based and must be made of CaF_2 for TEA-based gaseous detectors.

THRESHOLD CHERENKOV DETECTORS

The simplest type of Cherenkov detector just detects the presence of Cherenkov radiation. This can be particularly useful for a hadronic beam, where particle momenta have been selected in a suitable magnet system. In this case the momentum is known and singular, and a Cherenkov detector with a radiator with a suitably chosen index of refraction can provide discrimination of the flavour of the particles in the beam.

For a suitable choice of the index of refraction, a Cherenkov signal is evidence for a pion, while the absence of a signal would indicate a kaon.

A rough estimate of the incoming particle's velocity can be obtained from measuring the magnitude of the light signal, as the number of emitted photons is proportional to $\sin^2 \theta_{Ch}$.

RING IMAGING CHERENKOV DETECTORS

A direct measurement of the velocity can be achieved if the angle of individual Cherenkov photons can be measured. Typically, this will involve the measurement of as many photons as possible to reduce the error on the angle measurement for individual photons. Depending on the geometry and optics this will often, but not necessarily, result in a circular or elliptical detection pattern.

Reconstruction of the paths of several photons can be used to reconstruct the Cherenkov cone and thus supply information about the track parameters of the incoming particle.

From the relation for the Cherenkov angle $\cos \theta_{Ch} = (n\beta)^{-1}$ it follows that $d\beta/d\theta_{Ch} = \beta \tan \theta_{Ch}$, and thus for the resolution of the velocity measurement with a single angle measurement

$$\frac{\sigma_\beta}{\beta} = \tan \theta_{Ch}\, \sigma_{\theta_{Ch}}.$$

Contributions to the angle measurement error $\sigma_{\theta_{Ch}}$ come from chromatic dispersion, the photon position detection error (usually dominated by the detector segmentation), uncertainties in the knowledge of the emission point and imperfections of optical surfaces used to reflect the photons. Most of these errors are uncorrelated for different photons and as in state-of-the-art imaging Cherenkov detectors typically a few tens of photons are detected for each charged particle, the single-photon resolution can be (assuming that the angle measurement errors are the same for all photons) divided by $\sqrt{N}$, with N the number of detected photons. Typical event angle measurement errors $\sigma_{\theta_{Ch}}/\sqrt{N}$ are between 0.1 and 5 mrad.

The distributions for these errors usually do not have long tails (as opposed to charged particle dE/dx).

In an imaging Cherenkov detector the signal is given by the number of photons on the ring or similar pattern and thus the separation power is given by

$$n_{1,2} = \frac{\Delta \sin^2 \theta_{Ch}}{\sigma_{\sin^2 \theta_{Ch}}/\sqrt{N}} = \frac{\left(m_1^2 - m_2^2\right)c^2}{2k_r p^2},$$

$\sin^2 \theta_{Ch} = 1 - \dfrac{1}{n^2} - \left(\dfrac{mc}{pn}\right)^2$

and

$\sigma_{\sin^2 \theta_{Ch}} = 2 \sin \theta_{Ch} \cos \theta_{Ch}\, \sigma_{\theta_{Ch}}.$

with

$$k_r = n^2 \sin \theta_{Ch} \cos \theta_{Ch} \frac{\sigma_{\theta_{Ch}}}{\sqrt{N}},$$

k_r is often called the 'RICH detector constant'.

where $\sigma_{\sin^2 \theta_{Ch}}$ and $\sigma_{\theta_{Ch}}$ are the measurement errors for a single photon for $\sin^2 \theta_{Ch}$ and θ_{Ch}, respectively. The factor k_r is often approximated for relativistic particles ($\beta \simeq 1$), for example as $k_r \simeq \tan \theta_{Ch} \sigma_{\theta_{Ch}}/\sqrt{N}$ [510], or as $k_r \simeq \sqrt{n^2 - 1}\, \sigma_{\theta_{Ch}}/\sqrt{N}$ [505].

For an angle measurement error of 2 mrad and a radiator with $n = 1.5$, $k_r \simeq 2.2 \times 10^{-3}$ and the π-K separation power at 4 GeV is about 3.1.

Two main geometries are used to achieve an image of the photons on the detector planes.

Typically, but not necessarily the images are rings.

Proximity focusing Cherenkov detectors

In this geometry the photons are emitted over a limited distance of the path of the primary charged particle. Typically, this uses a limited thickness of solid or liquid radiator, to yield a large enough number of photons. The limited depth of the radiator provides a constraint on the point of emission for the photons. For a radiator depth d and a much larger proximity gap l between the radiator and the detector the error on the angle measurement is given by

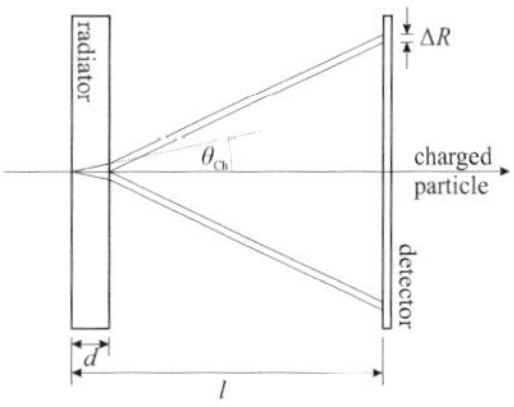

$$\frac{\sigma_{\theta_{Ch}}}{\theta_{Ch}} = \frac{\Delta R}{R} \simeq \frac{d}{l}.$$

Figure 12.8: Schematics of a proximity focusing Cherenkov detector.

A special case of proximity focusing are large volume neutrino experiments, where the medium is typically water. Here the limitation of the emission points is due to the limited distance (compared to the size of the tank) the leptons created in charged-current interactions travel before their energy is used up and they come to a stop. As there is no information on the track parameters of the incoming particle, the direction must be inferred from the shape of the image. Photon arrival times due to time-of-flight can support the determination of the direction of the lepton. Identification of the lepton type can be obtained from the sharpness of the image, as electrons initiate electromagnetic showers, with charged secondaries slightly off-axis from the originating electron, and thus the Cherenkov image is more blurred than for muons.

Low energy muons will have a very short range compared to the dimensions of the detector and will produce sharp rings as shown in the figure. Higher energy muons can have a significant range (for example, 2 GeV muons have a range of ~ 10 m) and therefore the inside of the ring will tend to be filled in by Cherenkov radiation from the muon as it propagates through the detector towards the detection surface. However, there will still be a sharp outer edge and the pattern can still be efficiently distinguished from that caused by electron showers.

Figure 12.9: Example event displays from the Super-Kamiokande detector [297] (© Deutsche Physikalische Gesellschaft. Reproduced by permission of IOP Publishing. CC BY-NC-SA). Left: electron event, right: muon event. The event displays show the unwrapped instrumented surface (11,000 photomultiplier tubes) of the cylindrical water tank (height and diameter ~ 40 m).

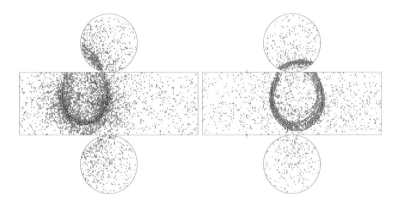

Mirror-focusing Cherenkov detectors

Alternatively, the photons can be focused by concave mirrors, with the detectors in the focusing plane of the mirror. Parallel photons are focused to the same detector location, and the Cherenkov cone onto a ring. In practical applications, the mirrors are segmented to allow for planar detector geometries, and to limit the depth of the Cherenkov detector system.

Hence the acronym RICH, for 'Ring Imaging Cherenkov' detector.

One advantage of the mirror-focused system is that there is no fundamental limit on the depth of the radiator, and thus this geometry is compatible with low-density radiators.

To optimise separation performance over a wide momentum range mirror-focusing RICH systems often comprise more than one radiator material. An example of a combined proximity- and mirror-focusing RICH system has been the DELPHI RICH, which used a liquid C_6F_{14} radiator

This is true for parabolic mirrors, for spherical mirrors the focus will not be in a point, and variations in the emission point will result in an error on the angle measurement.

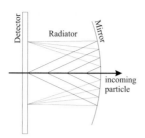

Figure 12.10: Schematics of a mirror-focusing Cherenkov detector. The detector plane can also be outside of the acceptance for incoming charged particle tracks.

At collider experiments RICH detectors usually have a mirror tilt to project the image outside of the detector, which introduces non-perfect focusing that results in chromatic aberration.

in a proximity focusing geometry and a C_5F_{12} gaseous radiator with mirror focusing. The photons were detected in a CH_4/TMAE gaseous drift chamber with a long (1.6 m) drift.

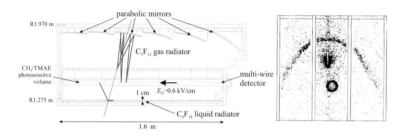

Figure 12.11: DELPHI barrel RICH. Left: Quadrant view (modified from [69]). Right: Photoelectron scatter plot of 200 superimposed events in a prototype (beam at 18° relative to the normal at $z = 142$ cm) [69]. The large parabolic pattern is from the liquid radiator, the small circle from the gaseous radiator. Other features are from quartz windows of the photo-sensitive region. The ionisation electrons are drifted along the length of the detector to the end on the right, where they are detected by a multi-wire detector.

The LHCb RICH system consists of two detectors, RICH1 and RICH2. In both detectors a mirror system consisting of a spherical mirror and a plane mirror deflects photons out of the acceptance for charged particles and focuses them onto a photon detector system consisting of hybrid photodiodes (HPDs). RICH1 uses aerogel (only in run 1) and C_4F_{10} as radiators, CF_4 is used in RICH2.

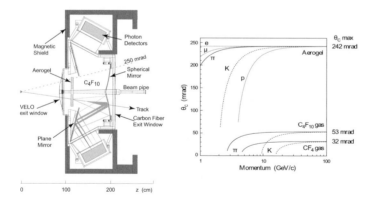

The aerogel radiator was removed after run 1 as its ability to provide PID below the Cherenkov threshold for kaons for the C_4F_{10} radiator was limited by the challenges to reconstruct the photons in the high track density environment, while its removal improves the performance of the gas radiator, as it removes the obstruction for photons upstream of the aerogel radiator [397]. In 2022 the HPDs have been replaced with multi-anode PMTs.

Figure 12.12: LHCb RICH1 [332]. Left: Side view cross-section. Right: Cherenkov angle versus particle momentum for the LHCb RICH radiators. Indices for refraction are $n = 1.03$ for aerogel, $n = 1.0014$ for C_4F_{10}, and $n = 1.0005$ for CF_4 (in RICH2, not shown).

12.5 TRANSITION RADIATION DETECTORS

As discussed in section 2.9, transition radiation in the X-ray regime has an intensity that is proportional to γ, and thus a measurement of this intensity, together with a measurement of the momentum, can be used to determine the mass of the charged particle. It is evident that this works best for ultra-relativistic ($\gamma \gg 1$) particles. Up to high momenta

For good summaries of transition radiation detectors see [59, 505].

$(\sim 100\ GeV/c)$ only $e^{\pm}$ will emit significant TR intensity, so that the typical application of TR detection is to distinguish $e^{\pm}$ from other charged particles in that energy regime. The commonly used figure-of-merit for a transition radiation detector (TRD) is the efficiency for mis-tagging a pion as an electron for a given electron identification rate (typically 90%).

TR has also been used for the identification of pions in hadron beams above 100 GeV/c [209, 127].

RADIATORS FOR TRANSITION RADIATION DETECTORS

The number of photons emitted at a single interface is $\mathcal{O}(z^2\alpha)$, where ze is the charge of the incoming particle, and thus a large number of interfaces is needed to obtain a detectable signal. At the same time, attenuation needs to be limited. The dominant photo-absorption mechanism for photons with about 10 keV is by the photo-electric effect, with an absorption length that scales like Z^{-m} with m between 3 and 5, whereas the radiated energy increases only with $\omega_p \propto \sqrt{Z}$. Thus, it is necessary to use a radiator material with a low Z. In addition, the material should be structurally stable, so that the interface surfaces are maintained. Early TRDs used foils from low-Z metals like lithium or beryllium. However, these metals are difficult to handle and more recent TRDs use polypropylene, $(C_3H_6)_n$, or Mylar. These materials can be used as foils, foams or fibre mats.

See section 2.1.

Z is the atomic number of the material.

Lithium is flammable, and beryllium is toxic.

Mylar is a trade name for a polyester film made from stretched polyethylene terephthalate (PET).

PHOTON DETECTION FOR TRANSITION RADIATION

The typical energy of photons that escape a radiator stack is a few 10 keV, as the lower energy photons are more strongly absorbed in the radiator material. The two challenges for the photon detector in TRDs are the low interaction cross-section for photons with this energy, and the small emission angle of the TR, which usually makes it impossible to separate the charged particle energy loss of the incoming particle from the energy deposition of TR photons. The energy deposited by TR photons is typically significantly larger than the local average charged particle energy loss, but energy transfers in the tail of the straggling distribution (δ-electrons) can create signals of similar size, and thus fake a TR photon.

The spectrum tends to be harder for thicker stacks.

To maximise TR photon absorption the detector material should have a high Z, and the most common choice of photon detectors are gaseous detectors with xenon as the main gas component. As usual, the noble gas xenon needs to be combined with a suitable quencher gas like CH_4 or CO_2. Two approaches can be taken for the design of a TRD. Either, the detector consists of thin elements, which allows for a fast response in a high-rate experiment, or it is deep, which can provide better segmentation beneficial in experiments with high track densities. In both cases it is common to include the position information for the charged particle in the general tracking of the experiment.

Xenon gas is also the detector material with the highest relativistic rise of the charged particle energy loss ($\sim$75%), which further increases the signal from charged particle loss for a particle with a large relativistic γ-factor.

The first approach was taken in the ATLAS TRT, where the detection elements are straw tubes made of polyimide with a diameter of 4 mm, operated with $Xe/CO_2/O_2$ 70/27/3. There are 52,544 straws in the barrel, and 122,880 in the endcaps. The straw tubes are embedded in the radiator that is made of polypropylene-polyethylene fiber mats. The absorption length for the lowest energy photons of interest (5 keV) in the radiator material is about 17 mm. Hits are identified with two different thresholds, a low threshold used for tracking (at $\sim$15% of the average signal expected for a minimum-ionising particle, about 250 eV), and a high threshold for the identification of TR photons (at about 6 keV).

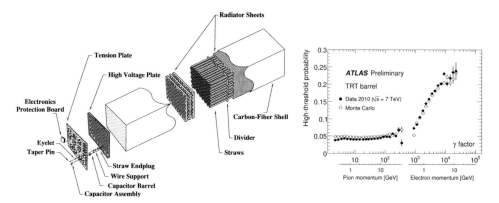

Figure 12.13: ATLAS TRT. Left: Isometric view of a barrel module [103]. Right: Probability of a TRT high-threshold hit for pions and electrons as a function of γ in 7 TeV collision events [142]. For the identification of particles the information from the whole TRT, comprising a large number of straws ($\mathcal{O}(30)$), is combined. The efficiency for electron identification of the TRT depends on the threshold of TRT hits required, but this is usually set to achieve ∼90% efficiency for electron identification.

The second approach has been taken by the ALICE collaboration. The ALICE TRD consists of a radiator comprising a carbon-fibre-laminated Rohacell layer and polypropylene fibre mats with a total thickness of 48 mm, and a gaseous detector filled with Xe/CO$_2$ 85/15, which consists of a drift section of 30 mm thickness, and a 7 mm thick multi-wire proportional chamber section with pad readout. In this detector particles with $\gamma > 1000$ will produce about 1.45 X-ray photons in the energy range of 1–30 keV from TR. The highest probability to be absorbed in the gas occurs when the photons enter the active volume (corresponding to the largest drift time), decaying exponentially on a scale given by the absorption length.

Rohacell® is the trade name for a closed-cell rigid expanded foam plastic based on polymethacrylimide (PMI).

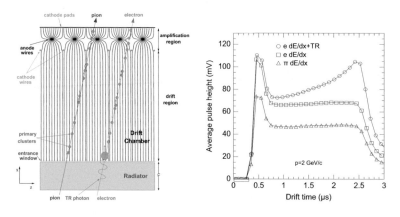

Figure 12.14: ALICE TRD [59]. Left: Schematic cross-sectional view of a detector module. Right: Average pulse height as a function of drift time for pions (triangles), electrons without a radiator (squares) and electrons with a radiator (circles), all with $p = 2$ GeV/c momentum.

For the identification of TR photons two different strategies can again be used [227]. Either the total charge deposited in the detector for the passage of the charged particle (dE/dx+TR) is measured after amplification by a charge sensitive amplifier and digitisation in an ADC ('Q-method'), or the number of clusters exceeding a threshold well above the average charged particle energy loss is counted ('N-method'). The second approach has the advantage that the number of such clusters follows a

Poisson distribution with a less prominent tail than of the straggling distribution of the energy loss of the charged particle, but it can be affected by space charge effects and diffusion of the ionisation charge during its drift in the detector gas. The two different methods require different optimisations of the readout electronics and the detector design.

In TRDs with gaseous detectors, the position resolution is usually insufficient to separate the photons from the charged track and thus a direct measurement of the emission angle is impossible. However, the better position resolution in semiconductor detectors can be used to to resolve the angle, and thus to improve the PID performance (see for example [50]). The separation can be enhanced by a magnetic field between the radiator and the detector.

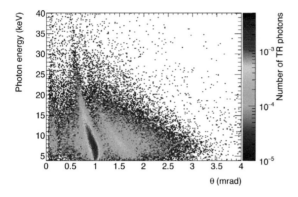

Figure 12.15: Photon energy versus reconstructed production angle obtained with a polypropylene radiator (180 foils with thickness 15.5 μm and 222 μm air gaps) for 20 GeV/c electrons recorded with a 480 μm thick p-in-n silicon pixel sensor [50].

PERFORMANCE OF TRANSITION RADIATION DETECTORS

The design choices for a TRD can be affected by many secondary considerations (speed, segmentation, performance as a tracking device, use in trigger systems, etc.), but, as experience shows, the design parameter which effectively drives the performance for its primary function, particle identification based on the detection of TR, is the depth of the TR detection system. Pion fake rates below 1% for electron identification rates of 90% have been achieved with detectors with a depth of several tens of cm.

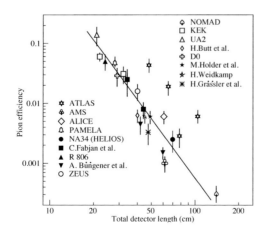

Figure 12.16: Pion efficiency measured (or predicted) for different TRDs as a function of the detector length for a fixed electron efficiency of 90% [505].

12.6 COMPLEMENTARITY AND COMBINATION OF PARTICLE IDENTIFICATION TECHNIQUES

From the discussion in this chapter so far it should be clear that the different dedicated particle identification detectors perform differently in different kinematic regions. No technique works well for all particles and momenta, so that the choice of PID system depends very much on the specific experiment. Some of the PID detectors can also serve additional functions like contributions to tracking, triggering, and/or pattern recognition.

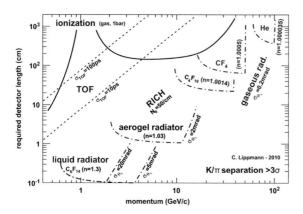

Figure 12.17: Approximate minimum detector length required to achieve a K/π separation of $n_{K,\pi} = 3$ with three different PID techniques [339]. For the energy loss technique a gaseous detector is assumed. For the TOF technique, the detector length represents the particle flight path over which the time-of-flight is measured. For the Cherenkov technique only the radiator thickness is shown. The thicknesses of an expansion gap and of the photon detectors have to be added.

The different approaches are uncorrelated, and thus the combination of the results, if such systems are present in an experiment, can improve the overall PID performance.

The ALICE experiment at the LHC incorporates all the PID technologies discussed here [42].

Figure 12.18: Particle identification by simultaneous dE/dx and TOF measurement in the momentum range 5–6 GeV/c for central Pb-Pb collisions in the NA49 experiment [24].

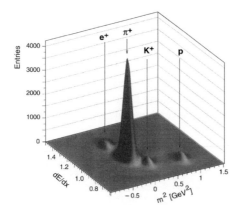

12.7 ELECTRON IDENTIFICATION

Identification of electrons in the presence of hadrons will use the very different shower profiles of electrons and charged hadrons that we considered in chapter 11. Electron identification is particularly challenging in a hadron collider like LHC because of the very large cross-section of hadrons from QCD jets. We will therefore focus our discussion on electron identification at hadron colliders, but the principles could be applied to other environments.

We will focus on identifying electrons at high values of transverse momentum (p_T) because of the very large hadron backgrounds at low p_T.

The key differences between electrons and hadron showers are the short longitudinal shower profile and the narrow transverse shower profile of the former. Therefore, if we have a calorimeter with fine granularity, we can use the lateral and the longitudinal shower profile to provide powerful electron/hadron separation.

Additional suppression of this type of hadron background can be achieved using transition radiation detectors (see section 12.5).

As we saw in chapter 11 the shower profile from photons is very similar to that from electrons. Therefore, a calorimeter alone cannot provide powerful rejection against backgrounds from $\pi^0 \to \gamma\gamma$. This type of background can be rejected by requiring electron candidates to have a good matching between the direction and momentum of the track reconstructed in the inner tracking system, and the energy and location of the shower in the electromagnetic calorimeter.

Another background arises from photon conversions in the detector material upstream of the calorimeter. If the photon conversion is sufficiently asymmetric the energy of either the e^- or e^+ can be large enough that the reconstructed track momentum is compatible with the measured calorimeter energy. This type of background can be suppressed by reconstructing the tracks of both the e^- and e^+ or by vetoing electron candidates which do not have a hit in the first layer of the tracking detector.

This provides one motivation for trying to minimise the material (in units of radiation lengths) in tracking detectors.

In a hadron collider experiment it is often necessary to distinguish 'prompt' electrons from electrons originating in semi-leptonic decays of heavy flavour hadrons (b and c hadrons). In general, there will be no other high energy particles close to a prompt electron. The electrons from heavy flavour decays on the other hand arise from a b or c jet and therefore will invariably be accompanied by high energy hadrons. We can therefore strongly suppress the background from electrons from heavy flavour decays by 'isolation' cuts. The cut will typically require a low number of charged hadrons and/or energy in the calorimeter close to the electron. We can reduce the background from photon conversions in the detector by requiring that the candidate electron track has a hit in the first layer of the vertex detector (see exercise 4).

In a hadron collider experiment (e.g. LHC) the cross-section for hadronic jets is many orders of magnitude larger than that for the production of prompt electrons from processes like $W \to e\nu_e$ and $Z \to ee$.

These general principles will apply to all collider experiments but the details will differ. We will consider electron identification in the ATLAS detector as an interesting example [91] because of the fine lateral and longitudinal sampling of this electromagnetic calorimeter.

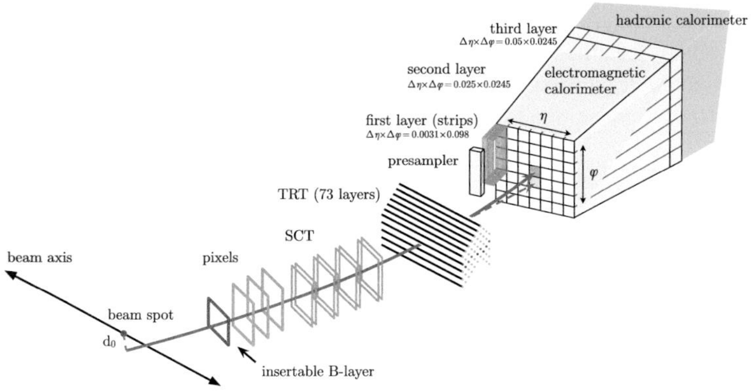

Figure 12.19: Schematic of the ATLAS tracking detectors and calorimeters used for electron identification. The dashed track before the presampler shows the path of a photon that has resulted from electron bremsstrahlung [91].

The electron reconstruction and identification consists of four principle steps:

1. *Seed Cluster.* The energy in the electromagnetic calorimeter in the different layers is summed within 'towers' of size $\Delta\eta \times \Delta\phi = 0.025 \times 0.025$. A 'sliding window' algorithm is then used to find clusters of 3×5 towers in η and ϕ. The sliding window algorithm works by summing the transverse energy ($\sum E_T$) in 3×5 towers and moving the centre of the window over the calorimeter to look for clusters with $\sum E_T > 2.5$ GeV.

   η is a polar angle variable defined by eq. (2.55) and ϕ is the azimuthal angle.

2. *Track matching.* A reconstructed track should point to the centre of the electron cluster within $|\eta_{cluster} - \eta_{track}| < 0.05$ and $-0.1 < \phi_{cluster} - \phi_{track} < 0.05$. The backgrounds from photon conversions are suppressed by requiring a minimum number of hits on the track and by requiring a hit in the innermost pixel layer.

   The width is larger in ϕ to accommodate the effects of electron bremsstrahlung.

3. *Electron identification.* Several calorimeter and track variables are used to efficiently identify genuine electrons and reject backgrounds from QCD jets. These will be discussed in more detail below.

4. *Electron isolation.* Prompt electrons (e.g. from processes like $Z \to e^+e^-$) can be separated from electrons from decays of b or c hadrons using energy around the electron cluster.

There are several variables which contribute to the electron identification stage. Some of these variables are described to illustrate how the general principles of electron identification are implemented in practice.

The details will change for different experiments. Here it will be described for the case of the ATLAS detector.

- *Longitudinal shower profile.* The ratio of the transverse energy in the electromagnetic calorimeter cluster to the transverse energy in the corresponding cells of the hadronic calorimeter.

- *Transverse shower profile.* The energy-weighted width in the second layer of the electromagnetic calorimeter is defined in terms of the energy in cell i, E_i and the pseudorapidity η_i as $\sqrt{\left(\sum_i E_i \eta_i^2\right)/\sum_i E_i - \left(\left(\sum E_i \eta^i\right)/\sum E_i\right)^2}$. A similar variable, R_ϕ is defined as the ratio of energy in a region of 3×3 cells over that of the energy in 3×7 cells (centred on the electron cluster).

- *Track quality.* The number of hits in the silicon tracker are used to suppress photon conversions, and the transverse impact parameter divided by the uncertainty, $d_0/\sigma(d_0)$, to suppress electrons from heavy flavour decays.

- *Transition radiation.* The information from the transition radiation tracker (TRT) can be used to further suppress backgrounds from charged hadrons.

- *Track-calorimeter matching.* The precise matching of the track-projected impact point at the calorimeter and the calorimeter cluster centre-of-gravity provides rejection against all backgrounds apart from photon conversions. The matching between the measured calorimeter energy E and the momentum p of the matched track provides further background rejection.

Some of these variables can be used in a simple selection algorithm. Other variables can provide some statistical separation between the electron signal and background, but do not show a clean separation between signal and background. Therefore the analysis combines many variables but as there are many variables a simple series of selections on each variable ('cuts') will tend to be inefficient.

If sufficiently tight cuts are placed on many different variables in order to achieve a high background rejection then the overall efficiency will be low.

A better approach for electron identification is a likelihood method[1]. The likelihood for signal electrons (background) for a given event is defined as

$$L_{S(B)}(\vec{x}) = \prod_{i=1}^{n} P_{S(B),i}(x_i), \qquad (12.5)$$

where the vector $\vec{x}$ represents the n variables used and $P_{S(B)}(x_i)$ are the probability distribution functions (PDF) for signal (background) for the individual variables evaluated at the measured value x_i. The detector performance varies strongly with η and electron transverse energy E_T. Therefore, the signal and background PDFs are evaluated in bins of η and E_T. The discriminant to separate electrons from backgrounds is then defined by

The values used in the selection (cuts) are also optimised for different η and E_T bins.

$$d_L = L_s/(L_S + L_B). \qquad (12.6)$$

For presentation purposes, a scaled variable is used that is defined as $d'_L = \tau \ln(d_L^{-1} - 1)$.

While no one variable provides very good discrimination between the electron signal and the background, the combined likelihood function does achieve very good separation. This is illustrated in Figure 12.20, which shows the distribution for signal and background for one such variable, R_ϕ and the likelihood discriminator d'_L.

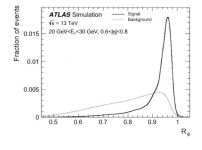 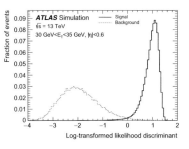

Figure 12.20: Distribution of electron signal and background for the variable R_ϕ (left) and the scaled discriminating variable d'_L (right) [91].

A cut is then made on a particular value of d'_L and the electron efficiency can be traded off against background rejection.

Finally, if we want to identify electrons from prompt processes (e.g. $Z \to e^+e^-$) as opposed to semi-leptonic decays of heavy flavours,

The electron selection can be adapted to identify photons by reversing the requirement that there is a charged track matching the cluster in the electromagnetic calorimeter. Alternatively, photons can be identified with clusters in the electromagnetic calorimeter that match an identified photon conversion in the tracking detector.

---

[1]This provides a significant improvement on the simple cut analysis. However, it does not allow for the significant correlations between the variables. An ideal algorithm would use a multi-dimensional likelihood, which combined all the variables. Consider the simple case of a likelihood based on two variables for each event, x_i, y_i. Ignoring the correlations between the two variables, the likelihood for signal and background would be $L = P_{S(B)}(x_i)P_{S(B)}(y_i)$. If we allow for the correlations between x and y using a joint probability distribution function $P(x, y)$ the likelihood for signal (background) would be $L = P_{S(B)}(x_i, y_i)$. However, this would be computationally impractical when the number of dimensions is large. An alternative approach is to use a machine learning algorithm like neural nets to take into account these correlations.

we can apply 'isolation' selections. The basic idea is to measure the transverse energy within a cone around the direction of the electron but excluding the electron energy itself. The cone is defined in (η, ϕ) space by $\Delta R = \sqrt{(\Delta \eta)^2 + (\Delta \phi)^2}$, where $\Delta \eta$ ($\Delta \phi$) is the difference in η (ϕ) between the edge of the cone and electron η (ϕ). The isolation can be computed using the calorimeter transverse energy or the sum of the transverse momentum of the charged tracks inside the cone.

Typically, the cone size is $\Delta R = 0.2$.

EFFICIENCY MEASUREMENT

The efficiency for the electron identification selections can be estimated from Monte Carlo simulations. The efficiency varies with electron η and transverse energy E_T. In order to minimise systematic uncertainties associated with inaccuracies in the simulation, data-driven methods are used to correct the simulation. This is done using a 'tag and probe' method, similar to that used for measuring tracking efficiencies for muons (see section 10.1).

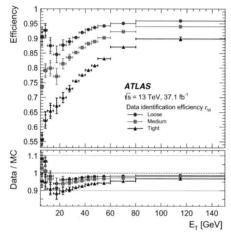

Figure 12.21: Data-driven measurements of electron identification efficiency as a function of electron transverse energy E_T for three values of the discriminating variable, called 'loose', 'medium' and 'tight' for the ATLAS experiment [91]. The bottom panel shows the ratio between the data-driven and the Monte Carlo simulation estimates.

12.8   MUON IDENTIFICATION

Muons lose their energy by ionisation rather than through bremsstrahlung or hadronic interactions. Therefore high energy muons will only lose a small amount of energy in the calorimeters and we can identify muons by matching tracks in dedicated position detectors behind the calorimeter ('muon chambers') to a track measured in the inner tracking detectors (before the calorimeter). In addition, the minimum-ionising signature of muons in the calorimeters can also be used as additional information on the path of a muon if calorimeter signals compatible with a minimum-ionising energy deposition can be linked to the track candidate from the tracker.

Only very high energy muons ($\gtrsim 1$ TeV) have significant energy loss by bremsstrahlung. The cross-section for muon-nucleus hadronic interactions is negligible for practical purposes.

Different matching strategies in terms of how the hypothesis is seeded and which information from other systems is included, and different matching criteria, are often employed concurrently. The different algorithms help to maximise coverage and achieve sensitivity in areas where one of the subsystems performs poorer. While these general principles are valid for all detector systems, the details depend on the magnetic field system and the tracking detectors used.

The different algorithms provide a trade-off between efficiency for identifying genuine muons versus the probability of misidentifying a hadron as a muon and can be selected for the needs of the specific physics analysis using the muon candidates.

If the calorimeter is sufficiently deep in units of hadronic interaction lengths, the probability of a charged hadron not interacting in the calorimeter will be negligible. However, there can be particles emerging from the hadronic calorimeter from the tail of the hadronic shower and these hadrons can potentially create fake muon signatures (this background is called 'punch-through').

Another source of background muons is the decay in flight of charged pions and kaons. The semi-leptonic decays, $\pi \to \mu \nu_\mu$ or $K \to \mu \nu_\mu$ can occur before the calorimeter and therefore result in genuine muons entering the calorimeter. This background can be suppressed by identifying kinks in the tracks in the inner tracking system. The cases in which the kink cannot be identified because the pion or kaon are parallel with the muon, will result in a residual background. As with electrons, we can reject muons from heavy flavour decays using isolation requirements.

The efficiency for the isolation selection can be measured in collider experiments using the same tag-and-probe method as discussed for muons (see section 10.1). The probability for identifying a genuine muon inside the acceptance in ATLAS and CMS is in the range 95% to 99%, and the probability of misidentifying a charged pion as a muon is in the range 0.1% to 0.2%, depending on how tight the muon selection is.

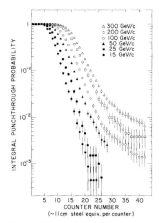

Figure 12.22: The integral probability distributions for hadronic shower punch-through as a function of calorimeter depth for hadrons of different energies, measured with the CCFR experiment [369]. One counter corresponds to 11 cm of steel, about $2/3 \lambda_{int}$.

The corresponding probability for kaons is slightly larger, in the range 0.3% to 0.5%.

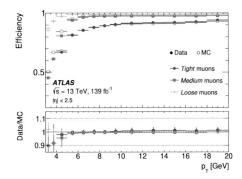

Figure 12.23: Measured muon reconstruction and identification efficiencies for different criteria in ATLAS [92]. The efficiencies have been measured in $J/\Psi \to \mu\mu$ using the tag-and-probe technique.

As for electrons, we can suppress the contribution from muons from heavy flavour decays by isolation requirements.

MUON MOMENTUM RESOLUTION

After the identification of a muon, it is the momentum of the muon that is usually of interest. As for any charged particle this is determined from the curvature of the tracks in a magnetic field. To understand the design considerations of muon detection systems, it is instructive to look at the design choices and performance of the muon systems in CMS and ATLAS.

In the CMS experiment, muons are detected in the central silicon tracker and in a dedicated muon system, which is located in four stations within the iron return yoke, which closes the field lines for the central tracking solenoid. The precision position information in the barrel section is provided by square gaseous drift tube (DT) detectors, while cathode strip chambers (CSCs) are used in the endcaps. Both systems are supplemented by resistive plate chambers (RPCs), which provide trigger and precise timing information, but with lower spatial precision.

For muons from Z or W decays there will in general be no other particles with large p_T at angles close to the muons. However, muons from heavy flavour decay such as $b \to c\mu\nu$ will tend to have other high p_T particles at angles close to the muon from the decays of the charm quark. See [202] for an example of the measurement of the efficiency for isolation selection.

The magnetic field strength inside the central solenoid is 3.8 T.

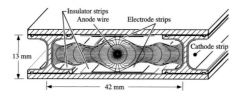

Figure 12.24: Layout of a DT cell with electric field lines of the drift field superimposed [203].

ATLAS uses a smaller superconducting solenoid with a magnetic field of 2 T for the inner tracking volume and a system of discrete coils to create a large air core toroid. There are 8 coils each in the barrel and the two endcaps, with a rotation of 22.5° between barrel and endcaps. The electromagnetic and hadronic calorimeters are between the tracker solenoid and the muon system toroids. For the precision position measurement in the central part, pressurised drift tubes (monitored drift tubes, MDTs) are used, together with RPCs for triggering, while in the forward direction, CSCs for precision and thin gap chambers (TGCs) for trigger measurements are used. Due to the higher pressure and the drift geometry, the single wire resolution of the MDTs is better than for the DTs in CMS (about 80 μm compared to 200–300 μm).

The large air-core toroid for ATLAS was chosen at the time of the planning of the LHC experiments to achieve a good standalone performance of the muon system, to add robustness in case the inner tracking system would not perform as well as planned in the unexplored regime of high energy and high luminosity at the LHC.

This concern turned out to be unnecessary, the inner tracking detectors for both ATLAS and CMS turned out to have excellent performance.

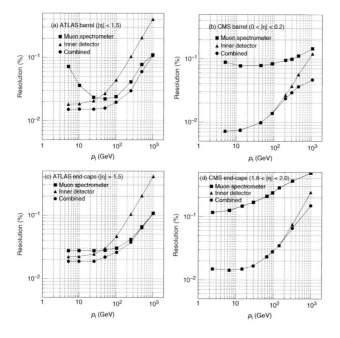

Figure 12.25: Relative momentum resolution as a function of the muon transverse momentum showing the stand-alone resolution of the muon systems and the inner trackers, and their combined resolution [417] (© Deutsche Physikalische Gesellschaft. Reproduced by permission of IOP Publishing. CC BY-NC-SA). (a) and (c) ATLAS for $|\eta| < 1.5$ and $|\eta| > 1.5$, respectively. (b) and (d) CMS for $|\eta| < 0.2$ and $1.8 < |\eta| < 2.0$, respectively.

The momentum resolution obtained for the ATLAS muon spectrometer demonstrates that the goal of good standalone momentum resolution was achieved. In the barrel for very low momenta the energy resolution is affected by energy loss fluctuations in the calorimeter, but above ~10 GeV/c the resolution improves due to the low material in the air-core field, until the resolution degrades again at high momenta, due to the sagitta measurement error.

In comparison, the standalone momentum resolution of CMS is worse, due to multiple scattering in the iron yoke. However, combined with the inner tracker the performance is better than for ATLAS, except for large momenta in the endcaps, because of the much larger value of BL^2 (see section 10.4). As the performance of the muon system in CMS is limited by multiple scattering, the demands for single wire resolution and chamber alignment of the precision detectors are much relaxed.

The effects of multiple scattering on muon momentum resolution is addressed in exercise 3.

The material also affects the trigger. In the case of an air-core magnet, the momentum resolution of the trigger chambers is usually sufficient to provide an efficient threshold up to relatively high p_T. For an iron core, information from the inner tracking detectors is needed to provide a sufficiently sharp threshold. A disadvantage of the muon system being embedded in the return flux to the central solenoid field is that the field has the opposite direction to that in the tracker, so the tracks curve in the opposite direction. This will bend back the trajectories of low momentum muons from hadron decays, so that they can point to the vertex and occasionally fake a high momentum muon.

12.9 TAU IDENTIFICATION

The identification of tau leptons is much more challenging than that of electrons or muons as in general purpose collider experiments the tau will decay in the tracking volume. The 3-body leptonic decays $\tau \to l \nu_\tau \overline{\nu}_l$ will in general result in electrons or muons with lower energy. This makes it harder to identify it in the presence of a very large hadronic background at low energy. In addition, there will be a background from genuine isolated electrons from $W \to e \nu_e$ and $W \to \mu \nu_\mu$ decays.

The decay also involves a neutrino.

Therefore, tau identification focuses on the hadronic decay modes. The largest background is from jets of hadrons from QCD processes. Several criteria can be used to distinguish this from tau decays.

- The lateral dimension of the jet of hadrons from a tau decay at high energy will be limited and in general less wide than that of the background.

- The hadronic tau decays will result in fewer charged hadrons than is typical in QCD jets.

- Genuine tau decays will have a net charge of ± 1, therefore only candidates with a net charge of ± 1 of the reconstructed final state are used [204].

The CMS tau identification uses the hadronic decay modes listed in Table 12.1.

Decay mode	Resonance	$\mathscr{BR}$ (%)
$\tau^- \to h^- \nu_\tau$		11.5
$\tau^- \to h^- \pi^0 \nu_\tau$	$\rho(770)$	25.9
$\tau^- \to h^- \pi^0 \pi^0 \nu_\tau$	$a_1(1260)$	9.5
$\tau^- \to h^- h^- h^+ \pi^0 \nu_\tau$	$a_1(1260)$	9.8

Table 12.1: Decay modes and branching ratios for tau decays used in CMS tau identification. $h^\pm$ refers to hadrons ($\pi^\pm$ or $K^\pm$). Where relevant, the table gives intermediate resonances [204].

The CMS tau reconstruction combines the information from the reconstructed charged tracks associated with a jet with electron and photon candidates within a window of $\Delta\eta \times \Delta\phi$. The total vector momentum and energy in this region (called strips) is calculated. In the first version of the procedure a fixed window size of 0.05×0.2 is used. Charged hadrons from tau decays interacting in the detector material can lead to hadrons detected outside this window. Similarly, the effects of pair production, multiple scattering and the large magnetic field can result in electrons or photons detected outside this window. These energy deposits will decrease the efficiency of isolation requirements. Conversely, at higher values of p_T of the tau, the energy will tend to be contained within a smaller cone. Therefore, in the more sophisticated version of the analysis, the fixed window algorithm is replaced with a dynamic window which depends on the p_T of the tau [204].

The next step in the analysis is to reconstruct the invariant mass of the charged hadrons using the momentum measured in the strip tracker, m_τ^h (for the decay modes other than $\tau^- \to h^- \nu_\tau$). The value of m_τ^h should be compatible within errors with the masses of the resonances listed in Table 12.1.

The identification of genuine taus while rejecting backgrounds from QCD jets is optimised using a Multi-Variate Analysis (MVA). There are two types of variables used [205]:

Multi-Variate Analysis uses machine learning to allow for the correlations between variables to optimise the classification of signal/background.

1. *Lifetime.* The finite lifetime of the tau (2.91×10^{-13} s) is used to define several variables that provide some discrimination against QCD backgrounds. The simplest one uses the transverse impact parameter d_0 (see section 10.6) of the 'leading' (i.e. highest value of p_T) hadron in the tau candidate final state and its estimated uncertainty σ_{d_0}. The 'significance' is defined by d_0/σ_{d_0}.

2. *Isolation.* Genuine tau decays should have a small total momentum in the tracker and energy in the calorimeter in the region around the reconstructed tau. An isolation cone size of $R = 0.5$ is used. In a high-luminosity collider like LHC, there are many primary interactions within the same bunch crossing. These will occur over a range of longitudinal coordinates z. The pattern recognition will therefore reconstruct several primary vertices, with each track associated to a particular vertex. In order to minimise pile-up effects, only tracks compatible with the 'hard scattering' vertex are used. For the neutral particles, a statistical subtraction of the energy due to pile-up is made.

The cone is defined in η, ϕ space as $R = \sqrt{(\Delta\eta)^2 + (\Delta\phi)^2}$.

i.e. the vertex with high transverse momentum tracks associated to it.

The performance of the tau identification in CMS is illustrated in Figure 12.26.

Additional MVAs are used to separate genuine taus from backgrounds from muons and electrons. The efficiencies shown in Figure 12.26 are based on Monte Carlo simulations. Data-driven corrections are determined using the tag-and-probe methodology. The procedure uses the signal from $Z \to \tau^+\tau^-$ with one tau decaying to a muon and the other tau decaying hadronically. The muon provides the 'tag' and the identification efficiency is measured for the hadronic tau decay mode. A 'visible' mass, m_{vis} is defined as the invariant mass of the muon and the tau. Distributions of m_{vis} are fitted to the expected peak from $Z \to \tau^+\tau^-$ and backgrounds.

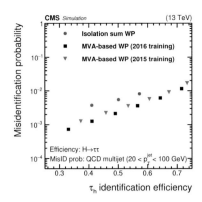

Figure 12.26: CMS tau identification efficiency for taus from a Higgs boson decay $H \rightarrow \tau^+\tau^-$ versus efficiency for falsely identifying a QCD jet as a tau [204].

12.10 MISSING E_T

In a collider detector, the probability of a neutrino interacting in the detector is practically zero. If a neutrino is created in the collision, this results in an imbalance of the measured momentum in the event, which can be used to infer the presence of the neutrino (or an exotic, weakly interacting, electrically neutral particle).

Missing transverse energy is a marker for many beyond the standard model physics searches.

In a collider experiment, a significant number of particles will travel inside or close to the beam pipe, where they avoid detection. Therefore it is not practical to measure the component of the total momentum in the beam direction. Also, typically the detector system which provides the most complete coverage of the solid angle, and can also supply measurements of neutral particles, are the calorimeters, which measure the energy of incoming particles. With sufficiently fine segmentation of the calorimeter, we can assign a direction of the energy flow from the collision point. Hence, we use 'missing transverse energy' to determine the imbalance in the detected particles, which can be used to identify the presence of neutrino(s) and get information on the kinematic properties.

This method cannot identify if there is more than one neutrino in the event.

In a simplified picture, we assume that the energy E_i is measured in calorimeter cell i and the line from the centre of the detector to the centre of the cell is at a polar angle θ_i and azimuthal angle ϕ_i. We then define the missing transverse energy 2-D vector as

$$E_\mathrm{T}^x = -\sum_i E_i \sin\theta_i \cos\phi_i, \text{ and } E_\mathrm{T}^y = -\sum_i E_i \sin\theta_i \sin\phi_i. \tag{12.7}$$

If there are identified muon(s) in the event, we need to add their transverse momentum to that determined by the calorimeter.

The magnitude of the missing transverse energy is then defined by

Sometimes the symbol $\not{E}_\mathrm{T}$ is used for the missing transverse energy.

$$E_\mathrm{T}^\mathrm{miss} = \sqrt{\left(E_\mathrm{T}^x\right)^2 + \left(E_\mathrm{T}^y\right)^2}.$$

In a real detector, there will always be a non-zero value of $E_\mathrm{T}^\mathrm{miss}$ from fluctuations in the energy measurements. The resolution in the measurement of $E_\mathrm{T}^\mathrm{miss}$ is found to scale approximately as

$$\sigma\left(E_\mathrm{T}^\mathrm{miss}\right) \propto \sqrt{\sum_i E_i \sin\theta_i}. \tag{12.8}$$

In addition to this source of background from measurement errors, there will be a background to 'prompt' neutrinos from heavy flavour decays (e.g. semi-leptonic decays of b-hadrons).

In the very sophisticated LHC detectors, the resolution can be improved by considering the momentum flow of different objects. The missing transverse energy in the x-direction (with an equivalent definition in y) is [93],

$$E_T^x = - \sum_{i \in \text{hard}} p_i^x c - \sum_{i \in \text{soft}} p_i^x c,$$

where the 'hard' terms refer to the reconstructed electrons, photons, muons, taus and hadronic jets. The 'soft' terms refer to the tracks from the primary vertex that were not assigned to any of the hard objects.

For example, for electrons the dedicated calibration and reconstruction procedures will give more precise measurements of the electron p_T than simply using the uncorrected energies in the calorimeter (see section 11.7). Similar procedures improve the resolution for the other hard objects.

Care has to be taken to avoid using the same calorimeter cells in multiple objects. There is a signal ambiguity resolution algorithm that assigns cells to not more than one object [93].

12.11 NEUTRINO FLAVOUR IDENTIFICATION

In neutrino experiments it is often important to be able to identify the flavour of a neutrino responsible for an interaction. It is not possible to identify the neutrino flavour in a neutral current process but in a charged current process lepton flavour conservation ensures that the flavour of the neutrino is transferred to the charged lepton. Therefore the problem of neutrino flavour identification is transferred to that of identifying the flavour of the outgoing charged lepton (e, μ or τ).

This is obviously essential for any experiment studying neutrino oscillations.

See Figure 2.38.

An example of a neutrino charged current interaction would be $\nu_e X \to e^- Y$, where X and Y represent some hadronic states.

The techniques used for identification of the charged leptons will vary depending on the detector technology used for the neutrino detector. For example, the Super-Kamiokande (SK) experiment uses a very large volume of water (approximately 50,000 tonnes) viewed by 13,000 photomultiplier tubes (PMTs). The μ and e from the neutrino interaction will in general be above threshold for the emission of Cherenkov radiation, which is then detected by the PMTs. Therefore both relativistic electrons and muons will result in a ring of 'hits' in the PMTs at an angle to the direction of the electron or muon. However, electrons will start electromagnetic showers in the water and this will then result in emission at a range of angles. Therefore the electron rings will tend to be fuzzier than the sharp muon rings. This information is used quantitatively by constructing a likelihood [294]

For a more detailed discussion of SK see section 14.2.

See section 12.4.

See Figure 12.9.

$$L(\Gamma, \theta) = \prod_j^{\text{unhit}} P_j(\text{unhit}|\mu_j) \times$$

$$\times \prod_i^{\text{hit}} \left\{ [1 - P_i(\text{unhit}|\mu_i)] f_q(q_i|\mu_i) f(t_i|\Gamma, \theta) \right\}. \quad (12.9)$$

Γ is the event hypothesis (e.g. electron- or muon-like) and θ represents the kinematic properties. $\mu_i = \mu_i(\Gamma, \theta)$ is the expected number of photoelectrons produced by the ith PMT given the hypothesis. P is the probability that it does or does not register a hit given the fitting hypothesis. The likelihood considers all the PMTs (the ones that are hit and and the ones that are not hit). For the hit PMTs, the measured charge and time are used. $f_q(q_i|\mu_i)$ is the probability of observing a charge q_i for a predicted charge μ_i and similarly $f(t_i|\Gamma, \theta)$ is the probability for a hit to occur at time t_i given the event hypothesis Γ and the kinematic properties.

In iron/scintillator calorimeters such as MINOS the neutrino flavour identification uses the shape of the energy deposition in the calorimeters. The electrons from ν_e will result in short electromagnetic showers,

For a more detailed discussion of MINOS see section 14.2.

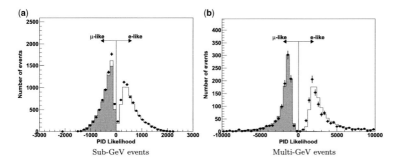

Figure 12.27: Electron/muon separation in the SK experiment [294]. The figures show the likelihood distributions for electrons and muons. The left (right) plot is for sub-GeV (multi-GeV) atmospheric neutrinos. The points with error bars are data and the histograms are from Monte Carlo simulations of ν_e and ν_μ interactions.

whereas ν_μ will result in muons that will propagate long distances in the calorimeter without showering. The momentum of the muon can be determined from its curvature in a magnetic field.

If the energy is sufficiently large the muons will exit the calorimeter but lower energy muons will 'range out', i.e. stop in the calorimeter. For these muons, the energy can be estimated from the measured track range.

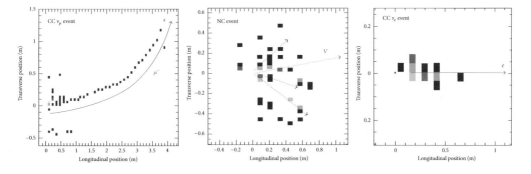

Figure 12.28: Neutrino interaction topologies observed in the MINOS detectors [236]. Left: A CC ν_μ interaction (muon track). Middle: A NC interaction (nuclear recoil). Right: A CC ν_e interaction (electron shower). Each coloured rectangle represents an excited scintillator strip, the colour indicating the amount of light: purple and blue are low light levels, through to orange and red for the highest light levels.

The identification of ν_τ is much more difficult due to the added challenge of identifying and reconstructing the tau from the charged current interaction. This was achieved in the DONUT experiment [314]. The observation was based on the use of photographic emulsion (see section 6.1), in which the decays of the τ could be detected as either kinks in the track ('1-prong') or vertices with one charged track turning into an odd number of charged tracks. Owing to the very small neutrino cross-sections, a very large volume of emulsion was required. It would not have been feasible to scan the full volume of emulsion for candidate vertices. Therefore external detectors were used to record charged tracks emerging from the emulsion and thus identify locations of interest where the emulsion was scanned and the tracks from the vertex measured. The emulsion targets consisted of a stack made of layers of 1 mm thick stainless steel sheets interleaved with emulsion plates made of 100 μm thick emulsion layers on each side of an 800 μm thick plastic base. The length of exposure for each target was set by the track density from muons, with a limit of 10^5 cm^{-2} that was reached within about one month.

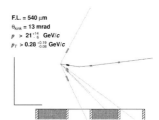

F.L. = 540 μm
θ_{kink} = 13 mrad
p > 21$^{+14}_{-6}$ GeV/c
p_T > 0.28$^{+0.19}_{-0.08}$ GeV/c

Figure 12.29: ν_τ CC interaction event in the DONUT experiment [314]. The neutrinos are incident from the left. The scale is given by the perpendicular lines with the vertical line representing 0.1 mm and the horizontal 1.0 mm. The target material is shown by the bar at the bottom of each part of the figure representing steel (grey), emulsion (hatched) and plastic (no shading).

12.12 JET TAGGING

The most important type of jet-tagging is identifying heavy flavours, i.e. hadrons that contain b or c quarks. We will focus here on b-tagging because the longer lifetime of b-hadrons makes them easier to identify. There are two possible signatures we can use to distinguish b-jets from light quark or gluon jets:

- The semi-leptonic decay mode of the b-hadrons can result in muons or electrons with high transverse momentum with respect to the direction of the b-hadron.

- The relatively long lifetime of the b-hadrons ($\mathcal{O}(1\text{ ps})$) allows to reconstruct their decay vertices, or to detect tracks that do not point back to the primary vertex.

The first technique is limited in efficiency by the magnitude of the branching ratios of the semi-leptonic decay modes of b-hadrons. In practice it is much easier to identify a muon inside a jet rather than an electron, because it can be detected in isolation in the muon system. On the other hand, it is difficult to distinguish in the calorimeter the electron shower from the other particles in the jet. It is therefore usually only the decays with muons that are used. The key signature is that of a muon with a direction close to the jet axis but with a significant component of momentum perpendicular to the jet axis.

BR($B^\pm \to l\nu_l X$) = $(10.99\pm0.28)\%$ and BR($B^0 \to l\nu_l X$) = $(10.33\pm0.28)\%$, where these branching ratios are averaged over electron and muon decay modes [505].

The second technique is based on the significant flight path of a b-hadron before it decays. Therefore, high energy b-hadrons can have decay distances of several mm. In most cases for collider detectors, the decays will occur inside the beam pipe and we have to reconstruct the decay from the charged tracks measured by the vertex detector surrounding the beam pipe. If a secondary vertex can be reconstructed that is significantly displaced from the primary vertex, this provides very powerful separation between B hadrons and light-quark hadrons. However, even if a secondary vertex cannot be reconstructed, powerful b-tagging can still be performed, as there will be tracks resulting from the b-hadron decay that do not point back to the primary vertex. A signed impact parameter is defined in the r-ϕ view such that genuine decays will have a positive value but the result of random measurement errors will result in a positive or negative value.

For B^0 mesons, $c\tau = 456$ μm [505] (τ is the lifetime of the meson).

A similar definition can be made in the longitudinal plane. A per-track significance is defined as $S_{d0} = d_0/\sigma_{d_0}$, where d_0 is the impact parameter and σ_{d_0} is its uncertainty. At large positive values of S_{d0} there is a clear signal for long-lived hadrons from b-quarks.

One relatively simple approach to b-tagging is to define a log-likelihood [94]

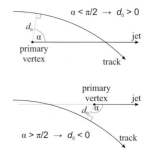

$$L = \sum_i \log\left(\frac{p_i^b}{p_i^u}\right), \tag{12.10}$$

Figure 12.30: Definition of signed transverse impact parameter d_0 (after [370]). The sign of the impact parameter is based on the angle between the jet and the line between the primary vertex and the point of closest approach of the track. Top: positive sign, bottom: negative sign.

where the sum runs over all the tracks associated to a particular jet and p_i^b (p_i^u) is the probability for a track from a b-jet (light-quark jet) to result in the given value of S_{d0}. The estimated values of p_i^b and p_i^u depend on the measured value of d_0 but also the particle kinematics as this determines the uncertainty σ_{d_0}.

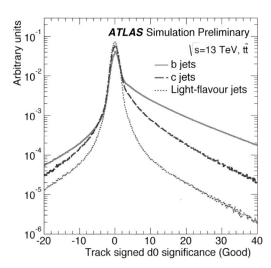

Figure 12.31: Simulated distributions of S_{d0} for tracks from b-jets, c-jets and light quark jets in ATLAS [94].

An alternative approach combines the measurements of the impact parameter in the transverse and longitudinal planes. A third method uses reconstructed secondary vertices. The simple log-likelihood approach ignores the strong correlations between tracks from the same jet. These correlations can be 'learnt' by machine learning algorithms, so they can produce superior performance compared to the simpler log-likelihood approaches.

In principle, the likelihood could be extended to a very large number of variables to accommodate this but this is not computationally practical.

The 'low-level' taggers are broadly based on two different approaches [94].

- *Impact parameter significance.* These use the large impact parameters (either in 2D or 3D) of charged particles from the decays of *b*-hadrons. In order to maximise the separation power between signal and background these algorithms use the estimated 'significance' for an impact parameter d_0 as d_0/σ_{d_0}.

- *Use of secondary vertices.* In one version this approach, the JETFITTER algorithm [95] tries to reconstruct the full *b*-hadron to *c*-hadron decay chain.

The performance can be improved by combining the results of the low-level taggers in so called high-level taggers. As the information between the low-level taggers is correlated, this is done using machine learning methods like Boosted Decision Trees or Neural Nets. The performance of *b*-tagging techniques are illustrated in the plot of *b*-tagging efficiency versus rejection of light quark jets (the inverse of the efficiency). The best *b*-jet detection efficiency versus rejection of light quark jets is achieved with the high-level taggers.

A rejection factor against light-quark jets of $\sim$1000 can be achieved at a *b*-tagging efficiency of 70%.

In addition to fake b-tags, physics analysis will also need to account for genuine B hadrons from gluon splitting $g \to b\bar{b}$.

The efficiencies calculated from Monte Carlo simulations need to be corrected for inaccuracies in the simulation. This is done by measuring the *b*-tagging efficiency in a sample of $t\bar{t}$ (with $t \to bW$ and $W \to e\nu_e$ or $W \to \mu\nu_\mu$) events and then defining scale factors as the ratio of efficiency determined from data to that from the Monte Carlo simulation.

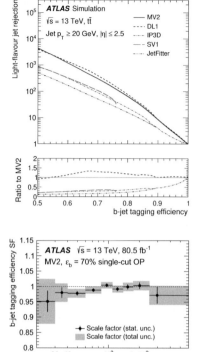

Figure 12.32: Light-quark flavour rejection ratio versus b-jet tagging efficiency in ATLAS for three low-level taggers (IP3D, SV1 and JetFitter) and two high-level taggers (MV2 and DL1) [94].

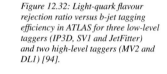

Figure 12.33: The scale factor between the b-tagging efficiency in ATLAS measured for data compared to Monte Carlo simulation for the MV2 algorithm as a function of jet transverse momentum [94].

We will now consider the LHCb Vertex Locator (VELO). The function of the VELO is slightly different to that of the central detectors (e.g. ATLAS or CMS). The VELO is used to not just to tag the presence of b-hadrons but to fully reconstruct the vertices. This allows for the full reconstruction of b and c hadrons from secondary and tertiary vertices. In addition, the study of CP violation and oscillations requires excellent time resolution for the secondary and tertiary vertices.

The LHCb experiment is a single-arm spectrometer at the LHC. It is optimised for reconstruction of b-hadrons in the forward direction. The VELO uses similar technology, silicon microstrips, to that of ATLAS and CMS but the layout is very different [3].

In order to survive the very high fluences expected for the VELO silicon, the detectors uses n^+-in-n and n^+-in-p sensors, rather than p^+-in-n.

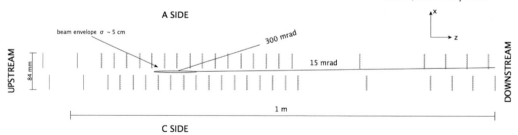

Figure 12.34: Layout of the silicon detectors in the LHCb VELO [3]. The VELO divides in x into two halves for retraction during LHC beam injection.

In operation, the silicon is moved within 7 mm of the beam line. However, this location would receive too high a fluence during beam injection before acceleration, when beam losses are much larger. Therefore it is retracted from the beam line before there are stable beams.

In order to optimise the space point resolution, analogue readout is used. Using analogue readout the baseline variations (pedestals) are subtracted from each channel in data taking. Analogue readout allows for the suppression of 'common mode' noise, which can arise from coherent noise across a sensor. The mean baseline level of all the ADC values corresponding to a sensor (excluding genuine hits) is calculated and this level is then used to correct all the ADC values for the sensor.

The analogue signals from the front end ASIC are transferred to a repeater board outside the VELO detector and then sent via 60 m of twisted pair cables to the counting room where they are digitised.

The main advantage of analogue readout is the improved space point resolution for a given strip width, compared to a binary readout. Neighbouring strip hits (defined as ADC values) above a given threshold after baseline subtraction are clustered together. The charge (given by the ADC value) centroid is used to determine the hit location on the sensor.

This threshold is set at six times the measured noise value.

The position resolution can be determined by measuring residuals of the hit location compared to a fitted track. The resolution increases with strip pitch. It is also very sensitive to the projected angle between the track and the sensor, as this determines the amount of charge sharing.

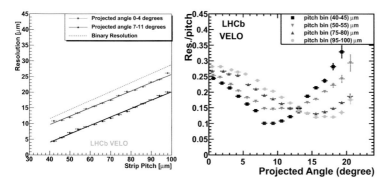

Figure 12.35: Cluster resolution in the LHCb VELO [3]. Left: As a function of strip pitch for different projection angles. Right: Resolution divided by pitch as function of the track-projected angle for four different strip pitches.

The measured cluster centroids are used to fit tracks. The tracks are then extrapolated back to the primary vertex to determine the impact parameter. There are two contributions to the resolution of the impact parameter d_0, multiple scattering and point measurement uncertainties. The measured impact parameter resolution can be parameterised (see exercise 2) as a function of transverse momentum, p_T as

$$\sigma_{d_0} = \frac{A}{p_T} + B. \tag{12.11}$$

The parameter A depends on the detector thickness before the first layer in units of radiation length, and B depends on the detector resolution.

The VELO strip detector has been replaced for LHC Run 2 with a hybrid pixel detector. This provides superior resolution, higher radiation tolerance and faster readout. This allows the detector to be moved closer to the beam pipe and therefore increases the acceptance [514].

In addition to using impact parameter measurements to identify decays of b and c hadrons, a vital requirement for LHCb is a full reconstruction of the decay chains and a precise measurement of the time of flight from the primary vertex to the decay vertex. The decay time in the rest frame of the particle is given by $t = mL/p$, where m is the mass of the

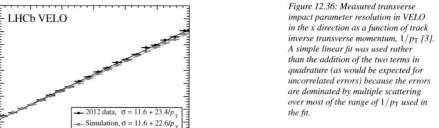

Figure 12.36: Measured transverse impact parameter resolution in VELO in the x direction as a function of track inverse transverse momentum, $1/p_T$ [3]. A simple linear fit was used rather than the addition of the two terms in quadrature (as would be expected for uncorrelated errors) because the errors are dominated by multiple scattering over most of the range of $1/p_T$ used in the fit.

decaying particle, L the distance from the primary to the decay vertex and p is the momentum in the LHCb frame. The resolution of the time measurement depends on the particular decay topology. Results obtained for $B_s^0 \rightarrow J/\Psi\phi \rightarrow \mu^+\mu^-K^+K^-$ are shown in Figure 12.37.

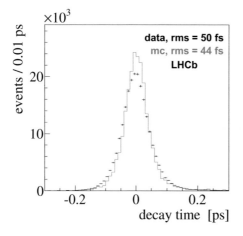

Figure 12.37: Measured decay time resolution in VELO for the topology $B_s^0 \rightarrow J/\Psi\phi \rightarrow \mu^+\mu^-K^+K^-$ [3]. The contribution from genuine decays is removed before making this plot to avoid a spurious tail at positive times. The resolution quoted is based on a fit to the negative side of the distribution.

Key lessons from this chapter

- Identification of particles provides an important kinematic constraint, as it identifies the mass of the particle unambiguously, and it identifies the flavour of the particle.

- Dedicated particle identification detectors provide a measurement of the velocity of the particle, in the form of either the relativistic β or γ factors. Combined with a momentum measurement this allows for reconstruction of the mass.

- Dedicated particle identification technologies comprise, in order of increasing $\beta\gamma$ of the sensitive momentum region, time-of-flight, dE/dx, Cherenkov radiation and transition radiation.

- The separation power of TOF is limited by the distance between the timing detectors, and their time resolution.

- Separation by dE/dx is easy for low energies, but becomes hard for minimum-ionising momenta and above. Due to the density effect, above minimum-ionising only gaseous detectors provide enough differentiation in dE/dx. The main challenge is the straggling of the energy loss, and one way to reduce its effects is the use of the truncated mean of a number of individual energy loss measurements. Alternatively, cluster-counting is less affected by straggling.

- The most powerful Cherenkov detectors measure the opening angle of the Cherenkov cone for a measurement of β. The challenge is that continuous tracking of the photons is not possible. In practise, the endpoint of the photon trajectory is measured, and the start point is inferred from other means (either proximity or mirror focusing). To maximise the range of sensitivity RICH system typically comprise more than one radiator, with different indices of refraction.

- The main challenge for the detection of transition radiation is the small probability for the emission of a photon on one interface, so that a large number of interfaces is required. To maximise photon collection in the detectors low-Z radiators and high-Z detectors are used. TR is sensitive to the relativistic γ factor, and thus it is primarily useful in experiments for electron identification.

- Other particles (electrons, muons, jets, tau and heavy flavour hadrons) can be identified from the characteristic signatures and combination of the tracking and calorimeter systems in a standard particle physics experiment. For muons this relies on filtering of hadrons in the calorimeters, for electrons on the typical dimensions of electromagnetic showers. In all cases matching of tracks and showers in different subsystems and isolation cuts are powerful means to suppress backgrounds.

- The flavour of neutrinos in neutrino experiments can be inferred from the flavour of the lepton produced in charged current reactions.

EXERCISES

1. *TOF separation power.*

   Starting from $m = p/(\beta\gamma) = p\sqrt{1/\beta^2 - 1}$ and $\beta = L/(tc)$, derive eq. (12.2). Derive eq. (12.3).

2. *Errors on vertex reconstruction.*

   Consider a cylindrical vertex detector with detector layers at a radius r_1 and r_2. The resolution of the detectors are σ_1 and σ_2.

   a) Ignoring the effects of the magnetic field and multiple scattering, what is the uncertainty on the measured impact parameter d_0?

   b) For the same geometry, ignore the effects of the finite measurement errors and consider the effects of multiple scatting. The material in the beam pipe and the first detector layer is x in units of radiation length. What is the uncertainty on the measured impact parameter d_0 as a function of transverse momentum p_T?

3. *Measurement in a muon spectrometer.*

   Consider a magnetised iron muon spectrometer. The magnetic field is $\vec{B}$ and the track length in the spectrometer is l. The momentum of a unit charged particle transverse to the magnetic field is p_T.

   a) What is the change in p_T from the start to the end of the trajectory in the magnetic field (ignoring the effects of multiple scattering)? You can assume that the deflection of the particle is small and that it has a speed of $\beta = 1$.

   b) What is the uncertainty in the angular deflection of the particle in traversing the iron, due to multiple scatting? What is the corresponding uncertainty in p_T?

   c) What is the resulting fractional resolution in p_T arising from multiple scattering alone? Evaluate this for a muon spectrometer with $l = 2$ m and $B = 2$ T. Comment on the performance of such a spectrometer, compared to the resolution achieved by ATLAS and CMS.

4. *B^0 and vertex layers.*

   Consider the decays of a B^0 hadron in a collider detector with a cylindrical vertex detector. The first detector layer is at a radius $r = 3$ cm. The lifetime of the B^0 is $\tau = 1$ ps. What fraction of the B^0 hadrons decay before the first layer for transverse momentum values of 5, 10 and 50 GeV/c?

5. *Momentum resolution in a solenoid.*

   Consider a tracking detector inside a solenoid with an axial magnetic field $B = 4$ T and a radius $R = 2$ m. The resolution of the space points at the start, middle and end of the tracks is 100 μm.

   Neglecting multiple scattering calculate the resolution for the measurement of p_T as a function of p_T for tracks that are perpendicular to the axial magnetic field. Compare this resolution with that of an iron core spectrometer (question 3).

6. *e/π separation using a TRT.*

   Consider a TRT with 30 straw chambers for the readout. The efficiency for getting a high threshold hit in one straw is 20% for electrons and 5% for pions. Electrons are selected if there are N_{min} high threshold hits out of 30. For the purposes of this question assume that the hits are uncorrelated.

   a) What value should be used for N_{min} if we want an electron efficiency $> 90\%$?

   b) What would the resulting pion efficiency be for this value of N_{min}?

   c) Compare your answer with that given for ATLAS in Figure 12.16 and discuss the origin of any discrepancies.

13 Triggers

13.1 INTRODUCTION

In general, in particle physics interesting events and the response they produce in detectors are instantaneous, or they have a very short duration given by decay lifetimes. This duration can be increased by the finite response time of the detectors, but is still very short compared to the operational time of the experiment. Other events, from interesting physics or backgrounds, will take place before and after, and thus, in general, the experiment will need to identify the occurrence of specific events in time and record the information from its detectors for this instance. This identification is provided by the trigger.

An exception are emulsion neutrino detectors (see section 6.1), which can operate with very long exposure times due to the low local event density. However, even these detectors are paired with electronic (and triggered) detectors, that indicate where the interesting physics event took place.

In experiments searching for rare physics, the rate of triggered events will be low, but signal events still need identification among backgrounds. In many other cases, the combined rate of signal and background events, and the amount of data recorded for each event, makes readout of all events occurring in the experiment prohibitive, and the task of the trigger in these experiments is to reduce the rate of events selectively to a manageable level, while significantly enhancing the fraction of potential signal events in the process.

For any measurements of (relative) rates, for example for measurements of cross-sections, the (relative) trigger efficiencies must be known, and the error on this knowledge is often a significant contribution to the error of the actual measurement. Events that are rejected by a trigger are lost, so great care has to be used in designing triggers and understanding any additional bias they introduce.

There are many different types of triggers that are used by different experiments. This chapter will not aim to be comprehensive, rather we start with discussing some basic principles, then look at the trigger challenges at more complex experiments, and finally study some examples for triggers in different particle physics experiments.

13.2 BASIC TRIGGER CONCEPTS

COINCIDENCES

Coincidences are a way to identify the hits belonging to a specific event topology. In coincidences the correlation of hits is done in the time domain, whereas a spatial correlation can be achieved by the position of the included detectors. Coincidences are usually used for fairly large detection elements (for example scintillator paddles).

Coincidences can also be used to obtain clean start/stop signals for time measurements (for example of drifttimes or TOF).

Coincidences are particularly useful when the detectors included are producing signals with a significant fake rate, for example due to noise or thermally induced signals in detectors with large internal gain (for example photomultipliers).

In coincidence circuits two or more binary signals that are expected for the observation of an event are required to occur in time with each other for a positive decision that the event has occurred. If we have two binary

DOI: 10.1201/9781003287674-13

input signals, with duration T_1 and T_2 and signal rates (including fakes) R_1 and R_2, respectively, then the rate of fake coincidences is

$$R_c^{\text{fake}} = R_1 R_2 (T_1 + T_2).\qquad(13.1)$$

Usually $T_1 + T_2 \ll (R_{1,2})^{-1}$ and thus the fake rate of the coincidence will be much smaller than the fake rates of the inputs.

Both input signals will have their inherent inefficiency from the sensor itself and the comparator decision in the conversion to a binary signal. If the efficiencies for the two signals are ε_1 and ε_2, respectively, then the combined efficiency is

$$\varepsilon_c = \varepsilon_1 \varepsilon_2.\qquad(13.2)$$

If the input fake rates are high, then the duration of the signals must be small, and the coincidence will require careful adjustment of the time of arrival of the two input signals. Including further inputs will improve the fake rejection.

Coincidences can also include inverted inputs, if the absence of a signal in that detector is part of the desired event topology. For example, if we wish to record the response of a detector to K^+ in a beam consisting of K^+ and π^+ we could add an upstream threshold Cherenkov detector. If the radiator is chosen appropriately for the momentum of the beam particles, the K^+ would be below threshold for Cherenkov radiation but the π^+ would be above threshold. We could then use the logical signal from the Cherenkov counter as a veto.

TRIGGER SIGNATURES AND TURN-ON

More generally, a binary trigger decision is based on an analogue measurement that is compared to a trigger threshold value. The analogue input can be simply the signal height in a single or multiple detector cells, or it can be kinematic information gathered from a detector system (like the momentum measured in a magnetic spectrometer, or the missing E_T in a collider experiment). Both efficiency and fake rate will depend on the threshold, with the rejection of backgrounds improving, but signal efficiency decreasing, for increasing threshold.

To minimise the processing power required for the trigger, and to achieve a fast decision time, the reconstruction of kinematic event properties is usually done at a relatively coarse level, with a limited resolution. As a result, the trigger will not switch its response exactly at the threshold value, but the transition from rejection to full acceptance will occur gradually as the real value of the kinematic trigger variable increases. This behaviour can be illustrated using trigger 'turn-on curves', which show the efficiency as a function of the true value of the kinematic trigger variable (which, for example, can be obtained in the more accurate offline analysis). A sharp turn-on facilitates the understanding of the trigger efficiency. In addition, the background processes usually have a steeply falling spectrum as a function of the variable used for the trigger decision (e.g. p_T). Improving the resolution allows to reduce the threshold and thus can significantly reduce the fraction of background events accepted.

The logical operation can be performed by an AND gate. For simple applications, for example in test beams, standardised modular electronics are in wide use, for example based on the Nuclear Instrumentation Module (NIM) standard [213, 329]. For more complex topologies field-programmable gate arrays (FPGAs) are available.

This assumes that the two signals arrive at the same time, which can be achieved with appropriate delays.

This is called an 'anti-coincidence'.

Coincidences have played an important role in early experimental particle physics using detectors with electronic readout, leading to the Nobel prize for Walther Bothe 'for the coincidence method and his discoveries made therewith'. See for example [156].

This is referred to as the trigger 'signature'.

Another reason for degraded resolution at the trigger level can be the lack of availability of accurate calibration or alignment data during the data taking.

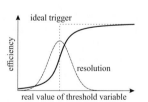

Figure 13.1: Schematic trigger turn-on curve.

13.3 COMPLEX TRIGGERS

TRIGGER TYPES

It is obvious that in general a trigger should have a high efficiency for recognising signal events, and at the same time reject a large fraction of background events. However, often more complex trigger capabilities are required. Different physics studies can require datasets obtained with more than one type of trigger, and some of the experimental bandwidth must usually be reserved for secondary datasets needed for instrumental and physics background studies, detector, tagging and trigger efficiency measurements, calibrations (for example for energy scales), etc. In a complex detector like at LHC there will be several trigger streams targeting different physics analyses. The trigger algorithms and thresholds at the different trigger levels will need to be tuned so as to maximise the physics potential given the constraint of the available bandwidth. One simple way to record data with unbiased triggers is pre-scaling, where a random fraction of the events is accepted, but at a rate which is reduced, so that it does not significantly obstruct the collection of signal data.

Another important way that pre-scaling is used is for triggers on high p_T jets. The threshold for an un-prescaled jet trigger will need to be very large in order to keep the accepted event rate at a manageable level. However, several lower thresholds can be used if the prescaling factors are decreased. Therefore data is recorded over a larger range of p_T and is available for analysis. For example, the cross-section can be measured over a much wider range of p_T than that which would have been possible using only an un-prescaled trigger.

Trigger efficiency is defined as the fraction of accepted events divided by the real events. In practice the latter corresponds to the events accepted in a physics offline analysis.

Robustness can be improved by using redundant triggers, i.e. triggers targeting the same type of physics events, but using different inputs.

Since trigger rate vary with the instantaneous luminosity, dynamic pre-scaling is sometimes used (pre-scaling fractions are decreased as the luminosity decreases during a run).

These pre-scaled triggers can be very useful for detector studies, e.g. to measure the trigger efficiency for a higher threshold trigger.

TIMING

The finite time that it takes to come to a trigger decision by hardware or software becomes relevant when the event rate in an experiment is high. We call the time used to form the trigger decision and distribute the acceptance signal to the individual readout channels the 'latency'.

After an event it usually takes some finite time until the detector is fully ready to record the next. This time is called the 'deadtime'. It can comprise detector-related time constants, for example to complete the drift of charges, or the delayed relaxation of excited states, and readout-related contributions from digitisation, and signal and trigger processing. In high-rate experiments the former is usually accepted as 'pile-up', as long as the data from individual events can be separated in the offline analysis.

For a trigger rate R_t and a deadtime T_d the fraction of recorded events is given by

$$\frac{R_{rec}}{R_t} = \frac{1}{1 + R_t T_d}. \tag{13.3}$$

This is maximised ($R_{rec} \sim R_t$), if the deadtime and the trigger rate can be kept small ($R_t T_d \ll 1$). In principle we can create a deadtimeless trigger at level 1 (L1) by using local pipeline memories that store the front-end data locally for the duration of the trigger latency. The data used for the hardware trigger is read out into pipelined processors. These processes will make a trigger decision after a fixed time. The trigger decision is provided to the front-end electronics after a fixed latency. Accepted events are

In accelerator experiments there are also more extended periods of operational deadtime, for example the time needed to fill the accelerator, where no collisions take place.

This allows for data to be stored in the pipeline memories with no deadtime but there will still be deadtime in the readout of the data from the pipelines. For LHC experiments, the pipelines are implemented in the front-end ASICs that read out the detector data. The finite bandwidth of the readout of the data from the buffers will result in deadtime if the L1 trigger rate is too high. In the case of ATLAS, preventative deadtime is introduced so that the buffers do not become full [96].

transferred to buffers for subsequent readout. The data corresponding to non-triggered events will be overwritten with new data.

For the LHC experiments the latency is typically 3 μs.

MULTI-LEVEL TRIGGERS

In complex high-rate, high-background experiments like at the LHC the data rates cannot be reduced to an acceptable level without the loss of too much efficiency for the signal events. The strategy is therefore to split the trigger decision into several levels (typically 2 to 4), with progressively lower trigger rates, allowing for progressively longer processing times. These higher level triggers are implemented in software, allowing more powerful algorithms to be used. Part of each trigger level is a local buffer that stores the data locally during the time of the trigger decision at this level. Trigger rates, latency and buffer size must be carefully matched, so that the readout bandwidth is fully used, while avoiding buffer over-flows. The maximum accept rate for each trigger level is set by the rate with which data can be transferred to, and processed at the next level.

Because the first trigger level encompasses fast and simple decisions, and requires local buffering, first level triggers are usually implemented in hardware, whereas higher level triggers typically run on farms of standard commodity computers.

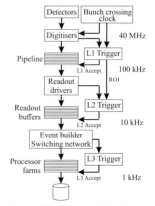

Figure 13.2: Typical multi-level readout and trigger structure. Due to increased processing power modern multi-level triggers are typically two-level triggers.

Field-programmable gate arrays (FPGAs) are the most powerful devices for use in level 1 trigger processing, as they allow for easy modification of trigger algorithms.

TRIGGERS AT COLLIDER EXPERIMENTS

In experiments at accelerators, the time structure of the bunches in the beam is usually used as a first indication when to activate the readout of the experiment.

At collider experiments the primary targets for triggers are usually high-p_T muons, electrons or photons, high energy jets of hadrons or missing transverse energy. Combinations of shower shapes in the calorimeters and tracking information are used to select electrons and photons. Particles penetrating the calorimeter and giving hits in the muon chambers are used for muon triggers. In addition, isolation criteria can be used for leptons to reject backgrounds from QCD jets. Multi-object triggers (for example combinations of leptons) can reduce trigger rates.

Very different trigger signatures are needed for detectors targeting heavy quark (b or c) production or in heavy ion experiments.

Pipelined level 1 triggers operate with a fixed latency, which excludes, for example, the use of iterative algorithms. Level 1 trigger operations are thus usually either simple arithmetic operations or functions using memory lookup tables. However, the complexity of algorithms that can be performed at level 1 has increased over time and for example the ATLAS L1 trigger can perform mass cuts. To cope with the higher luminosity in LHC run 3 a new custom electronics processing system was built that enables the calorimeter data to be available with a tenfold increase of granularity, which allows the use of more sophisticated identification algorithms [97].

The numbering of the trigger levels is not consistent, with some experiments labelling the first trigger level as 'level 1', while other call this 'level 0'.

Higher trigger levels can reconstruct tracks and perform matching of information in different sub-detectors. They can also reconstruct event kinematics for effective mass and topology cuts. The complexity of higher level triggers has greatly increased, due to the wide availability of high-performance processors (graphics processing units, GPUs) and parallel operation. Level 2 trigger processing is generally still limited in execution

For example matching of tracks in the tracker and the muon system, or of tracks and clusters in the calorimeter.

time. One way to achieve this is by dividing processing into 'Regions of Interest' (RoIs) determined by the Level 1 trigger data.

The RoI approach will be discussed later in this chapter in the context of ATLAS and CMS triggers.

The identification of the correct bunch-crossing by the L1 trigger is essential in order to initiate the readout of the data from the pipelines corresponding to the correct bunch crossing(s). Bunch-crossing identification can be challenging in large high-rate experiments as in the LHC, where the bunch-crossing frequency is 25 ns, much shorter than the signal arrival times and duration in many detectors, and even the time-of-flight of particles through the detector.

13.4 TRIGGERS AT LHC

One of the most challenging environments for triggering are the experiments at the LHC. At $\sqrt{s} = 14$ TeV the total cross-section for pp interactions is $\sigma \approx 100$ mb or about 100 million events per second at an instantaneous luminosity of $\mathscr{L} = 10^{33}$ cm^{-2}s^{-1}. The difficulty is illustrated by considering the case of Higgs boson production. The total Higgs boson cross-section $\sigma(H) = 54$ pb and thus the event rate for events containing Higgs is 10^{-9} times smaller than the total collision rate.

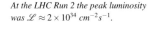

At the LHC Run 2 the peak luminosity was $\mathscr{L} \approx 2 \times 10^{34}$ cm^{-2}s^{-1}.

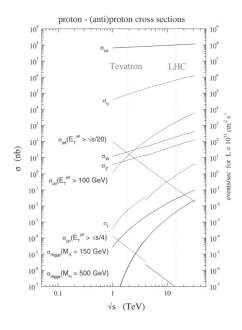

Figure 13.3: Standard model cross-sections and event rates at hadron colliders [178].

Furthermore, the branching ratios for the more easily identifiable decay modes of the Higgs are very small, e.g. $BR(H \rightarrow \gamma\gamma) = 2.27 \times 10^{-4}$. Therefore, the LHC has to operate at very high luminosity to achieve a useful rate for these type of rare events, which results in an extremely high rate of pp collisions. It is unfeasible to store the data from all these collisions and therefore very powerful trigger systems are required to reduce the rate sufficiently to allow the data to be permanently stored. At the same time the triggers need to have a high efficiency for detecting these rare events.

An additional complication for triggers at LHC is that there are collisions every 25 ns when bunches of protons collide. The length of time required to form a trigger decision is much longer than this. To avoid very large deadtime, pipelined triggers are used.

The size of the general purpose LHC detectors is such that it takes a highly relativistic particle more than one bunch crossing (25 ns) from the vertex to the outer parts of the experiment.

In general the triggers need to reduce the readout rate to a level that is acceptable based on the bandwidth available. Therefore, as the luminosity increases, there is a need to adjust the triggers to keep the rate manageable. This can be done by a combination of increasing the trigger thresholds and by tightening the selection procedures used to select events. In addition, multi-object triggers can be used for some physics channels and these can use lower thresholds than single object triggers.

For example for $H \to \gamma\gamma$ a di-photon trigger can be used with lower p_T thresholds than for single-photon triggers.

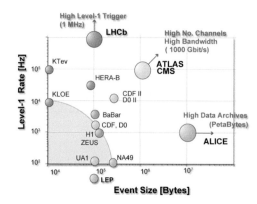

Figure 13.4: Raw data transfers from Level 1 trigger in LHC experiments and a comparison with earlier experiments [391].

In LHCb the amount of data per event is smaller than in the general-purpose experiments (ATLAS and CMS), and this allows for a much larger L0 trigger rate. Until long shutdown 2 (LS2, 2018) this was about 1 MHz. The L0 trigger was followed by a two-level HLT, running on an Event Filter Farm (EFF) of 52,000 logical CPU cores [450]. The upgrade of the LHCb detector uses a pure software trigger which will be discussed below.

The ALICE experiment [42] at the LHC is designed for the study of heavy-ion collisions. The bunch-crossing rate for these type of collisions and the instantaneous luminosity are lower and thus the trigger rates are lower [321], but the very large track multiplicity results in very large event data sizes.

The bunch-crossing interval for heavy ion collisions at the LHC is 125 ns.

In the following we will consider two examples of the multi-level approach for electron and muon triggers at the general purpose collider experiments. We will then look at the upgraded LHCb trigger system in which all the data is read out at 40 MHz and pure software triggers are used.

ELECTRON/PHOTON TRIGGER IN ATLAS

There are important physics channels that involve high transverse momentum electrons and photons (e.g. $pp \to t\bar{t}$, $t \to b e \nu_e$ and $pp \to H \to \gamma\gamma$).

The ATLAS e/γ trigger consists of two levels. At Level 1 (L1) a pipelined hardware trigger is used to reduce the rate from the bunch crossing rate of 40 MHz to $\sim$ 20 kHz [98]. The L1 e/γ trigger only uses calorimeter information. The trigger is pipelined, so that decisions are made every 25 ns with a latency of 2.5 μs. In order to minimise the

This rate is $\sim$ 20% of the total L1 trigger bandwidth.

bandwidth required for the trigger, reduced granularity is used at L1 (see Figure 13.5).

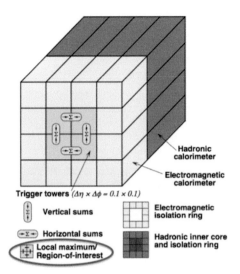

The coarse granularity calorimeter trigger data is read out and processed in dedicated pipelined processors. A 'sliding window' algorithm is used, in which the transverse energy in a region of 2×2 neighbouring trigger towers is calculated. The region is stepped over the calorimeter in the η and ϕ directions to find local maxima of the energy in the window. As the signal involves isolated electrons and photons, powerful rejection of backgrounds is achieved using cuts on the energy in the ring of 12 cells surrounding the 2×2 cluster in the EM and hadronic calorimeters. Additional rejection against backgrounds is obtained by cuts on the hadronic calorimeter energy behind the 2×2 EM cluster.

The backgrounds arise from QCD jets which have a very high rate, so that even if a low fraction of these events is misidentified as electrons the result is a significant accepted background rate.

The events are read out after an L1 trigger, so that the High Level Trigger (HLT) can be a software trigger using the full granularity of the detector. The L1 e/γ trigger seeds a Region of Interest (RoI) centred on the 2×2 window. In order to reduce CPU time the HLT e/γ trigger algorithms only look for electrons and photons within the RoIs defined by the L1 e/γ trigger. To optimise the CPU requirement, simple algorithms are first applied in order to reject some of the backgrounds. For electron candidates with $p_T > 15$ GeV/c, simple cuts on variables like the ratio of hadronic to electromagnetic energy in the cluster. This stage uses fast track reconstruction inside the RoI to perform track-cluster matching cuts [98].

After these fast algorithms have passed an event with electron candidates, more sophisticated algorithms are used. The cluster reconstruction and calibration used a similar likelihood method to that used in the offline analysis (see the discussion of eq. (12.6)).

For electron candidates with $p_T < 15$ GeV/c the background rates are higher and more sophisticated algorithms based on neural networks are used to reduce the rates. This procedure uses slightly more CPU time than the simple cuts algorithm but improved the background rejection by a factor of 1.5–1.6 [98].

The overall electron efficiency can be factorised into the product of trigger efficiencies and offline efficiency, $\varepsilon_{\text{all}} = \varepsilon_{\text{trig}}\varepsilon_{\text{offline}}$. The trigger efficiency is defined as $\varepsilon_{\text{trig}} \equiv N(\text{trig})/N(\text{offline})$, where $N(\text{trig})$ and $N(\text{offline})$ are the number of electrons identified at the trigger and offline level, respectively. The same type of 'tag and probe' analysis is used to measure the electron trigger efficiency as used to measure the offline efficiency (see section 12.7).

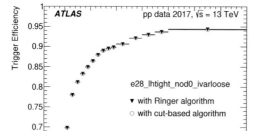

Figure 13.6: Measured electron trigger efficiency in ATLAS as a function of electron transverse energy (E_T) for two different algorithms (see text for details) [98].

The data show the characteristic turn-on behaviour with a low efficiency at very low p_T and gradually rising to a plateau at high p_T. The plateau is below 100% because of the need to reject a large fraction of the background. The efficiency varies across the detector as it is lower in the regions in which there is more material in front of the calorimeter. The effect of pile-up also reduces the trigger efficiency. Therefore the trigger efficiency is also measured as a function of η and the number of pile-up vertices, μ [98].

A similar data-driven approach is used to determine the photon trigger efficiency using samples of $Z \to e^+ e^- \gamma$ in which the tag is given by the $e^+ e^-$ pair with an invariant mass below that of the Z and an invariant mass of the three-body system compatible with that of the Z [98].

MUON TRIGGER IN CMS

CMS uses hardware triggers to reduce the bunch crossing rate of 40 MHz to 100 kHz. The rate is reduced to about 1 kHz by the HLT which used software triggers running on a CPU 'farm' with about 30,000 cores [206]. As there are several triggers for the different interesting final states, one trigger can only use a fraction of the total bandwidth, which is about 15% in the case of the muon triggers.

For an overview of the CMS muon system see section 12.8.

The level 1 muon trigger uses 'trigger primitives' from the different stations, which are track segments in the different layers with an estimate of the angular coordinates, θ and ϕ from the different layers of each subsystem. Different hardware trigger systems are used for the three chamber types. For the case of the fast RPCs, the track segments are determined by a hardware pattern comparator. For the case of the slower DTs, local track segments are reconstructed by custom electronics on the chambers. This electronics also identifies the bunch crossing. The track segments are processed by FPGAs, which synchronise the data and send it via fast optical links to the counting room.

The maximum drift time of the DTs of $\mathcal{O}(400\ ns)$ is much longer than the interval between bunch crossings of 25 ns.

In the counting room, the global muon trigger (GMT) matches track segments in the different layers and determines track quality and estimates the transverse momentum (p_T). It uses a combination of look-up tables and FPGAs. The value of p_T is estimated from the deflection in ϕ due to the solenoidal magnetic field. The GMT combines muon candidate tracks from three different algorithms. It resolves ambiguities with overlapping tracks by keeping the best quality tracks.

The resolution in p_T is limited by multiple scattering for most of the interesting range of p_T (see section 12.8).

About 20–25% of the L1 muon triggers do not correspond to genuine muons. Therefore the High Level Trigger (HLT) uses a more sophisticated algorithm to reject most of this background. The first step is to reconstruct standalone muons from the muon chambers alone (L2 muons). The vertex constraint is added to improve the resolution of the p_T measurement.

This is implemented in software and so can use essentially the same algorithms as used for offline muon reconstruction. The L2 muons are then used to define a region of interest for the track reconstruction in the inner tracker. By matching the muon candidates to the tracker, the p_T resolution is greatly improved, thus allowing further reduction in trigger rates. The much greater precision of the inner tracker is then used to reduce the trigger rate by applying a sharper threshold in p_T.

The CPU time required for the complete reconstruction of the full inner tracker would be too long for the HLT operating using a single processor. Parallel processing is achieved by providing each processor with separate events.

Many physics channels involve isolated muons, whereas the muons from backgrounds like π/K decays in flight will typically result in other particles in close proximity to the muon. Therefore powerful background rejection can be achieved with isolation cuts. The total p_T of additional tracks and calorimeter clusters within a cone of $\Delta R = 0.3$ is evaluated and used to reject non-isolated muons. The power of the isolation requirement is illustrated in LHC Run 2, for which the trigger threshold for the isolated muon trigger was $p_T = 24$ GeV/c, whereas the non-isolated trigger used a threshold of $p_T = 50$ GeV/c [207].

$\Delta R = \sqrt{(\Delta\eta)^2 + (\Delta\phi)^2}$, where $\Delta\eta$ is the difference in η between the tracks/clusters and the muon track, with an equivalent definition for $\Delta\phi$.

In a similar methodology as used for electron triggers, the muon trigger efficiency is measured relative to that of selected offline muons. The same type of 'tag and probe' analysis is used to measure the efficiency from data.

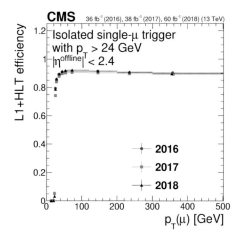

Figure 13.7: Measured muon trigger efficiency in CMS from LHC Run 2 as a function of muon p_T [206].

LHCB SOFTWARE TRIGGER

The principle triggers for the LHCb experiment target events with hadrons containing b or c quarks. The very large cross-sections (see Figure 13.3) make it very challenging to design efficient triggers at high luminosity. The triggers for the hadronic decays of b and c hadrons are particularly challenging as the identification requires the precision tracking to identify the particles that do not point to the primary vertex.

For the LHC Runs 1 and 2 the LHCb experiment used a conventional hardware trigger (L0) and a software selection (HLT). In this period the peak luminosity was limited to 4×10^{32} cm^{-2}s^{-1}. One crucial development was splitting the HLT into two levels (HLT1 and HLT2) [4]. HLT1 performs partial event reconstruction that allows selections to reduce the rate to 110 kHz. These events are stored on a large disk buffer for long

The rate of the L0 trigger was limited to 1.1 MHz.

enough for the updated alignment and calibrations to be determined. This allows the CPU farm to apply the final alignment and calibration procedures. The HLT2 (running on the CPU farm) can then make selections that are the same as those that would be made in conventional offline analysis. This allows for sharp trigger thresholds to be implemented without loss of efficiency and this reduces the rate by an additional order of magnitude.

For Run 3, the target luminosity for LHCb was increased to 2×10^{33} cm^{-2}s^{-1} [333]. Using the L0 trigger would have required raising the trigger thresholds and would have resulted in a large loss of interesting physics events. Therefore, for LHC Run 3, a major additional improvement was the change to a full software trigger [334].

This corresponds to an average number of inelastic interactions per bunch crossing of 7.6.

All events are read out from the detector at the bunch-crossing rate of 40 MHz and the trigger is performed by the two-level HLT. This requires the use of 8800 Versatile Link transmitters (optical links) using a data rate of 4.8 Gbps. The Versatile Links are based on a CERN developed custom radiation hard optical transceiver [476]. The rates at the different stages are summarised in Figure 13.8.

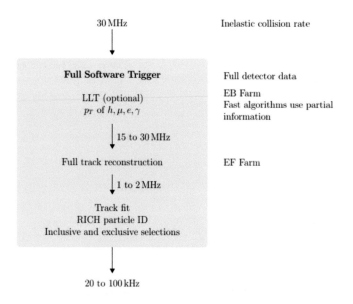

Figure 13.8: Estimated data rates for the LHCb all-software trigger. The 30 MHz input rate is averaged over the filled and empty bunches in the LHC. The Low Level Trigger (LLT) is an optional feature to implement simple selections on individual object p_T in a similar way to that used in a hardware trigger [334]. As for the trigger in Run 1 and 2, the data output from HLT1 is stored on disk long enough to apply the final alignment and calibrations.

The HLT1 runs on GPUs [5] and can accept an input data rate of 40 Tbit/s using about 500 GPU cards. HLT1 can perform pattern recognition and identify particles as hadrons or muons. Crucially for the LHCb application it can identify decay vertices of long-lived particles. The HLT2 software trigger requires the use of a dedicated CPU farm with up to 4000 nodes. This allow for a processing time per event of 13 ms. The average event size is 100 kB and accepting events for storage at 20 kHz, results in an output bandwidth of 2 GB/s [334].

The concept of a pure software trigger was also considered for the ATLAS and CMS experiments. These experiments contain many more channels than LHCb and therefore the costs would have been very high. In addition, the radiation levels in the pixel detectors in these experiments are too high for the operation of optical links and therefore copper cables are used to transfer the data to a lower radiation region where the optical transceivers can be placed. The very large number of these electrical links that would have been required by a pure software trigger would result in too large a mass of copper.

13.5 TRIGGERS FOR RARE DECAY EXPERIMENTS

The trigger requirements for rare decay experiments are very different to that from collider experiments. In general, the event rates will be very much lower. We will consider the LZ experiment as an example (see section 8.2). The rate of signal events is of course extremely small and possibly zero. However, in order to control the backgrounds it is essential to acquire suitable data samples. For example, events outside a fiducial volume can be used to extrapolate the backgrounds into the signal region. For the LZ experiment the estimated background rate is 40 Hz [34]. Higher rates are achieved with dedicated calibration sources such as LEDs for which the rate will be up to 150 Hz. For these relatively low rates the data can be continually digitised and read into a circular buffer without the need for a hardware trigger. Data extractors are used to detect interesting events and compress the data which is then read out to temporary buffers. The Event Builders then assemble all the data from each event for offline analysis [34].

Data extractors and event builders are implemented in software.

TRIGGERS FOR NEUTRINO DETECTORS

The triggers required for accelerator and non-accelerator neutrino experiments are different. We will consider the triggers used for the MINOS experiment (see section 14.2) as an example of an accelerator neutrino detector. The neutrino beam has a 1 μs 'beam spill' structure, with a repetition time of 2 s [374].

The time structure arises from the proton accelerator used to create the neutrino beams.

The pixels of each PMT are digitised and the pedestal is subtracted. Within the beam spill time window, all hits above a threshold are read out to a trigger processor. The trigger processor implements a simple algorithm for the far detector to select physics events for permanent storage. The hits are time-ordered and clusters of hits within about 200 ns are defined. The primary selection algorithm uses clusters within a window of 200 ns to identify candidate events. The main trigger algorithm for MINOS requires that at least n planes out of any group of m contiguous planes should have hits within the time window. The instantaneous rate of neutrino interactions at the 'far detector' is less than 200 Hz, which is much lower than the rate of PMT hits from radioactivity and dark currents.

The far detector is located 730 km from the accelerator in order to study neutrino oscillations.
Other algorithms are used for calibration and background studies.

Typical values were $n = 4$ and $m = 16$.

The Super-Kamiokande experiment (see section 14.2) is an example for a non-accelerator based experiment. It originally had a hardware trigger, which triggered a readout of the experiment when the number of PMT hits out of the about 13,000 PMTs in the experiment within a time-window of 200 ns exceeded a threshold. Depending on the energy threshold the rate of triggers was between 10 to 1000 Hz. The signals from the PMTs were dominated by the dark counts rate, which was about 4 kHz.

PMT dark currents are mostly made up of signals due to thermionic electron emission in the PMT.

However, to improve the physics reach of the experiment, and to allow for the search for relic supernova neutrinos the energy threshold had to be lowered, resulting in higher trigger rates, and the time window had to be increased. As a consequence, the trigger windows would have overlapped and the detection of the physical event would have been made much more difficult. Therefore the readout of the experiment was changed to continuous readout at a rate of 60 kHz with a software event selection in a farm of standard commodity PCs [507]. The total data rate this system can process is around 800 MB/s. To verify correct operation of the trigger, high rate tests, simulating data from a supernova burst by flashing LED light in the water tank (2.5M flashes in 10 s) were performed.

For this search the time-window had to be increased to more than 200 µs to suppress the most prominent background from atmospheric neutrinos, which can be rejected by detecting the 2.2 MeV gamma ray from capturing the recoil neutron from the interaction of the atmospheric neutrino. The longer time window is necessary because of the time until capture.

Key lessons from this chapter

- Coincidences can be used to identify geometrical correlations of signals that are representing interesting event topologies. At the same time, coincidences can reduce fake rates from individual detectors.

- Trigger decisions are binary decisions typically based on a continuous variable. A sharp turn-on of the trigger as a function of the continuous variable is beneficial.

- In a typical particle physics experiments several types of trigger are employed (different physics targets, calibration data, redundancy). The total rate of triggers must be compatible with the capabilities of the data acquisition system. Triggers with a rate exceeding this must be pre-scaled.

- To reduce readout deadtime, data is buffered locally, usually in a pipeline memory.

- In experiments with a high rate of background events, multi-level triggers are employed, to match at each level the trigger rate, the latency and the local buffering capacity.

- Advances in network technology and processing power of standard commodity PCs enable the move towards pure software triggers.

EXERCISES

1. *Coincidence triggers.*

   An experiment uses N_{PMT} photomultiplier tubes (PMTs) for a trigger. The dark rate in each PMT is R. The trigger is defined by the requirement that there should be at least N_{hits} within a time interval of ΔT. Explaining any assumptions you make, what is the rate of triggers from coincidences of the background?

2. *Trigger purity.*

   A secondary beam at an accelerator is produced by collisions of a proton beam with a target. This beam has a fixed momentum and is known to consist of 90% π^+ and 10% K^+. A threshold Cherenkov counter is used to select K^+. The efficiency for the counter to trigger on π^+ (K^+) is 99% (1%). How can the signal from the Cherenkov detector be used to select K^+? What would the purity (fraction of K^+) be for the triggered events?

3. *Muon lifetime measurement.*

   A scintillator setup consisting of three scintillators is used to measure the lifetime of cosmic muons with a TDC. The cosmic muon comes to a stop in S_1 and decays to an electron (and two undetectable neutrinos).

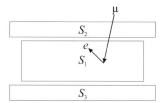

a) Find logical expressions describing coincidences providing a robust start and stop signal for the TDC.

b) If the rate of cosmic muons is about 1 per second, and the hit rate from the PMTs is 1 kHz (mostly due to thermal noise), how long would the duration of the scintillator signals need to be to keep the rate of fake coincidences at an acceptable level?

c) Why is it sensible to give the inverted signal used in the anti-coincidence a longer duration?

4. *Trigger rates.*

a) Derive eq. (13.3).

b) Consider a trigger system using N buffers to store the data while a trigger processor determines whether to keep or reject the event. Let the mean arrival rate of triggers be λ and the mean trigger processing time be μ. The arrival times of the events and the processing times are distributed exponentially. Write down the probability that there was one event in the queue at time $t + dt$ if there was one event in the queue at time t. Hence show that for equilibrium for the case $n \neq 1$

$$\mu P_{n+1} + \lambda P_{n-1} - (\lambda + \mu)P_n = 0,$$

and for $n = 1$

$$\lambda P_0 = \mu P_1.$$

c) Using the results from b) show that $P_n = \rho^N P_0$, where $\rho = \lambda/\mu$.

d) The buffer is full and there is deadtime if $n = N$. Solve for P_N (i.e. the deadtime fraction) by requiring that $\sum_0^N P_n = 1$.

e) For the case $\rho = 0.3$ what is the deadtime fraction with no buffers. How many buffers would be required to reduce the deadtime fraction to be $<5\%$?

14 Detector systems and applications

In this chapter we will pull together all that we have discussed in previous chapters, and see how different detector technologies can work together to combine into powerful experiments that are optimised for physics measurements, but can also suppress backgrounds efficiently. We will also discuss some of the challenges the environment of the experiments and the integration of the different systems impose.

14.1 COLLIDER DETECTORS

Modern colliders and their associated experiments constitute a large investment, financially and in time and effort. To maximise the scientific output, general purpose collider experiments are designed for a broad range of tasks, which also gives the capability for internal cross-checks and complementary and mutually supporting measurements. These detector systems are designed to have good resolution for the kinematic properties of electrons, photons, muons and jets, as well as missing transverse energy. In addition, there is a need to identify jets originating from b or c hadrons (see section 12.12) and τ (see section 12.9), and detached vertices from the decay of short-lived particles.

More specialised collider detectors target more specific physics, e.g. LHCb which is optimised for heavy flavour physics (b and c-hadrons).

Considering the different amounts of material in different detector technologies, the effect of upstream material on their measurement, the different typical channel density and their cost per unit area, there is an obvious sequence of detector technologies a particle will be presented with as it makes its way from the interaction point (IP) through the collider experiment.

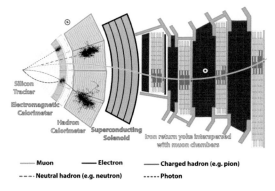

Figure 14.1: A slice of the CMS detector, showing schematically the response to different particles [115].

The innermost detector layers typically have the task to reconstruct primary and secondary vertices. As some short-lived particles, for example mesons containing c or b quarks, or taus have short proper lifetimes ($c\tau < 1$ mm), these layers should be as close as possible to the primary interaction. However, the radius of these vertex detector layers is normally limited by the beam pipe, the track-density resulting in a

The LHCb detector is an exception in that the VELO detector can be inserted inside the beam pipe

DOI: 10.1201/9781003287674-14

high hit-occupancy and the high background particle fluences close to the beam, which cause radiation damage in the vertex detectors. To achieve the goal of good vertex reconstruction in very high track densities these detectors must be highly segmented, making semiconductor pixel detectors the obvious choice for these detectors. The detectors should also contain very little scattering material. Radiation-hard techniques are a must for hadron colliders, for example 3D sensors.

The beam pipe might also require a thermal insulation layer to allow for bake-out of the beam pipe to remove surface contaminants that could degrade the beam pipe vacuum. The degradation of the vacuum occurs when these surface contaminants are hit by stray beam particles.

The vertex layer is then surrounded by further tracking layers. The combined tracking detector measures the momentum of the charged particles in a magnetic field. The standard magnet configuration for tracking detectors is a solenoid, as it gives good uniform p_T resolution. To achieve a good momentum resolution, a high magnetic field and a long lever arm are required. In hadron collider experiments, where event rates and track densities are high, fast and accurate position detectors are required, making currently silicon strip detectors the most capable detection technology. However, progress in MAPS detectors and other semiconductor technologies will bring pixelated readout into the tracker in future experiments. In experiments where fast response is not a concern, like in heavy-ion or lepton colliders, gaseous TPCs are used, as they provide a large number of space points, often with dE/dx information, with minimal scattering material in the tracking volume, albeit with a poorer position resolution per point.

LHCb [332] differs from the layout of a standard collider experiment, as it is optimised for the detection of particles in the forward direction and thus it does not have a solenoidal spectrometer, but a dipole.

There is a trade-off in cost and performance which will determine at which radius there is a switch between pixel and strip detectors.

Depending on the physics priorities of the experiment, either as part of the tracking detector or as dedicated layers just after the tracker, particle identification detectors (see chapter 12) can be employed.

Outside of the tracker, the calorimeters are placed, usually divided into an inner electromagnetic part, and an outer hadronic calorimeter. They are the key detector system to detect neutral particles, and they are critical for the determination of the missing transverse energy. To achieve the latter, the calorimeters have to extend up to a high pseudorapidity, typically $|\eta| \sim 5$. In practice it is very difficult to extend the coverage beyond this range because the calorimeters have to be outside the beam pipe.

The differences between electromagnetic and hadronic calorimeters can be in the construction (homogeneous vs sampling), choice of absorber Z and/or segmentation. Particles don't distinguish between them. They start a shower by the interactions they undergo in the material presented. Typically about half the energy of an incident hadron is deposited in the electromagnetic calorimeter.

A key design consideration in every collider experiment is the position of the solenoid coil for the central tracking system. If it is placed before the calorimeter its material will degrade the energy resolution. One way to compensate for this is the use of a pre-shower detector. If an electron interacted in the material upstream of the calorimeter it will result in a large signal in the pre-shower detector which can be used to compensate for the energy lost in the passive material in front of it.

Another motivation for a finely grained pre-shower detector is the identification of single photons in the presence of backgrounds from $\pi^0 \to \gamma\gamma$ which is crucial for the study of the Higgs decay $H \to \gamma\gamma$.

Placement of the calorimeter inside of the solenoid coil increases the size and cost of the magnet. To limit this, the calorimeters have to be very dense. In addition, their readout has to work in the strong magnetic field of the central tracker.

In any case, the field lines of the magnet must be closed. To prevent excessive stray fields, closure of the field lines is usually accomplished by an iron return yoke, which can be used as the absorber of the hadronic calorimeter (as in ATLAS [85]), or for a solenoidal bending field for the muon system (as in CMS [196]).

The muon detection system is located outside the calorimeters. Due to the large size and the moderate track densities even at hadron colliders, gaseous detectors are used. Precision layers are complemented by trigger layers that are faster but have a coarser granularity.

Triggering is a critical issue for every collider detector, but in particular for experiments at hadron colliders (see chapter 13).

SERVICES

The goal for the overall layout of a typical 4π collider detector is a compact integration of all the subsystems while maintaining full hermeticity. One key consideration in this are the services that are required to operate the detectors, as they constitute significant dead material, and the services for the inner layers need to penetrate the outer layers.

The aim is to have a solid angle coverage as close to 4π as possible. Therefore, general purpose collider detectors are sometimes referred to as 4π detectors.

Typical services in a collider detector are:

- *High voltages* for drift and amplification fields, bias voltages, etc. The voltages typically range from about 100 V for semiconductor detectors to a few kV for wire chambers, with currents that are typically low ($<\mu A$). It is typically sufficient to segment the supply for larger sections of the detector, with internal distribution, and the typical number of cables for a system is between 10^2 and a few 10^3.

- *Electrical power* for front end-electronics, with a typical supply voltage between 1 and 5 V DC. In particular if the front-end electronics has to provide digitisation and buffering, supply currents can be very high (several kA), requiring large conductors. For high power inner systems like semiconductor trackers improved powering techniques are thus developed, which provide the power with reduced current by increasing the voltage in the supply cables. Two strategies are pursued: Serial powering or DC-DC conversion. Because of the high power in each channel and the need to decouple individual modules, typically high granularity of the supply is required (10^3 to 10^4 cables per system).

In serial powering (see for example [265]) loads are connected in series like the bulbs in a Christmas tree lighting system. In DC-DC conversion systems (see for example [26]) the voltage is reduced by storing the input energy temporarily in a coil or capacitor, and then switching to release that energy to the output at a different voltage (see exercise 3).

- *Gases* for gaseous detectors or environmental gas supply. These are supplied through tubes and hoses, the size of which limits the number to 10 to a few 100 per system, typically with further distribution inside the detector.

- *Cooling.* There are two main needs for cooling in a detector. First, all the heat generated by detectors and front-end electronics needs to be removed from the detectors. This can be a substantial amount (several 10 kW for typical semiconductor systems at LHC). For high-power systems this requires liquid or even evaporative cooling (see section 9.6), whereas in low power detectors, for example at lepton colliders, gas cooling might be sufficient.

In addition to heat removal capacity (power) there might also be requirements for the temperature of the coolant (for example to maintain thermal stability for radiation-damaged semiconductors).

  The second need for cooling is the supply of liquids to cryogenic detectors and superconducting magnets. These are typically cryogenic liquids like LN_2 or LAr.

  Cooling lines typically need to be thermally insulated to prevent condensation or formation of ice. Their size limits their number to typically 10 to a few 100 per system, with further distribution inside the detector.

- *Detector control system* (DCS), sometimes also called 'slow control') for monitoring of detector system parameters (T, V, I, etc.) and control (setting parameters in the front-end electronics). These tasks are typically (but not necessarily) slow and the signals are low voltage/low current. Often these systems are multiplexed, so that a few 10 to 100 lines are sufficient. For safety-critical tasks interlock lines (monitoring and control) are needed. For high reliability these might require individual hardware lines, which can result in significant numbers (several 1000) in large systems.

About DCS for LHC experiments, see for example [448].

- *Data readout.* Highly segmented detector systems like pixel detectors can have up to several 10^9 channels. For such large channel numbers on-detector data reduction is required (zero-suppression, multiplexing, etc.). Electrical lines have limited bandwidth and relatively high mass, and thus optical links are often used at speeds of several hundred Mb/s to a few Gb/s. As the power of data lines is low, electrical read-out lines can be small, but many are needed (10^3 to 10^5). Optical lines can be fewer after multiplexing, but they are more delicate, and need proper protection. The opto-electronic conversion introduces additional complexity to the readout system.

For a review of opto-electronics in the LHC experiments see [476].

Opto-electronics components are generally not sufficiently radiation-hard for the use at the inner radii of LHC experiments.

RADIATION ENVIRONMENT AND ACTIVATION

In particular at hadron colliders the radiation environment is a major design driver. The radiation fields in the LHC experiments are dominantly produced by beam collisions, with contributions from beam-gas interactions and other machine losses. They are strongest around the beam pipe, in particular close to magnets and collimators, but there is also a more uniform component due to neutrons.

The neutrons arise predominantly from spallation in the calorimeters. The energy of the neutrons can be reduced by elastic scattering to an energy below threshold for causing radiation damage. In practise this is achieved with polyethylene moderators.

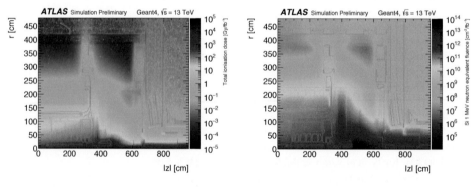

Figure 14.2: Total ionising dose in Gy(Si) (left) and 1 MeV neutron equivalent fluence in silicon (right) from GEANT4 simulations of the ATLAS detector, shown for a quadrant of the detector (beampipe along the horizontal axis) [101]. Note the increased radiation levels close to the beam pipe, leakage along the service gap between the barrel and the endcaps (at $z \simeq 350$ cm), and increased hadron fluences due to the showers in the dense material of the endcap calorimeter (370 cm $\lesssim z \lesssim 600$ cm).

Effects of radiation backgrounds are

- *Increased particle rates in the detector.* This leads to increased occupancy and deadtime. The former creates challenges for pattern recognition and separation of signals, while the latter limits data taking rates.

- *Radiation damage to detectors*. In semiconductor detectors this leads to bulk damage and consequently increased leakage currents, to type inversion and charge trapping which leads to a loss of signal (see section 9.11). In wire chambers this leads to creation of polymers, and thus increased leakage current and loss of gain (see section 7.7). In scintillation detectors it leads to loss of transparency and thus signal (see section 5.6).

- *Damage to front-end electronics*. This can be minimised using modern deep sub-micron technologies and appropriate design techniques.

- *Radiation damage to materials*. The very high radiation levels at hadron colliders can also also cause very significant effects on passive material. For example all plastics will deteriorate with radiation damage and it is essential to select an appropriate plastic for a given application. Similarly some adhesives are much more radiation tolerant than others. Another example is the increase in attenuation in optical fibres due to ionising radiation. Custom radiation-tolerant fibres have been developed and are used in very high radiation environments.

  Data on radiation hardness of different materials has been compiled in a series of CERN technical reports [471, 470, 134, 340].

  Commercial products are not designed to be radiation tolerant. Companies can change processes in subtle ways which affect the radiation tolerance. It is thus important to perform tests to verify radiation hardness, also during production on batch samples.

- *Single event effects (SEEs)* in electronics (see section 3.9).

- *Activation of detector components*. This makes personnel access for maintenance or replacement of parts progressively more challenging with time. This, together with the compact designs of detectors and services, and the need for environmental barriers, can make access to inner parts of detectors at hadron colliders prohibitively difficult. Highly reliable technologies and redundancy are therefore a major design requirement.

  Heavy metals can undergo radioactive transmutation to radioactive isotopes. Therefore the use of metals like gold and silver should be minimised.

Dealing with the consequences of the radiation environment often requires compromises in the detector design. Sometimes this involves a direct trade-off between the performance of the detector and its successful long term operation in the radiation environment. The demands of designing a detector that can sustain radiation damage can drive requirements for the environment, like for cooling of silicon detectors, and can influence the choice of detector technology and segmentation.

For example limiting ageing effects in wire chambers will require reducing the gas gain.

Due to the high importance of the radiation effects in collider detectors at the LHC, reliable predictions of the radiation fields have been an essential prerequisite for the design of these detectors. As the dominant source of the radiation is from beam-beam interactions, the starting point of these predictions are Monte Carlo event generators such as PYTHIA8 [458, 459] and DPMJET-III [430]. The particles originating from the proton–proton collisions interact with the material in the detector and the accelerator (collimators, focusing magnets, etc.), causing electromagnetic and hadronic showers which give rise to the complex radiation fields. For the prediction of these, Monte Carlo transport software like GEANT4 [29, 44, 45] or FLUKA [120, 151] are being used.

For a detailed review of radiation environments, their prediction, and the effects of radiation for LHC detectors see [219].

14.2   NEUTRINO EXPERIMENTS

For experiments in neutrino physics there are very different types of detectors, depending on the energy of the neutrinos to be studied. In general, the weak interaction neutrino cross-sections are very small, so large detectors are required. For experiments studying neutrino oscillations over very long lengths it is necessary to use a 'near' detector (i.e. close to the production target) to measure the flux and a 'far' detector to measure neutrinos after they have oscillated. The rates in the near detector will be large but the rates in the far detector will be very low, so in general very different detector types will be used for near and far detectors. The far detector must be capable of identifying the neutrino flavour to be sensitive to neutrino oscillations.

For low energy neutrinos, there will be significant backgrounds from radioactivity and cosmic rays, and therefore similar precautions will be required to those used in the experiments searching for rare events (see section 14.3).

The detection of low energy (MeV) anti-neutrinos uses inverse β decay, $\bar{\nu}_e p \to e^+ n$. The positron will annihilate with an electron in the detector, $e^+ e^- \to 2\gamma$, and the γs can be detected in a scintillator. The neutron will be captured by a nucleus leading to an excited state that decays radiatively to the ground state. This process occurs on a longer time scale (in the order of μs), and therefore the signal for an $\bar{\nu}_e$ interaction is a prompt signal, followed by a delayed signal.

Requiring the capture signal can be used to give a very big reduction in the radioactive backgrounds and this technique was used in the discovery of the $\bar{\nu}_e$, where the neutron was captured by cadmium nuclei in the scintillator solution of the detector [421].

A modern example of this approach is provided by the Daya Bay experiment [226]. To capture the neutrons, gadolinium is added to the liquid scintillator. This has two advantages:

This experiment measured the $\bar{\nu}_e$ fluxes at different distances from a nuclear reactor, in order to study neutrino oscillations.

- the large cross-section for neutron capture by Gd reduces the timescale for this process and thus reduces the backgrounds from random coincidences;

- the radiative decay of the excited Gd nuclei results in an 8 MeV γ signal, which is a higher energy than that produced by radioactive backgrounds.

The Gd-liquid scintillator is contained in a transparent acrylic vessel to allow the scintillation light to be viewed by photomultipliers. The liquid scintillator must be chemically compatible with acrylic and stable over a period of years. Linear alkyl benzene (LAB) was chosen as the scintillator because of its good scintillation yield and compatibility with acrylic.

Acrylic has lower radioactivity than glass.

Earlier experiments had problems with the scintillator turning yellow. The resulting attenuation length decrease rendered the detectors inoperable.

The chemical structure of LAB is $C_6H_5C_nH_{2n+1}$, with n in the range from 10 to 13.

WATER CHERENKOV NEUTRINO DETECTORS

A different approach to the detection of neutrinos is to use very large water Cherenkov detectors. These are very large vessels filled with pure water and viewed with an array of photomultipliers. The experiments are typically located in an underground cavern to provide shielding against cosmic rays.

The Super-Kamiokande (SK) experiment [254] uses 50,000 tons of ultra-pure water in a cylindrical steel vessel. The Cherenkov light from

An even larger detector using these principles (HyperK) is planned for the study of CP violation in neutrino oscillations [344].

relativistic particles is detected by an array of PMTs. As discussed in section 12.4, the pattern of hits in this array is given by the intersection of the Cherenkov cone with the detector array.

The volume of the SK experiment is divided into an on Outer Detector (OD) and an Inner Detector (ID). The ID and OD are optically separated by a stainless steel framework. The ID is the actual fiducial volume of the experiment. It is instrumented with 11,146 hemispherical PMTs with 50 cm diameter. This provides an effective area coverage with photocathode surfaces of 40%. The purpose of the OD is to detect and veto incoming charged particles. It is instrumented with a smaller number of PMTs. As there is a lower area coverage of the PMTs in the OD two techniques are used to increase the light collection efficiency. Firstly, wavelength shifter (WLS) plates are used to shift the UV light to the blue/green region, for which the quantum efficiency of the PMTs is higher. Secondly, a reflective liner is used on the outer walls, which spreads the light over several PMTs. This degrades pattern recognition, but the increased light yield improves the energy resolution.

In 2002, during filling of the tank with water, one PMT imploded, which then caused a shock wave that damaged further PMTs and created a cascade which resulted in over half the PMTs being destroyed. These PMTs were replaced and protection was introduced to prevent such an incident in the future.

As the OD is used as a veto detector, this is an overall performance improvement.

In order to minimise radioactive backgrounds and to ensure a very long attenuation length for the Cherenkov light, the water is purified in multiple purification systems. For calibration a variety of systems are used. These include a system with a laser and CCDs to monitor the transparency of the water. A xenon light is used to measure the relative gains of the PMTs. A nickel source is used to produce 9 MeV photons from thermal neutron capture on nickel using the reaction ^{58}Ni$(n, \gamma)^{59}$Ni [11]. The activity of the source is such that the probability of detecting two photons is less than 1%. The resulting spectrum shows the clear peak from single photoelectrons. Finally, a low energy linear electron accelerator is used to inject electrons of precisely known energy.

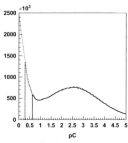

Figure 14.3: *Charge signal spectrum (single photoelectron peak at about 2.5 pC) measured with the Super-Kamiokande PMTs using the ^{58}Ni γ source [11].*

In addition to the array of hardware calibration systems, in-situ calibration procedures were used based on through-going muons, stopping muons and the resulting decay electrons ($\mu^- \to e^- \nu_\mu \bar{\nu}_e$). The energy resolution for events with a single muon ring was around 2% for events with a visible energy around 1 GeV [294]. The energy resolution for low energy electron neutrinos is [10]

$$\sigma_E/E = -0.123/(E/\text{MeV}) + 0.376/\sqrt{E/\text{MeV}} + 0.0349.$$

The first term which represents the noise has an unphysical negative value but this is an artefact of the fit.

TRACKING CALORIMETERS

For higher energy (multi-GeV) neutrinos produced from accelerators, detectors with a large amount of material are required to obtain a significant signal rate. Therefore a typical approach has been to use an iron/scintillator calorimeter. In such a detector, electrons from $\nu_e X \to e^- Y$ can be identified from the electromagnetic showers in the calorimeter. The muons from $\nu_\mu X \to \mu^- Y$ can be identified by penetrating tracks that do not shower. In addition, if the iron is magnetised, the momentum of the muons can be reconstructed from the curvature of the tracks.

We will consider the MINOS experiment [371] as a modern example of this approach. The MINOS detector used alternating planes of scintillating strips and 2.54 cm of steel. The scintillating strips were 4.1 cm wide, 1 cm deep and up to 8 m in length. In order to minimise the effect

This is required for long baseline neutrino oscillation experiments.

For these type of experiments, the calorimeter serves as the target and the detector.

If the muon stops inside the detector, the energy can be more precisely determined from the range.

See also section 12.11.

of attenuation for these very long strips, the scintillating strips were read out by wavelength shifting (WLS) fibres. The steel was magnetised by toroidal coils, providing an average magnetic field of 1.42 T.

The resolution for the muon momentum measurements was dominated by multiple scattering, and was $\Delta p/p = 12\%$ for the typical muon momenta in the experiment. In addition, for low energy muons that stopped in the detector the momentum could be determined from the range of the track.

The WLS fibres absorb light around 420 nm and re-emit light at 470 nm, which greatly increases the attenuation length. Outside the active detector region the light is coupled to clear fibres as these have an even longer attenuation length.

The resolution for electromagnetic showers was limited by the thickness of the steel plates. Improved resolution could have been achieved with thinner plates, but for a fixed detector budget this would have resulted in a lower target mass and hence a lower number of signal events. The calorimeter energy resolution for electron showers was $\sigma_E/E = 24\%/\sqrt{E/\text{GeV}} \oplus 4\%/(E/\text{GeV})$ and for hadron shower measurements $\sigma_E/E = 56\%/\sqrt{E/\text{GeV}} \oplus 2\%$. The total mass for the 'far'-detector was 5400 tons.

A similar but smaller detector was located near to the accelerator neutrino source.

Several calibration systems were required to achieve this performance. These included a UV LED system to inject light into the WLS fibres. This was used to monitor the uniformity and stability of the PMTs and associated electronics. In addition to the electronic and optical calibration systems it was essential to make an accurate calibration of the energy scale for electromagnetic and hadronic showers. This was done using a smaller calorimeter made with the same steel and scintillator technology. This calorimeter was exposed to beams of low energy electrons, muons and pions to determine the absolute energy scale. The cross-calibration with the MINOS detectors was performed using signals from cosmic ray muons.

The exposure was performed at a CERN PS test beam.

A new approach to large mass neutrino experiments is using a liquid argon TPC as described for the DUNE experiment in section 8.1. This provides very good tracking as well as calorimetric information.

14.3 PARTICLE DETECTORS FOR RARE EVENTS

Examples for experiments targeting rare events are searches for dark matter or neutrinoless double β decay ($0\nu\beta\beta$). The general requirements for these type of experiments are:

Conventional particle detector searches for dark matter generally target WIMP dark matter. Weakly Interacting Massive Particles (WIMPs) are a candidate for the dark matter in the Universe. They would interact with ordinary matter with a strength similar to the weak interaction and would have masses of $\mathcal{O}(100~\text{GeV})$. Very different technologies are required to search for other possible types of dark matter.

- A large active mass and a long running time;

- High efficiency for detection of the signal event;

- Backgrounds must be low enough so that they do not obscure the search signal.

Sometimes it is possible to perform useful searches even if the background rates are larger than the expected signal rates, if the background rates can be accurately calculated. However, in this case the sensitivity will be limited by Poisson-statistical fluctuations in the number of background events. Thus the sensitivity will be proportional to $\sqrt{N_{\text{target}} \times t}$, where N_{target} is the number of target atoms and t is the duration of the experiment. On the other hand, if the background rates can be significantly lower than the signal, the sensitivity will be proportional to $N_{\text{target}} \times t$;

The first way backgrounds can be reduced is by placing the detectors in deep underground laboratories, which provides shielding against cosmic rays. Even with this shielding there will be a significant flux of muons. If the active detector is surrounded by a muon detection layer, then candidate events caused by incoming muons can be vetoed. The simplest method is to use large area scintillators coupled to photomultiplier tubes to detect the muons. Alternatively, if the active detector can provide 3D spatial information, then events near the outer surface of the detector can be vetoed.

In addition, great care must be taken to minimise the radioactivity of all materials used in the construction of the detector, including the cables. Other techniques for reducing backgrounds are more specific to the particular combination of the event topology of the search and the employed detector technology,

These experiments are extremely challenging in very different ways to collider and neutrino experiments and a very diverse range of approaches are being developed. A few examples will be considered here.

An example for a rare decay search is the Majorana experiment [52]. The characteristic signature searched for is a double-β decay without the emission of neutrinos, $^A_Z X \to \, _{Z+2}^{A} Y \beta \beta$. The standard model background is conventional double-β decay $^A_Z X \to \, _{Z+2}^{A} Y \beta \beta \, \bar{\nu}_e \bar{\nu}_e$. The Majorana experiment uses Germanium enriched in the isotope ^{76}Ge which undergoes double-β decay with a half life of $T_{1/2} = 1.78 \times 10^{21}$ years.

This experiment is designed to search for Majorana neutrinos (a neutrino that is identical to its anti-neutrino).

The detector must contain an appropriate isotope for which double-β decay is energetically possible.

Apart from eliminating cosmic rays and radioactive backgrounds, we need to separate the signal from conventional double-β decay. The combined energy of the $\beta\beta$ will be equal to the Q-value of the reaction for the signal, whereas it will show a broad distribution between 0 and the Q-value for the background. Therefore very powerful background reduction can be achieved if sufficiently good energy resolution can be obtained.

The Majorana detector is optimised for good energy resolution, which can be achieved by semiconductor detectors. The energy resolution at low energy of a Ge detector is limited by the electronic noise, therefore we want a large detector with a minimum capacitance (see section 3.4 for the discussion of how electronic noise varies with capacitance). This is achieved with a geometry called a point contact (PC) detector.

The point contact geometry has been originally proposed for n^+ contacts [346]. If we approximate the contact as a hemisphere of radius r then the capacitance is given by $C \simeq 2\pi \varepsilon_r \varepsilon_0 r$. For germanium and with a realistic value of r of 1 mm, this gives a capacitance of about 1 pF, which is much smaller than in conventional Ge gamma spectrometers.

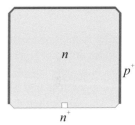

Figure 14.4: Schematic cross-section of a cylindrical point contact detector showing the electrical contacts.

If there is a uniform impurity concentration, then as the bias voltage is increased and the detector is depleted, there would be a radial electric field, such that electrons would drift towards the axis and not towards the n^+ electrode. The weighting field will only be large near this electrode, so only a small signal will be induced. We need a way to alter the electric field to give it an axial component. This can be achieved with an axial impurity gradient of the doping.

This can be achieved as a result of the segregation of impurities that tends to occur during crystal growth.

In the Majorana detector central p^+ contacts are used instead of n^+, for two reasons. First, the $\sim$0.5 mm thick n^+ layer generated by drifting lithium into the lattice covers most of the outer surface of the detector, resulting in lower sensitivity to very low energy minimum-ionising backgrounds (surface betas, x-rays, etc.). Second, with central n^+ contacts this geometry suffers from trapping of electrons as they travel over the long distances through the crystal to the n^+ electrode in that geometry, severely degrading the energy resolution [346]. p-type detectors suffer less from this effect.

The detected signal is corrected for the differential non-linearity. The energy scale is determined from the γ peaks from radioactive decays. The FWHM in a reduced size demonstrator for the Majorana experiment was 2.53 ± 0.08 keV at an energy of 2039 keV [52].

The demonstrator consisted of 58 high purity Ge detectors with 14.4 kg of natural Ge detectors and 29.7 kg of detectors enriched to $88.1 \pm 0.7\%$ ^{76}Ge.

Apart from the excellent energy resolution, detectors with this geometry can identify backgrounds with multiple interactions (for example multiple Compton electrons from γ backgrounds). As nearly all the signal will be generated near the p^+ electrode, where the weighting field is large, the signals from spatially separated electrons will result in several pulses, as opposed to the signal which should be a single pulse.

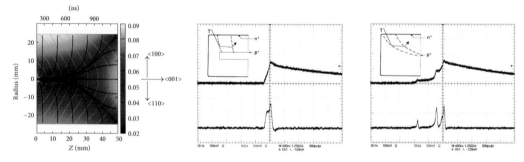

Figure 14.5: p-type point contact (PPC) detector. Left: Electric field intensity (color pattern), drift trajectories (black lines), and drift time isochrones (grey) in the PPCs for the Majorana demonstrator [12]. Middle and right: Effect of electrode geometry on the pulse shape measured with an oscilloscope for a multiple-site gamma interaction [111]. Middle: conventional geometry. Right: PPC detector. The top trace in each display corresponds to the preamplifier output, the bottom trace is the output of a timing filter amplifier (TFA) set at 10 ns differentiation and 10 ns integration time (essentially a bandpass filter).

In order to minimise external backgrounds the Majorana experiment also includes different layers of passive shielding. The innermost shielding layer uses ultra-pure copper. Additional shielding is provided by further layers of copper, lead and boroated polyethylene (used to moderate the neutrons, i.e. reduce their energy by elastic scattering on hydrogen atoms, and to absorb neutrons). The detector is surrounded by an active muon veto system.

The copper was produced underground to minimise cosmogenic activation.

The isotope ^{10}B, which has a natural abundance of 20%, has a large cross-section for neutron capture for low energy neutrons.

COMBINED READOUT

The signature of an interaction of one dark matter candidate, a weakly interacting particle (WIMP), with ordinary nuclei is the recoil of the nucleus. Typical background events in these experiments are caused by electrons and γs. Signal and background can be distinguished in a sub-K calorimeter if the calorimeter measurement is complemented by another measurement (scintillation or ionisation), because the yield for the complementary signal is significantly different for the same amount of heat deposited. Typically, less than 40% of the ionisation (in semiconductor detectors) and less than 10% of the scintillation light (in scintillating crystals) produced by electrons or photons is observed for nuclear recoils with the same heat deposit.

The energy resolution from these signals is considerably worse than for the heat measurement, as the energy required to produce an electron/hole pair or a scintillation photon is larger. For the same reason, the resolution of the measurement of ionisation charge is better than for scintillation.

The EDELWEISS experiment [68] uses 36 bolometers made of germanium, each with a mass between 820 and 890 g, operating at 18 mK. At this temperature the nuclear recoils cause an increase of temperature of roughly 10^{-7} K per keV. The temperature of each bolometer is measured with two NTD germanium sensors. Ionisation charge is read out by 200 nm thick Al electrodes that are deposited on the surface of the germanium crystal in the form of annular concentric rings 150 μm wide with a 2 mm pitch. The germanium crystals are biased by a small potential difference applied to electrodes at opposite faces of the crystals. The signal is measured by the integrated charge induced by the drifting electron-ion pairs. Consecutive electrode rings are operated at different bias voltages to separate the volume close to the surface of the crystal, for which the ionisation signal can be reduced by surface effects like charge trapping, from the bulk volume.

The lower Z value for silicon and germanium improves the sensitivity to low mass WIMPs compared to the use of xenon in the LZ detector (see section 8.2).

For a discussion of NTDs see section 11.2.

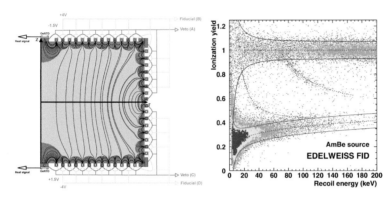

Figure 14.6: EDELWEISS III detector [68]. Left: Cross-section through a detector. Shown are the electrical connections with interleaved electrodes at different potential, and the drift lines of the electrons in the surface (dark) and bulk (light) regions. Right: Ionisation yield versus recoil energy for a neutron calibration using an AmBe source. The two red (blue) solid lines delimit the 90% C.L. nuclear (electron) recoil band. Purple dashed lines correspond to inelastic scattering of neutrons on the first (13.28 keV) or the third (68.75 keV) excited state of ^{73}Ge.

As the electrons and holes drift in the electric field they acquire energy from the field, which is ultimately converted into phonons. Therefore the measured phonon energy E_p is larger than the actual nuclear recoil energy E_r that we want to measure,

$$E_p = E_r + V_{bias}Q, \qquad (14.1)$$

where V_{bias} is the voltage difference of the electrodes on the two sides of the detectors and Q is the ionisation charge. For a large V_{bias}, the additional phonon signal will obscure the phonon energy from the nuclear recoil, and we will lose sensitivity.

An example for a combined heat/scintillation readout calorimeter is CRESST III [8]. Here the detector consists of $CaWO_4$ crystals that have a size of $20 \times 20 \times 10$ mm^3 and a mass of about 24 g each. The crystal is held by 'sticks' made of the same material. The temperature readout is provided by Transition Edge Sensors (TES). The CRESST-III detector used a thin tungsten film (thickness 200 nm), operated at around 15 mK. The very small signal can be read out by very sensitive Superconducting Quantum Interference Detectors (SQUID).

The scintillation light is detected by a bolometer, made of a 4 mm thick square silicon-on-sapphire wafer of 20 mm edge length, also held by $CaWO_4$ sticks and equipped with a TES.

In practice this means that V_{bias} should be as low as a few Volts. As the detector is operated at temperatures below 50 mK the intrinsic carrier concentration is extremely low, so the leakage currents in the silicon (germanium) crystals are sufficiently low that there is no need to use the strong reverse bias normally used in semiconductor detectors.

See also section 11.2.

The advantage of a measurement of the scintillation light is that the photon detector does not need to physically connect to the calorimeter, and thus does not influence its thermal properties.

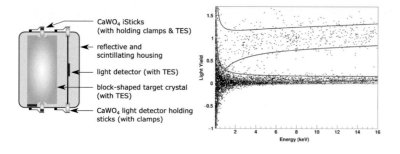

*Figure 14.7: CRESST III detector [8].
Left: Schematic of a detector module
(not to scale). Parts in blue are made
of CaWO$_4$, the TESs are sketched in
red. Right: Neutron calibration data in
the light yield versus energy plane. The
bands are fits for β/γ-events (blue),
nuclear recoils off oxygen (red), and
tungsten (green), where the respective
lines correspond to the upper and lower
90% boundaries of the respective band.*

14.4 PARTICLE DETECTORS IN SPACE

Particle detectors in space-borne experiments use the same detector technologies as the earth-bound experiments we have discussed so far. However, there are different integration and operational challenges:

- Detectors must withstand the stresses during launch;

- The power available to space-borne experiments is limited and the heat generated in the detector must be removed by a cooling system and radiated to space (which is at 2.7 K). Temperatures can vary significantly, depending on the amount of infrared radiation the experiment is exposed to (for example facing or opposing the earth).

- The experiment must not produce any stray magnetic field outside of the experimental volume;

- The data rate from the detector must be compatible with the bandwidth of the satellite's downlink;

- Access for repairs and maintenance is impossible, or at least extremely restricted. This also usually excludes replenishment of consumables and volatile components (gases or liquids).

- Exposure to neutral particles: In low earth orbits (LEO, between 200 and 2000 km height) atomic oxygen can be a major source of erosion;

- Exposure to charged particles: At about 400 km (orbit of the ISS), about 1% of the atmosphere is ionised. This fraction increases to 100% at geosynchronous altitudes (37,000 km). This plasma is strongly influenced by the earth magnetic field and local densities vary significantly, also with height. The plasma can charge up elements of the spacecraft.

  The Van Allen belts are regions where energetic ions and electrons experience long-term magnetic trapping. The energy of these trapped particles is greater than 30 keV and can reach hundreds of MeV. The intensity of the trapped radiation flux can reach up to $10^8 - 10^9$ cm^{-2}s^{-1} at a distance of about 2 R_E for electrons with $E_k > 0.5$ MeV and at about 3 R_E for $E_k > 0.1$ MeV protons [119]. R_E *is the earth radius (about 6371 km).*

  The total dose collected during a space flight is relatively small, typically in the order of 10 Gy(Si). The long term damage from radiation will thus be negligible. However, there can be Single Event

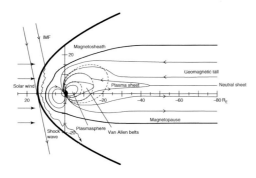

Figure 14.8: Earth's magnetosphere in the noon–midnight plane [400]. The dashed lines are the original dipole field. The solid lines are magnetic fields modified by external currents. IMF stands for interplanetary magnetic field, which is the component of the solar magnetic field that is dragged out from the solar corona by the solar wind flow to fill the solar system.

Effects (SEE) caused by single, heavily ionising particles depositing large energy in the electronics (see section 3.9).

SPACE-BORNE DETECTORS FOR CHARGED PARTICLES

A class of space-borne particle physics experiments measures composition, rates and energy spectra of the charged particle flux around earth, comprising a large range of ions. The central component of such an experiment is a magnetic spectrometer, typically instrumented with silicon detectors. An example for such an experiment is AMS-02 [30], which is part of the International Space Station (ISS).

AMS-02 consists of several detector systems. The central part is a magnetic spectrometer. To allow for an extended operation (>3 y) of the experiment, without the need to refill cryogenic liquids, a permanent magnet is used. It consists of 64 sectors, each made of 100 high-grade NdFeB blocks, assembled in a cylindrical shell 800 mm high with an inner diameter of 1115 mm. The segments are oriented in a 'magic ring' configuration to produce a dipole field of 0.14 T perpendicular to the axis at the centre of the magnet, while minimising any stray field (3–4×10^{-4} T anywhere at a distance larger than 2 m from the centre). The total weight of the magnet including the support structure is 2.2 t.

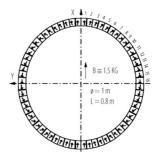

Figure 14.9: 'Magic ring' configuration [119]. In this configuration the fields from the magnetic elements in the ring add up to a vertical dipole field across the ring with little stray field outside.

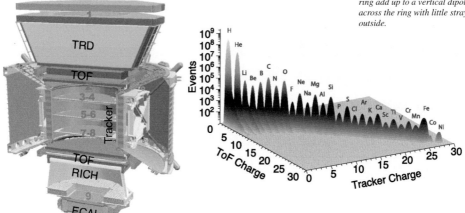

Figure 14.10: The AMS-02 experiment. Left: Schematics of the experiment [462]. Right: Charges reconstructed in the tracker and the TOF system [30].

Particle tracks are measured by nine layers of silicon detectors, which, after combination with the magnetic field map, yields the track rigidity $R = pc/Q$, where p is the momentum and Q the charge of the incoming particle. The silicon detectors do have analogue readout to allow for a measurement of the charge of the incoming particle (as the radius of the track in the spectrometer scales with Q^{-1}), but to provide the dynamic range to accommodate the large range of energy deposition for different ions, the pre-amplifier has been given a highly non-linear characteristics. The position is reconstructed from the distribution of charges in the hit strips. The position resolution depends on the inclination angle and, due to the non-linear response, on the charge of the incoming particle [55].

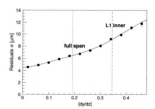

Figure 14.11: Position resolution in the AMS-02 silicon tracker as a function of the track inclination [55].

A Transition Radiation Detector (TRD) is located at the top of AMS-02. Its main purpose is to identify electrons and positrons by transition radiation while rejecting protons at a level of 10^{-3}. The TRD also provides an independent tracking capability, and determination of the charge value of the nuclei by measuring the rate of energy loss (dE/dx). It is made of 6 mm straw tubes with a Xe/CO_2 gas mixture, interleaved with a polyethylene/polypropylene fibre fleece radiator. The leak rate of the detector gas is sufficiently low to achieve >30 y of operation with the on-board gas supply (~55 kg).

Time-of-flight counters (TOF) consisting of two planes of scintillation counters with PMT readout above and two planes below the magnet provide a charged particle trigger to AMS, determine the direction and velocity of incoming particles, and measure the charge via dE/dx. The time resolution of each counter is 160 ps for $Z = 1$ nuclei. The timing resolution improves with increasing magnitude of the charge to a limit of 50 ps for $Z > 6$ nuclei. Anti-coincidence counter (ACC) surrounding the experiments veto events that could be contaminated by sideways going particles.

Below the spectrometer is a proximity-focusing RICH, with NaF crystal ($n = 1.33$) and aerogel ($n = 1.05$) radiators, an expansion volume with a depth of 470 mm, and segmented PMTs for photon detection.

The electromagnetic calorimeter [21] is made of 1 mm thick grooved lead foils interleaved with scintillating fibres with a lead/fibre/glue volume ratio of 1/0.57/0.15, read out with PMTs. The total thickness of the ECAL is 17 X_0. The resolution measured in test beams was $\sigma_E/E = (10.4 \pm 0.2)\%/\sqrt{E/\text{GeV}} \oplus (1.4 \pm 0.1)\%$.

The overall AMS-02 experiment measures $5 \times 4 \times 3$ m^3, and has a weight of 7.5 t.

SPACE-BORNE GAMMA RAY DETECTORS

An example for a space-born gamma ray telescope is the Fermi Gamma-Ray Space Telescope (FGST). It includes two instruments, the Large Area Telescope (LAT) [107], and the Gamma-ray Burst Monitor (GBM) [366]. It occupies an 565 km orbit with an inclination of 25.6 degrees.

The LAT is an imaging gamma-ray detector that detects photons with energies from 20 MeV to 300 GeV, with a field of view of about 20% of the sky. High-energy γ-rays cannot be optically focused. Thus, the LAT is a pair-conversion telescope, which consists of a precision converter-tracker and a calorimeter. The converter-tracker has 16 planes of high-Z

material (tungsten), in which γ-rays can convert to an e^+e^- pair, interleaved with 36 planes of silicon strip detectors, and spaced by carbon-fibre composite structures. The first twelve planes of tungsten are each $2.7\% \, X_0$ (0.095 mm) in thickness, while the final four are each $18\% \, X_0$ (0.72 mm) in thickness. The complete stack of tungsten converts about 63% of γ rays with an energy above 1 GeV (above 1 GeV the pair-conversion cross-section saturates, see section 2.1). The longitudinal separation is chosen to give good angular resolution. For photons with low energies, the hits in the first two layers following the conversion are the most relevant, as multiple scattering of the e^+e^- pair degrades the measurement of their position and direction. Also, the detector layers are located immediately behind the tungsten foils, so that only the multiple scattering in the first absorber layer is relevant. The lateral segmentation into strips with a pitch of 228 μm and the ratio of strip pitch to vertical spacing between tracker planes of 0.0071 has been chosen to achieve good direction reconstruction for high energy photons (>1 GeV).

The FGST-LAT calorimeter consists of bars of CsI crystals, read out by p-i-n diodes. The total vertical depth of the calorimeter is $8.6 \, X_0$ (for a total instrument depth of $10.1 \, X_0$). The depth is a compromise in shower containment against maximum permitted mass. The calorimeter is segmented into eight longitudinal segments and laterally into units with a width corresponding to one Molière radius. The fine segmentation is used to improve the energy measurement at high energies using a shower profile analysis.

An anticoincidence detector (ACD) made of plastic scintillator tiles read out with PMTs surrounds the experiment to reject charged particle backgrounds.

The FGST-LAT dimensions are $1.8 \times 1.8 \times 0.72$ m^3. The power required and the mass are 650 W and 2.8 t, respectively.

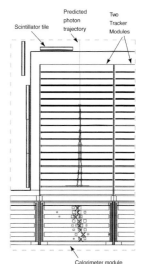

Figure 14.12: Reconstruction of a 470 MeV photon conversion in one of two FGST tracker towers operated together with calorimeter modules in a tagged-photon beam test at CERN [106].

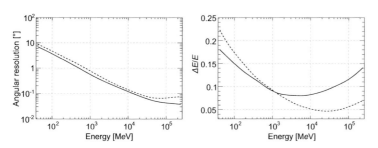

Figure 14.13: FGST-LAT performance as a function of photon energy at normal incidence (solid curve) and for incoming particles at an angle of 60° off-axis (dashed curve) for 68% containment [107]. Left: Angular resolution for conversions in the thin foil section of the tracker. Right: Energy resolution. The energy resolution is worse for very high energy because of shower leakage.

The primary role of the FGST-GBM is to complement the LAT with observations at lower energies ($\sim$8 keV to $\sim$40 MeV). It consists of 14 scintillation detectors (twelve NaI(Tl) crystals for the 8 keV to 1 MeV range and two BGO crystals with sensitivity from $\sim$200 keV to $\sim$40 MeV), and can detect gamma-ray bursts in that energy range across the whole of the sky not occluded by the Earth.

The NaI(Tl) crystal disks have a diameter of 5$''$ and a thickness of 0.5$''$. On the front face, there is a entrance window made of a 0.2 mm thick Be sheet and a 0.7 mm thick silicone layer, which defines the low energy threshold of 8 keV. The NaI(Tl) scintillators are distributed on the surface of the spacecraft pointing in different directions, so that the direction of a gamma ray burst can be determined from the relative count rates.

The combination of LAT and GBM provides the possibility to measure γ rays over seven orders of magnitude in energy.

The two BGO scintillators have a diameter and a length of 5″. They are positioned at opposite sides of the spacecraft so that any burst above the horizon will be visible to at least one of them. The scintillation light from the crystals is read out by PMTs, one each for the NaI(Tl) crystals and two each for the BGO crystals, to improve light collection and to provide redundancy.

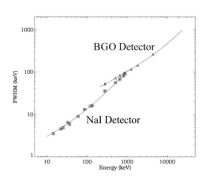

Figure 14.14: FGST-GBM [366]. Left: Mounting of crystals on spacecraft (0-11 NaI(Tl), 12-13 BGO). Right: Energy resolution for the two types of scintillators.

14.5   RELIABILITY

Large detector systems with many channels often need to operate for a long period of time. In cases like the inner detectors of collider experiments access is almost impossible because of the convoluted nested geometry in these experiments and activation of the detector material. Therefore, reliability is a major concern. Reliability is even more critical in space missions in which it is (almost) impossible to carry out any maintenance or repairs. Failures can occur at many levels, electrically or mechanically.

Much of the methodology of reliability engineering was developed by NASA [493].

A key strategy to achieve reliability is rigorous testing of all components during design, prototyping and construction of a detector. In general, at the development level components should be exposed to conditions more extreme than expected in operation to identify failure modes and if possible make changes to improve reliability. Typical stresses for electronics include temperature, and voltage or current. For HEP applications radiation damage is usually an important stress. These tests should also be repeated on a batch basis during production of the final components to check for variations in the production processes.

Performance verification during the design phase is referred to as 'quality assurance' (QA), and during the construction as 'quality control' (QC).

This methodology is called Highly Accelerated Lifetime Tests (HALT).

A complementary methodology, which aims to predict long-term performance of components, is Highly Accelerated Stress Test (HAST) in which the Mean Time To Failure (MTTF) is measured as a function of appropriate stress variables. Phenomenological models are then used to fit the data. The models can then be used to predict the lifetime at the operating conditions. A simple example of this approach are reliability studies of vertical cavity surface emitting lasers (VCSELs) used for data transmission from the detector, for which the MTTF scales with temperature T and current I according to the Arrhenius equation [472]

Testing of final production components at elevated stress levels is referred to as Highly Accelerated Stress Screen (HASS).

$$\frac{\mathrm{MTTF}(I_1, T_1)}{\mathrm{MTTF}(I_2, T_2)} = \left(\frac{I_2}{I_1}\right)^2 \frac{\exp\left(-E_\mathrm{A}/k_\mathrm{B}T_2\right)}{\exp\left(-E_\mathrm{A}/k_\mathrm{B}T_1\right)}, \qquad (14.2)$$

where E_A is the activation energy. The value of E_A can be determined from a fit to accelerated ageing tests at different currents and temperatures. A similar approach is used for other devices, in which accelerated ageing data is used to fit different phenomenological models, thus allowing an extrapolation of MTTF to the operating conditions.

A typical value would be $E_A \approx 1$ eV.

Another strategy to improve reliability is the use of 'redundancy' in which if one component fails then another one can perform the same function. Redundancy should always be used if human safety is involved or if a single point failure would be 'mission critical'. However, it is usually impractical to use redundancy for all channels of a large detector. Additional detector layers (e.g. in a tracking detector) can, in addition to sometimes improving system performance (as long as the additional material has no detrimental effects), also provide some measure of redundancy. Redundancy does not guarantee reliability, for example it does not achieve reliability if components or sub-system lifetimes are shorter than the required operation time of the detector.

Some small failure rates always have to be accepted and considered in the design of the experiment.

Newer Deep Sub-Micron (DSM) chip technologies have provided dramatic improvements in performance but at the cost of decreased reliability. One of the failure modes is due to electromigration [493]. The decrease in transistor feature size results in a decrease in the height and cross-section of the conductors. Therefore, there can be very large electron current densities which can ionise the atoms which then move in the electric field. This can result in voids in the conductor of the tracks connecting the components in the chip. These decrease the effective area of the conductor and can lead to catastrophic thermal failure due to high local current densities.

Other failure modes are Time-Dependent Dielectric Breakdown, Hot Carrier Injection (HCI) and Negative Bias Temperature Instability [493].

In general, the reliability of electronics can be improved by decreasing the operating currents/voltages and running at lower temperature. However, this might conflict with the requirements for channel density, noise performance etc.

In addition, all components and sub-systems need to be tested during and after production to check that they meet the specifications before being integrated into larger structures. Components can also be operated before installation for longer periods of time than required to validate the performance in an attempt to weed out 'infant mortality'.

At low temperatures the carrier mean free paths become longer and therefore a 'hot' carrier (i.e. a carrier with an energy larger than average) can acquire sufficient energy in an electric field to reach the gate oxide of a CMOS transistor. The build-up of charge can result in the transistor failing. This mechanism is called 'hot carrier injection' (HCI) [493]. HCI is more severe at lower temperatures and is thus a major issue for electronics in cryogenic detectors.

Other failure modes of electronic systems include cracks in thin conductors and connector failures. The desire to minimise material results in thinner, more fragile conductors and very dense small form factor connectors, which can be less robust.

A typical mechanical failure would be failure of a glue joint, or failure of a part due to unexpected loads, either due to the magnitude or the frequency of high-load events, or because of a manufacturing flaw. Similarly to electrical systems, mechanical properties also need to be understood and appropriate procedures need to be developed. Mechanical designs can also be guided by Finite Element Analysis (FEA). International engineering standards are a good guidance to help design systems with comfortable safety margins, although the requirements of particle physics experiments (for example the minimisation of scattering material) can sometimes justify a more aggressive approach. In this case extensive prototyping is required to validate the approach.

Ionising radiation can break chemical bonds and therefore reduce the strength of a glue.

This involves a combination of testing prototypes to a higher stress than expected during operation (and repeating these tests on a batch basis) and testing every part when this is practical.

No compromises can be made if personal safety is involved.

Detectors are usually assembled at room temperature and therefore, if they are operated at a different temperature, there will be thermal stresses. Hence, performance should be verified under thermal cycling during development and in production.

Space missions have to survive very high acceleration during launch, therefore all components must pass rigorous vibration testing.

Key lessons from this chapter

- Particle physics experiments usually rely on the whole range of detector techniques discussed in this book, to achieve mutually supporting, complementary and/or redundant measurements.

- The integration of these different detectors involves a challenging optimisation to achieve low material, coverage, compactness and low cost, requiring careful optimisation and compromises.

- LHC detectors are probably the most complex pieces of scientific equipment ever built. Very extreme challenges in terms of data rates and radiation damage needed to be understood.

- Neutrino detectors require very large fiducial mass, and the challenge is to develop powerful yet affordable detectors.

- Rare decay experiments have extreme requirements for controlling backgrounds like radioactivity.

- Modern space experiments like AMS-02 and FGST are similar in many ways to large terrestrial high energy physics detectors, with added challenges due to launch forces, environmental conditions, and the impossibility to service the detector.

- Detectors in large and complex experiments have very demanding reliability requirements.

EXERCISES

1. *Solenoidal magnetic fields and muon triggers.*

   Consider a cylindrical detector with a solenoid magnetic field of magnitude B. The flux of the solenoid is returned in a yoke made of iron. The solenoid has a radius R_1 and the flux return extends from R_1 to a radius of R_2,

   a) Explaining any assumptions you make, show that $\int_0^{R_2} B(r)\mathrm{d}r = 0$.

   b) Consider a charged particle created in a collision on the solenoid axis moving in the plane perpendicular to the B field. Write down an expression for the force on the particle and determine the resultant torque.

   c) What is the net change in angular momentum of the charged particle from creation to when it reaches a radius R_2?

   d) What are the implications for a possible muon trigger system based purely on the measurement of the muon trajectory outside the return yoke for this geometry?

2. *A rare decay search experiment.*

   An experiment is to be designed to look for the decay $p \to \mu^+ \pi^0$.

   a) Assuming that the partial decay rate for this decay mode is $1/(10^{32}\ \mathrm{y})$, make a rough estimate of the mass of protons required in order to have a measurable counting rate.

b) What would be the experimental signature of such a decay in a detector based on the Cherenkov effect? Suggest a suitable choice for the Cherenkov material, and give a basic outline of the overall detector. What sources of background will need to be considered? How could these backgrounds be reduced?

3. *Powering.*

A power supply system is required to deliver a current I_0 at a voltage of V_0 (between input and return lines) for N modules. The resistance of the cables carrying the current from the power supply to the modules is R.

Calculate the power dissipated in the cable(s) and the voltage drop along the cable(s) for the four cases below, explaining any approximations you use. Why do we need to minimise the power dissipation in the cables and the voltage drop along the cables? What are the advantages and disadvantages of each approach?

a) *Individual module powering.* Each module is powered by an independent power supply and cable.

b) *Parallel powering.* All modules are powered by one power supply and one cable.

c) *Serial powering.* One power supply is used to feed the current required by one module through one cable to the first module and all N modules are connected in series.

d) *DC-DC Converters.* A power supply outputs a voltage of NV_0 and is connected in parallel to all N modules. A DC-DC converter on each module transforms the voltage from NV_0 to V_0.

References

[1] M. Aaboud et al. "A measurement of material in the ATLAS tracker using secondary hadronic interactions in 7 TeV pp collisions". *JINST* 11.11 (2016), P11020. DOI: 10.1088/1748-0221/11/11/P11020.

[2] G. Aad et al. "ATLAS pixel detector electronics and sensors". *JINST* 3.07 (2008), P07007. DOI: 10.1088/1748-0221/3/07/P07007.

[3] R. Aaij et al. "Performance of the LHCb Vertex Locator". *JINST* 9.09 (2014), P09007. DOI: 10.1088/1748-0221/9/09/P09007.

[4] R. Aaij et al. "A comprehensive real-time analysis model at the LHCb experiment". *JINST* 14.04 (2019), P04006–P04006. DOI: 10.1088/1748-0221/14/04/p04006.

[5] R. Aaij et al. "Allen: A high-level trigger on GPUs for LHCb". *Computing and Software for Big Science* 4.1 (2020). DOI: 10.1007/s41781-020-00039-7.

[6] B. Abbott et al. "Production and integration of the ATLAS insertable B-Layer". *JINST* 13 (2018), T05008. DOI: 10.1088/1748-0221/13/05/T05008.

[7] D. Abbott et al. "Simple derivation of the thermal noise formula using window-limited Fourier transforms and other conundrums". *IEEE Trans. Educ.* 39.1 (1996), pp. 1–13. DOI: http://doi.org/10.1109/13.485226.

[8] A. Abdelhameed et al. "First results from the CRESST-III low-mass dark matter program". *Phys. Rev. D* 100.10 (2019), p. 102002. DOI: 10.1103/PhysRevD.100.102002.

[9] K. Abe et al. "Performance of the CRID at SLD". *NIMA* 343.1 (1994), pp. 74–86. DOI: 10.1016/0168-9002(94)90536-3.

[10] K. Abe et al. "Solar neutrino results in Super-Kamiokande-III". *Phys. Rev. D* 83.5 (2011), p. 052010. DOI: 10.1103/PhysRevD.83.052010.

[11] K. Abe et al. "Calibration of the Super-Kamiokande detector". *NIMA* 737 (2014), pp. 253–272. DOI: 10.1016/j.nima.2013.11.081.

[12] N. Abgrall et al. "The MAJORANA DEMONSTRATOR Neutrinoless Double-Beta Decay Experiment". *Adv. High Energy Phys.* 2014 (2014), p. 365432. DOI: 10.1155/2014/365432.

[13] H. Abramowicz et al. "The response and resolution of an iron-scintillator calorimeter for hadronic and electromagnetic showers between 10 GeV and 140 GeV". *NIM* 180.2 (1981), pp. 429–439. DOI: 10.1016/0029-554X(81)90083-5.

[14] R. Acciarri et al. "Design and construction of the MicroBooNE detector". *JINST* 12.02 (2017), P02017. DOI: 10.1088/1748-0221/12/02/P02017.

[15] D. Acosta et al. "Results of prototype studies for a spaghetti calorimeter". *NIMA* 294.1 (1990), pp. 193–210. DOI: 10.1016/0168-9002(90)91833-W.

[16] I. Adam et al. "The DIRC particle identification system for the BaBar experiment". *NIMA* 538.1 (2005), pp. 281–357. DOI: 10.1016/j.nima.2004.08.129.

[17] W. Adam et al. "The forward ring imaging Cherenkov detector of DELPHI". *NIMA* 338.2 (1994), pp. 284–309. DOI: 10.1016/0168-9002(94)91314-5.

[18] W. Adam et al. "Status of the R&D activity on diamond particle detectors". *NIMA* 511.1 (2003), pp. 124–131. DOI: 10.1016/S0168-9002(03)01777-7.

[19] T. Adams et al. "Beam test evaluation of electromagnetic calorimeter modules made from proton-damaged PbWO4 crystals". *JINST* 11.04 (2016), P04012. DOI: 10.1088/1748-0221/11/04/P04012.

[20] M. Adinolfi et al. "The KLOE electromagnetic calorimeter". *NIMA* 482.1 (2002), pp. 364–386. DOI: 10.1016/S0168-9002(01)01502-9.

[21] C. Adloff et al. "The AMS-02 lead-scintillating fibres Electromagnetic Calorimeter". *NIMA* 714 (2013), pp. 147–154. DOI: 10.1016/j.nima.2013.02.020.

[22] P. Adzicet et al. "Energy resolution of the barrel of the CMS Electromagnetic Calorimeter". *JINST* 2.04 (2007), P04004. DOI: 10.1088/1748-0221/2/04/P04004.

[23] S. Aefsky. "Alignment of the Muon Spectrometer in ATLAS". *Phys. Proc.* 37 (2012), pp. 51–56. DOI: 10.1016/j.phpro.2012.02.355.

[24] S. Afanasiev et al. "The NA49 large acceptance hadron detector". *NIMA* 430.2 (1999), pp. 210–244. DOI: 10.1016/S0168-9002(99)00239-9.

[25] A. Affolder et al. "CDF Central Outer Tracker". *NIMA* 526.3 (2004), pp. 249–299. DOI: 10.1016/j.nima.2004.02.020.

[26] A. Affolder et al. "DC-DC converters with reduced mass for trackers at the HL-LHC". *JINST* 6.11 (2011), p. C11035. DOI: 10.1088/1748-0221/6/11/C11035.

[27] A. Affolder et al. "Radiation damage in the LHCb vertex locator". *JINST* 8.08 (2013), P08002. DOI: 10.1088/1748-0221/8/08/P08002.

[28] G. Aglieri Rinella et al. "Performance Studies of Pixel Hybrid Photon Detectors for the LHCb RICH Counters". *Nucl. Phys. B Proc. Suppl.* 150 (2006), pp. 285–289. DOI: 10.1016/j.nuclphysbps.2005.01.247.

[29] S. Agostinelli et al. "Geant4—a simulation toolkit". *NIMA* 506.3 (2003), pp. 250–303. DOI: 10.1016/S0168-9002(03)01368-8.

[30] M. Aguilar et al. "The Alpha Magnetic Spectrometer (AMS) on the international space station: Part II — Results from the first seven years". *Phys. Rep.* 894 (2021), pp. 1–116. DOI: 10.1016/j.physrep.2020.09.003.

[31] M. Aharrouche et al. "Energy linearity and resolution of the ATLAS electromagnetic barrel calorimeter in an electron test-beam". *NIMA* 568.2 (2006), pp. 601–623. DOI: 10.1016/j.nima.2006.07.053.

[32] N. Akchurin et al. "Hadron and jet detection with a dual-readout calorimeter". *NIMA* 537.3 (2005), pp. 537–561. DOI: 10.1016/j.nima.2004.07.285.

[33] N. Akchurin et al. "Detection of electron showers in dual-readout crystal calorimeters". *NIMA* 686 (2012), pp. 125–135. DOI: 10.1016/j.nima.2012.04.092.

[34] D. Akerib et al. "The LUX-ZEPLIN (LZ) experiment". *NIMA* 953 (2020), p. 163047. DOI: 10.1016/j.nima.2019.163047.

[35] T. Akesson et al. "Aging studies for the ATLAS Transition Radiation Tracker (TRT)". *NIMA* 515.1 (2003), pp. 166–179. DOI: 10.1016/j.nima.2003.08.145.

[36] A. Akindinov et al. "The MRPC detector for the ALICE Time Of Flight System: Final Design and Performances". *Nucl. Phys. B - Proc. Supp.* 158 (2006), pp. 60–65. DOI: 10.1016/j.nuclphysbps.2006.07.035.

[37] A. Akindinov et al. "Performance of the ALICE Time-Of-Flight detector at the LHC". *Eur. Phys. J. Plus* 128.4 (2013), p. 44. DOI: 10.1140/epjp/i2013-13044-x.

[38] B. Al Atoum et al. "Electron transport in gaseous detectors with a Python-based Monte Carlo simulation code". *Comput. Phys. Commun.* 254 (2020), p. 107357. DOI: 10.1016/j.cpc.2020. 107357.

[39] M. Aleksa et al. "Results of the ATLAS solenoid magnetic field map". *J. Phys. Conf. Ser.* 110.9 (2008), p. 092018. DOI: 10.1088/1742-6596/110/9/092018.

[40] T. Alexopoulos et al. "A spark-resistant bulk-micromegas chamber for high-rate applications". *NIMA* 640.1 (2011), pp. 110–118. DOI: 10.1016/j.nima.2011.03.025.

[41] ALICE collaboration. "Performance of the ALICE experiment at the CERN LHC". *Int. J. Mod. Phys. A* 29.24 (2014), p. 1430044. DOI: 10.1142/S0217751X14300440.

[42] ALICE collaboration. "The ALICE experiment at the CERN LHC". *JINST* 3.08 (2008), S08002. DOI: 10.1088/1748-0221/3/08/S08002.

[43] G. Alkhazov. "Statistics of electron avalanches and ultimate resolution of proportional counters". *NIM* 89 (1970), pp. 155–165. DOI: 10.1016/0029-554X(70)90818-9.

[44] J. Allison et al. "Geant4 developments and applications". *IEEE TNS* 53.1 (2006), pp. 270–278. DOI: 10.1109/TNS.2006.869826.

[45] J. Allison et al. "Recent developments in Geant4". *NIMA* 835 (2016), pp. 186–225. DOI: 10.1016/j.nima.2016.06.125.

[46] W. Allison. *The flight of a relativistic charge in matter: Insights, calculations and practical applications of classical electromagnetism.* Springer Lecture Notes in Physics, 2023. ISBN: 978-3-031-23445-3. DOI: 10.1007/978-3-031-23446-0.

[47] W. Allison and J. Cobb. "Relativistic charged particle identification by energy loss". *Ann. Rev. Nucl. Part. Sci.* 30.1 (1980), pp. 253–298. DOI: 10.1146/annurev.ns.30.120180.001345.

[48] W. Allison and P. Wright. "The physics of charged particle identification: dE/dx, Cerenkov and transition radiation". *Experimental Techniques in High-Energy Nuclear and Particle Physics.* Ed. by T. Ferbel. World Scientific Publishing, 1991, pp. 371–417. ISBN: 9789810208684. DOI: 10.1142/1571.

[49] J. Alme et al. "The ALICE TPC, a large 3-dimensional tracking device with fast readout for ultra-high multiplicity events". *NIMA* 622.1 (2010), pp. 316–367. DOI: 10.1016/j.nima.2010.04.042.

[50] J. Alozy et al. "Studies of the spectral and angular distributions of transition radiation using a silicon pixel sensor on a Timepix3 chip". *NIMA* 961 (2020), p. 163681. DOI: 10.1016/j.nima.2020.163681.

[51] M. Altarelli et al. "Superconvergence and sum rules for the optical constants". *Phys. Rev. B* 6.12 (1972), pp. 4502–4509. DOI: 10.1103/PhysRevB.6.4502.

[52] S. Alvis et al. "Search for neutrinoless double-β decay in ^{76}Ge with 26 kg yr of exposure from the Majorana Demonstrator". *Phys. Rev. C* 100.2 (2019), p. 025501. DOI: 10.1103/PhysRevC.100.025501.

[53] U. Amaldi. "Fluctuations in calorimetry measurements". *Phys. Scr.* 23.4A (1981), p. 409. DOI: 10.1088/0031-8949/23/4A/012.

[54] S. Amato et al. *LHCb magnet: Technical Design Report.* LHCb. Geneva: CERN, 2000. URL: https://cds.cern.ch/record/424338.

[55] G. Ambrosi et al. "The spatial resolution of the silicon tracker of the Alpha Magnetic Spectrometer". *NIMA* 869 (2017), pp. 29–37. DOI: 10.1016/j.nima.2017.07.014.

[56] C. Amole et al. "Dark matter search results from the PICO-60 CF_3I bubble chamber". *Phys. Rev. D* 93.5 (2016), p. 052014. DOI: 10.1103/PhysRevD.93.052014.

[57] E. Anassontzis et al. "The Barrel Ring Imaging Cherenkov counter of DELPHI". *NIMA* 323.1 (1992), pp. 351–362. DOI: 10.1016/0168-9002(92)90315-U.

[58] A. Andresen et al. "Construction and beam test of the ZEUS forward and rear calorimeter". *NIMA* 309.1 (1991), pp. 101–142. DOI: 10.1016/0168-9002(91)90095-8.

[59] A. Andronic and J. Wessels. "Transition radiation detectors". *NIMA* 666 (2012), pp. 130–147. DOI: 10.1016/j.nima.2011.09.041.

[60] P. Anthony et al. "An accurate measurement of the Landau-Pomeranchuk-Migdal effect". *Phys. Rev. Lett.* 75 (1995), pp. 1949–1952. DOI: 10.1103/PhysRevLett.75.1949.

[61] R. Apfel. "The superheated drop detector". *NIM* 162.1 (1979), pp. 603–608. DOI: 10.1016/0029-554X(79)90735-3.

[62] J. Apostolakis et al. "An implementation of ionisation energy loss in very thin absorbers for the GEANT4 simulation package". *NIMA* 453.3 (2000), pp. 597–605. DOI: 10.1016/S0168-9002(00)00457-5.

[63] R.-D. Appuhn et al. "The H1 lead/scintillating-fibre calorimeter". *NIMA* 386.2 (1997), pp. 397–408. DOI: 10.1016/S0168-9002(96)01171-0.

[64] E. Aprile and T. Doke. "Liquid xenon detectors for particle physics and astrophysics". *Rev. Mod. Phys.* 82.3 (2010), pp. 2053–2097. DOI: 10.1103/RevModPhys.82.2053.

[65] E. Aprile et al. "Observation of anticorrelation between scintillation and ionization for MeV gamma rays in liquid xenon". *Phys. Rev. B* 76.1 (2007), p. 014115. DOI: 10.1103/PhysRevB.76.014115.

[66] Y. Arai et al. "Timing optimization of thin gap chambers for the use in the ATLAS muon endcap trigger". *NIMA* 367.1 (1995), pp. 398–401. DOI: 10.1016/0168-9002(95)00532-3.

[67] A. Ariga et al. "Nuclear emulsions". *Particle Physics Reference Library: Volume 2: Detectors for Particles and Radiation*. Ed. by C. Fabjan and H. Schopper. Springer International Publishing, 2020, pp. 383–438. ISBN: 978-3-030-35318-6. DOI: 10.1007/978-3-030-35318-6_9.

[68] E. Armengaud et al. "Performance of the EDELWEISS-III experiment for direct dark matter searches". *JINST* 12.08 (2017), P08010. DOI: 10.1088/1748-0221/12/08/P08010.

[69] R. Arnold et al. "A ring imaging Cherenkov detector, the DELPHI Barrel RICH Prototype: Part A: Experimental studies of the detection efficiency and the spatial resolution". *NIMA* 270.2 (1988), pp. 255–288. DOI: 10.1016/0168-9002(88)90695-X.

[70] N. Arora, J. Hauser, and D. Roulston. "Electron and hole mobilities in silicon as a function of concentration and temperature". *IEEE Trans. Electron Devices* 29.2 (1982), pp. 292–295. DOI: 10.1109/T-ED.1982.20698.

[71] X. Artru, G. Yodh, and G. Mennessier. "Practical theory of the multilayered transition radiation detector". *Phys. Rev. D* 12.5 (1975), pp. 1289–1306. DOI: 10.1103/PhysRevD.12.1289.

[72] M. Artuso et al. "Construction, pattern recognition and performance of the CLEO III LiF-TEA RICH detector". *NIMA* 502.1 (2003), pp. 91–100. DOI: 10.1016/S0168-9002(02)02162-9.

[73] N. Ashcroft and N. Mermin. *Solid state physics*. Fort Worth: Saunders College Publishing, 1976. ISBN: 978-0030839931.

[74] B. Assaf. Lecture notes on thermal noise. 2018. URL: https://badihassaf.files.wordpress.com/2018/03/lecture-notes-thermal-noise.pdf.

[75] A. Astbury et al. *A 4π Solid Angle Detector for the SPS Used as a Proton-Antiproton Collider at a Centre of Mass Energy of 540 GeV*. Geneva: CERN, 1978. URL: http://cds.cern.ch/record/319371.

[76] M. Atalla, E. Tannenbaum, and E. Scheibner. "Stabilization of silicon surfaces by thermally grown oxides". *The Bell System Technical Journal* 38.3 (1959), pp. 749–783. DOI: 10.1002/j.1538-7305.1959.tb03907.x.

[77] ATLAS collaboration. "Resolution of the ATLAS muon spectrometer monitored drift tubes in LHC Run 2". *JINST* 14.09 (2019), P09011–P09011. DOI: 10.1088/1748-0221/14/09/p09011.

[78] ATLAS collaboration. "Operation and performance of the ATLAS semiconductor tracker". *JINST* 9.08 (2014), P08009. DOI: 10.1088/1748-0221/9/08/P08009.

[79] ATLAS collaboration. "Performance of the ATLAS track reconstruction algorithms in dense environments in LHC Run 2". *Eur. Phys. J. C* 77 (2017), p. 673. DOI: 10.1140/epjc/s10052-017-5225-7.

[80] ATLAS collaboration. "Electron and photon performance measurements with the ATLAS detector using the 2015–2017 LHC proton-proton collision data". *JINST* 14.12 (2019), P12006. DOI: 10.1088/1748-0221/14/12/P12006.

[81] ATLAS collaboration. Public ATLAS luminosity results for Run-2 of the LHC. URL: https://twiki.cern.ch/twiki/bin/view/AtlasPublic/LuminosityPublicResultsRun2#2018_pp_Collisions.

[82] ATLAS collaboration. *Technical proposal: A High-Granularity Timing Detector for the ATLAS phase-II upgrade*. Geneva: CERN, 2018. DOI: 10.17181/CERN.CIUJ.KS4H.

[83] ATLAS collaboration. "Alignment of the ATLAS inner detector in Run 2". *Eur. Phys. J. C* 80.12 (2020), p. 1194. DOI: 10.1140/epjc/s10052-020-08700-6.

[84] ATLAS collaboration. ATLAS approved combined inner detector plots. URL: https://twiki.cern.ch/twiki/bin/view/AtlasPublic/ApprovedPlotsID.

[85] ATLAS collaboration. "The ATLAS experiment at the CERN Large Hadron Collider". *JINST* 3.08 (2008), S08003. DOI: 10.1088/1748-0221/3/08/S08003.

[86] ATLAS collaboration. "Drift time measurement in the ATLAS liquid argon electromagnetic calorimeter using cosmic muons". *Eur. Phys. J. C* 70.3 (2010), pp. 755–785. DOI: 10.1140/epjc/s10052-010-1403-6.

[87] ATLAS collaboration. "Electron and photon energy calibration with the ATLAS detector using LHC Run 1 data". *Eur. Phys. J. C* 10 (2014), p. 3071. DOI: 10.1140/epjc/s10052-014-3071-4.

[88] ATLAS collaboration. "Electron and photon energy calibration with the ATLAS detector using 2015–2016 LHC proton-proton collision data". *JINST* 14.03 (2019), P03017. DOI: 10.1088/1748-0221/14/03/P03017.

[89] ATLAS collaboration. "Jet energy scale and resolution measured in proton–proton collisions at $\sqrt{s} = 13$ TeV with the ATLAS detector". *Eur. Phys. J. C* 81.8 (2021), p. 689. DOI: 10.1140/epjc/s10052-021-09402-3.

[90] ATLAS collaboration. "Measurement of the top quark mass in the $t\bar{t} \to$ lepton+jets channel from $\sqrt{s} = 8$ TeV ATLAS data and combination with previous results". *Eur. Phys. J. C* 79.290 (2019). DOI: 10.1140/epjc/s10052-019-6757-9.

[91] ATLAS collaboration. "Electron reconstruction and identification in the ATLAS experiment using the 2015 and 2016 LHC proton–proton collision data at $\sqrt{s} = 13$ TeV". *Eur. Phys. J. C* 79.8 (2019), p. 639. DOI: 10.1140/epjc/s10052-019-7140-6.

[92] ATLAS collaboration. "Muon reconstruction and identification efficiency in ATLAS using the full Run 2 pp collision data set at $\sqrt{s} = 13$ TeV". *Eur. Phys. J. C* 81.7 (2021), p. 578. DOI: 10.1140/epjc/s10052-021-09233-2.

[93] ATLAS collaboration. "Performance of missing transverse momentum reconstruction with the ATLAS detector using proton–proton collisions at $\sqrt{s} = 13$ TeV". *Eur. Phys. J. C* 78 (2018), p. 903. DOI: 10.1140/epjc/s10052-018-6288-9.

[94] ATLAS collaboration. "ATLAS b-jet identification performance and efficiency measurement with $t\bar{t}$ in pp collisions at $\sqrt{s} = 13$ TeV." *Eur. Phys. J. C* 79.11 (2019), p. 970. DOI: 10.1140/epjc/s10052-019-7450-8.

[95] ATLAS collaboration. Topological b-hadron decay reconstruction and identification of b-jets with the JetFitter package in the ATLAS experiment at the LHC. 2018. URL: https://cds.cern.ch/record/2645405.

[96] ATLAS collaboration. "The ATLAS Level-1 Central Trigger". *J. Phys. Conf. Ser.* 331 (2011), p. 022041. DOI: 10.1088/1742-6596/331/2/022041.

[97] ATLAS collaboration. "The Phase-1 upgrade of the ATLAS first level calorimeter trigger". *JINST* 11 (2016), p. C01018. DOI: 10.1088/1748-0221/11/01/C01018.

[98] ATLAS collaboration. "Performance of electron and photon triggers in ATLAS during LHC Run 2". *Eur. Phys. J. C* 80 (2020), p. 47. DOI: 10.1140/epjc/s10052-019-7500-2.

[99] ATLAS collaboration. "The ATLAS electron and photon trigger". *J. Phys. Conf. Ser.* 1162 (2019), p. 012037. DOI: 10.1088/1742-6596/1162/1/012037.

[100] ATLAS Experiment - Public Results, $H \to 2e2\mu$ candidate event with 25 additional reconstructed primary vertices recorded in 2016 - HIGG-2016-25. ATLAS Experiment © 2022 CERN. URL: https://twiki.cern.ch/twiki/bin/view/AtlasPublic/EventDisplayRun2Physics.

[101] ATLAS Experiment Public Results - RadiationSimulationPublicResults. URL: https://twiki.cern.ch/twiki/bin/view/AtlasPublic/RadiationSimulationPublicResults.

[102] ATLAS TRT collaboration. "The ATLAS Transition Radiation Tracker (TRT) proportional drift tube: design and performance". *JINST* 3.02 (2008), P02013–P02013. DOI: 10.1088/1748-0221/3/02/p02013.

[103] ATLAS TRT collaboration. "The ATLAS TRT barrel detector". *JINST* 3.02 (2008), P02014. DOI: 10.1088/1748-0221/3/02/P02014.

[104] V. Atrazhev. "High mobility liquids as a result of competition between structural localization of electrons and reduction of polarization electron-atom attraction". *Proceedings of 1999 IEEE 13th International Conference on Dielectric Liquids*. 1999, pp. 13–17. DOI: 10.1109/ICDL.1999.798856.

[105] D. Attree et al. "The evaporative cooling system for the ATLAS inner detector". *JINST* 3.07 (2008), P07003–P07003. DOI: 10.1088/1748-0221/3/07/p07003.

[106] W. Atwood et al. "Design and initial tests of the Tracker-converter of the Gamma-ray Large Area Space Telescope". *Astropart. Phys.* 28.4 (2007), pp. 422–434. DOI: 10.1016/j.astropartphys.2007.08.010.

[107] W. Atwood et al. "The large area telescope on the Fermi Gamma-Ray Space Telescope Mission". *Astrophys. J.* 697.2 (2009), p. 1071. DOI: 10.1088/0004-637X/697/2/1071.

[108] B. Aubert et al. "Performance of a liquid argon electromagnetic calorimeter with an "accordion" geometry". *NIMA* 309.3 (1991), pp. 438–449. DOI: 10.1016/0168-9002(91)90247-N.

[109] B. Aubert et al. "Construction, assembly and tests of the ATLAS electromagnetic barrel calorimeter". *NIMA* 558.2 (2006), pp. 388–418. DOI: 10.1016/j.nima.2005.11.212.

[110] B. Baller. "Liquid argon TPC signal formation, signal processing and reconstruction techniques". *JINST* 12.07 (2017), P07010. DOI: 10.1088/1748-0221/12/07/P07010.

[111] P. Barbeau, J. Collar, and O. Tench. "Large-mass ultralow noise germanium detectors: performance and applications in neutrino and astroparticle physics". *J. Cosm. Astropart. Phys.* 2007.09 (2007), p. 009. DOI: 10.1088/1475-7516/2007/09/009.

[112] G. Barbiellini et al. "Energy resolution and longitudinal shower development in a Si/W electromagnetic calorimeter". *NIMA* 235.1 (1985), pp. 55–60. DOI: 10.1016/0168-9002(85)90245-1.

[113] J. Bardeen and W. Shockley. "Deformation potentials and mobilities in non-polar crystals". *Phys. Rev.* 80.1 (1950), pp. 72–80. DOI: 10.1103/PhysRev.80.72.

[114] V. Barnes et al. "Observation of a hyperon with strangeness minus three". *Phys. Rev. Lett.* 12.8 (1964), pp. 204–206. DOI: 10.1103/PhysRevLett.12.204.

[115] D. Barney. "CMS detector slice". 2016. URL: http://cds.cern.ch/record/2120661.

[116] R. Bates et al. "The effects of radiation on gallium arsenide radiation detectors". *NIMA* 395.1 (1997), pp. 54–59. DOI: 10.1016/S0168-9002(97)00628-1.

[117] G. Bathow, E. Freytag, and K. Tesch. "Measurements on 6.3 GeV electromagnetic cascades and cascade-produced neutrons". *Nucl. Phys. B* 2.6 (1967), pp. 669–689. DOI: 10.1016/0550-3213(67)90068-5.

[118] M. Battistin et al. "The thermosiphon cooling system of the ATLAS experiment at the CERN Large Hadron Collider". *Int. J. Chem. Reactor Eng.* 13.4 (2015), pp. 511–521. DOI: 10.1515/ijcre-2015-0022.

[119] R. Battiston. "Spaceborne experiments". *Particle Physics Reference Library: Volume 2: Detectors for Particles and Radiation.* Ed. by C. Fabjan and H. Schopper. Springer International Publishing, 2020, pp. 823–870. ISBN: 978-3-030-35318-6. DOI: 10.1007/978-3-030-35318-6_18.

[120] G. Battistoni et al. "Overview of the FLUKA code". *Ann. Nucl. Energy* 82 (2015), pp. 10–18. DOI: 10.1016/j.anucene.2014.11.007.

[121] G. Beck and G. Viehhauser. "Analytic model of thermal runaway in silicon detectors". *NIMA* 618.1 (2010), pp. 131–138. DOI: 10.1016/j.nima.2010.02.264.

[122] G. Beck et al. "Radiation-tolerant breakdown protection of silicon detectors using multiple floating guard rings". *NIMA* 396.1 (1997), pp. 214–227. DOI: 10.1016/S0168-9002(97)00749-3.

[123] G. Beck et al. "Thermo-electrical modelling of the ATLAS ITk Strip Detector". *NIMA* 969 (2020), p. 164023. DOI: 10.1016/j.nima.2020.164023.

[124] H. Becquerel. "Sur les radiations émises par phosphorescence". *Comptes Rendus Acad. Sci.* 122 (1896), pp. 420–421. URL: https://gallica.bnf.fr/ark:/12148/bpt6k30780/f422.item.

[125] H. Becquerel. On Radioactivity, a New Property of Matter. Nobel Lecture. 1903. URL: https://www.nobelprize.org/prizes/physics/1903/becquerel/lecture/.

[126] A. Bell. "Upon the electrical experiments to determine the location of the bullet in the body of the late President Garfield; and upon a successful form of induction balance for the painless detection of metallic masses in the human body". *Am. J. Sci.* s3-25.145 (1883), pp. 22–61. DOI: 10.2475/ajs.s3-25.145.22.

[127] N. Belyaev et al. "Development of transition radiation detectors for hadron identification at TeV energy scale". *J. Phys. Conf. Ser.* 1390.1 (2019), p. 012126. DOI: 10.1088/1742-6596/1390/1/012126.

[128] M. Berger et al. XCOM: Photon Cross Sections Database. DOI: 10.18434/T48G6X.

[129] G. Bertrand, M. Hamel, and F. Sguerra. "Current status on plastic scintillators modifications". *Chem. Eur. J* 20.48 (2014), pp. 15660–15685. DOI: 10.1002/chem.201404093.

[130] H. Bethe. "Zur Theorie des Durchgangs schneller Korpuskularstrahlen durch Materie". *Annalen der Physik* 397.3 (1930), pp. 325–400. DOI: 10.1002/andp.19303970303. English translation in *Selected Works of Hans A Bethe*. World Scientific Series in 20th Century Physics: Volume 18. World Scientific, 1997. ISBN: 978-981-02-2876-7. DOI: 10.1142/3295.

[131] H. Bethe. "Molière's theory of multiple scattering". *Phys. Rev.* 89.6 (1953), pp. 1256–1266. DOI: 10.1103/PhysRev.89.1256.

[132] H. Bethe and W. Heitler. "On the stopping of fast particles and on the creation of positive electrons". *Proc. R. Soc. Lond. A* 146 (1934), pp. 83–112. DOI: http://doi.org/10.1098/rspa.1934.0140.

[133] H. Bethe and R. Jackiw. *Intermediate quantum mechanics*. 3rd ed. CRC Press, 1986. ISBN: 9780429493645. DOI: 10.1201/9780429493645.

[134] P. Beynel, P. Maier, and H. Schönbacher. *Compilation of radiation damage test data. Index des résultats d'essais de radiorésistance*. CERN Yellow Reports. Geneva: CERN, 1982. DOI: 10.5170/CERN-1982-010.

[135] S. Bhasin et al. "Performance of a prototype TORCH time-of-flight detector". *NIMA* 1050 (2023), p. 168181. DOI: 10.1016/j.nima.2023.168181.

[136] S. Biagi. "A multiterm Boltzmann analysis of drift velocity, diffusion, gain and magnetic-field effects in argon-methane-water-vapour mixtures". *NIMA* 283.3 (1989), pp. 716–722. DOI: 10.1016/0168-9002(89)91446-0.

[137] S. Biagi. "Monte Carlo simulation of electron drift and diffusion in counting gases under the influence of electric and magnetic fields". *NIMA* 421.1 (1999), pp. 234–240. DOI: 10.1016/S0168-9002(98)01233-9.

[138] S. Biagi. Magboltz - transport of electrons in gas mixtures. URL: https://magboltz.web.cern.ch/magboltz/.

[139] H. Bichsel. "Straggling in thin silicon detectors". *Rev. Mod. Phys.* 60.3 (1988), pp. 663–699. DOI: 10.1103/RevModPhys.60.663.

[140] H. Bichsel and R. Saxon. "Comparison of calculational methods for straggling in thin absorbers". *Phys. Rev. A* 11.4 (1975), pp. 1286–1296. DOI: 10.1103/PhysRevA.11.1286.

[141] H. Bichsel and H. Schindler. "The interaction of radiation with matter". *Particle Physics Reference Library: Volume 2: Detectors for Particles and Radiation*. Ed. by C. Fabjan and H. Schopper. Springer International Publishing, 2020, pp. 5–44. ISBN: 978-3-030-35318-6. DOI: 10.1007/978-3-030-35318-6_2.

[142] A. Bingül. "The ATLAS TRT and its performance at LHC". *J. Phys. Conf. Ser.* 347.1 (2012), p. 012025. DOI: 10.1088/1742-6596/347/1/012025.

[143] M. Binkley et al. "Aging in large CDF tracking chambers". *NIMA* 515.1 (2003), pp. 53–59. DOI: 10.1016/j.nima.2003.08.130.

[144] J. Birks. "Scintillations from organic crystals: Specific fluorescence and relative response to different radiations". *Proc. Phys. Soc. A* 64.10 (1951), pp. 874–877. DOI: 10.1088/0370-1298/64/10/303.

[145] J. Birks. *The theory and practice of scintillation counting*. Pergamon Press, 1964. ISBN: 978-0-08-010472-0. DOI: 10.1016/C2013-0-01791-4.

[146] P. Blackett and G. Occhialini. "Photography of penetrating corpuscular radiation". *Nature* 130.3279 (1932), pp. 363–363. DOI: 10.1038/130363a0.

[147] V. Blanco Carballo et al. "GEMGrid: a wafer post-processed GEM-like radiation detector". *NIMA* 608.1 (2009), pp. 86–91. DOI: 10.1016/j.nima.2009.06.023.

[148] W. Blum, W. Riegler, and L. Rolandi. "Gas ionization by charged particles and by laser rays". *Particle Detection with Drift Chambers*. Berlin, Heidelberg: Springer, 2008, pp. 1–48. ISBN: 978-3-540-76684-1. DOI: 10.1007/978-3-540-76684-1_1.

[149] H. Blumgart and O. Yens. "Studies on the velocity of blood flow: I. The method utilized". *J. Clin. Invest.* 4.1 (1927), pp. 1–13. DOI: 10.1172/JCI100106.

[150] A. Bodek et al. "Observation of light below Cerenkov threshold in a 1.5 meter long integrating Cerenkov counter". *Z. Phys. C* 18.4 (1983), pp. 289–299. DOI: 10.1007/BF01573729.

[151] T. Böhlen et al. "The FLUKA code: Developments and challenges for high energy and medical applications". *Nuclear Data Sheets* 120 (2014), pp. 211–214. DOI: 10.1016/j.nds.2014.07.049.

[152] N. Bohr. "II. On the theory of the decrease of velocity of moving electrified particles on passing through matter". *The London, Edinburgh, and Dublin Philosophical Magazine and Journal of Science* 25.145 (1913), pp. 10–31. DOI: 10.1080/14786440108634305.

[153] R. Boie et al. "Second coordinate readout in drift chambers by timing of the electromagnetic wave propagating along the anode wire". *IEEE TNS* 28.1 (1981), pp. 471–477. DOI: 10.1109/TNS.1981.4331221.

[154] A. Bondar et al. "Cryogenic avalanche detectors based on gas electron multipliers". *NIMA* 524.1 (2004), pp. 130–141. DOI: 10.1016/j.nima.2004.01.060.

[155] W. Bonivento et al. "A complete simulation of a triple-GEM detector". *IEEE TNS* 49.4 (2002), pp. 1638–1643. DOI: 10.1109/TNS.2002.805170.

[156] L. Bonolis. "Walther Bothe and Bruno Rossi: The birth and development of coincidence methods in cosmic-ray physics". *American J, Phys.* 79.11 (2011), pp. 1133–1150. DOI: 10.1119/1.3619808.

[157] M. Bosetti et al. "Systematic investigation of the electromagnetic energy resolution on sampling frequency using silicon calorimeters". *NIMA* 345.2 (1994), pp. 244–249. DOI: 10.1016/0168-9002(94)90997-0.

[158] R. Bouclier et al. "Ageing of microstrip gas chambers: problems and solutions". *NIMA* 381.2 (1996), pp. 289–319. DOI: 10.1016/S0168-9002(96)00268-9.

[159] R. Bouclier et al. "High rate operation of micro-strip gas chambers". *IEEE TNS* 43.3 (1996), pp. 1220–1226. DOI: 10.1109/23.506667.

[160] G. Boyle et al. "Electron scattering and transport in liquid argon". *J. Chem. Phys.* 142.154507 (2015). DOI: 10.1063/1.4917258.

[161] W. Bragg and R. Kleeman. "LXXIV. On the ionization curves of radium". *The London, Edin-burgh, and Dublin Philosophical Magazine and Journal of Science* 8.48 (1904), pp. 726–738. DOI: 10.1080/14786440409463246.

[162] B. Bransden and C. Joachain. *Physics of Atoms and Molecules*. London & New York: Longman, 1982. ISBN: 9780582444010.

[163] A. Breskin. "CsI UV photocathodes: history and mystery". *NIMA* 371.1 (1996), pp. 116–136. DOI: 10.1016/0168-9002(95)01145-5.

[164] A. Bressan et al. "Two-dimensional readout of GEM detectors". *NIMA* 425.1 (1999), pp. 254–261. DOI: 10.1016/S0168-9002(98)01405-3.

[165] M. Bressler et al. "A buffer-free concept bubble chamber for PICO dark matter searches". *JINST* 14.08 (2019), P08019–P08019. DOI: 10.1088/1748-0221/14/08/p08019.

[166] C. Brizzolari et al. "Enhancement of the X-Arapuca photon detection device for the DUNE experiment". *JINST* 16.09 (2021), P09027. DOI: 10.1088/1748-0221/16/09/P09027.

[167] I. Bronic. "On a relation between the W value and the Fano factor". *J. Phys. B: Atomic, Molecular and Optical Physics* 25.8 (1992), pp. L215–L218. DOI: 10.1088/0953-4075/25/8/004.

[168] F. Brooks. "Organic scintillators". *Progress in Nuclear Physics* 5 (1956), p. 284.

[169] A. Bross and A. Pla-Dalmau. "Radiation damage of plastic scintillators". *IEEE TNS* 39.5 (1992), pp. 1199–1204. DOI: 10.1109/23.173178.

[170] "Bubble chamber: D meson production and decay". 1978. URL: https://cds.cern.ch/record/39469.

[171] "Bubble chambers" (1979). URL: https://cds.cern.ch/record/850624.

[172] W. Burton and B. Powell. "Fluorescence of tetraphenyl-butadiene in the vacuum ultraviolet". *Appl. Opt.* 12.1 (1973), pp. 87–89. DOI: 10.1364/AO.12.000087.

[173] A. Buzulutskov et al. "First results from cryogenic avalanche detectors based on gas electron multipliers". *IEEE TNS* 50.6 (2003), pp. 2491–2493. DOI: 10.1109/TNS.2003.820633.

[174] M. Cacciari, G. Salam, and G. Soyez. "The anti-kt jet clustering algorithm". *JHEP* 2008.04 (2008), p. 063. DOI: 10.1088/1126-6708/2008/04/063.

[175] P. Camarri et al. "Streamer suppression with SF6 in RPCs operated in avalanche mode". *NIMA* 414.2 (1998), pp. 317–324. DOI: 10.1016/S0168-9002(98)00576-2.

[176] F. Campabadal et al. "Beam tests of ATLAS SCT silicon strip detector modules". *NIMA* 538.1 (2005), pp. 384–407. DOI: 10.1016/j.nima.2004.08.133.

[177] F. Campabadal et al. "Design and performance of the ABCD3TA ASIC for readout of silicon strip detectors in the ATLAS semiconductor tracker". *NIMA* 552.3 (2005), pp. 292–328. DOI: 10.1016/j.nima.2005.07.002.

[178] J. Campbell, J. Huston, and W. Stirling. "Hard interactions of quarks and gluons: a primer for LHC physics". *Rep. Prog. Phys.* 70.1 (2006), p. 89. DOI: 10.1088/0034-4885/70/1/R02.

[179] M. Campbell et al. "GOSSIP: A vertex detector combining a thin gas layer as signal generator with a CMOS readout pixel array". *NIMA* 560.1 (2006), pp. 131–134. DOI: 10.1016/j.nima.2005.11.199.

[180] N. Campbell. "The study of discontinuous phenomena". *Proc. Camb. Phil. Soc.* 15 (1909), pp. 117–136. URL: https://archive.org/details/proceedingsofcam15190810camb/page/116/mode/2up.

[181] C. Canali et al. "Electron and hole drift velocity measurements in silicon and their empirical relation to electric field and temperature". *IEEE Trans. Electron Devices* 22.11 (1975), pp. 1045–1047. DOI: 10.1109/T-ED.1975.18267.

[182] M. Capeans. "Aging and materials: lessons for detectors and gas systems". *NIMA* 515.1 (2003), pp. 73–88. DOI: 10.1016/j.nima.2003.08.134.

[183] J.-F. Caron et al. "Improved particle identification using cluster counting in a full-length drift chamber prototype". *NIMA* 735 (2014), pp. 169–183. DOI: 10.1016/j.nima.2013.09.028.

[184] N. Cartiglia et al. "Design optimization of ultra-fast silicon detectors". *NIMA* 796 (2015), pp. 141–148. DOI: 10.1016/j.nima.2015.04.025.

[185] N. Cartiglia et al. "The 4D pixel challenge". *JINST* 11.12 (2016), p. C12016. DOI: 10.1088/1748-0221/11/12/C12016.

[186] M. Catanesi et al. "Hadron, electron and muon response of a uranium-scintillator calorimeter". *NIMA* 260.1 (1987), pp. 43–54. DOI: 10.1016/0168-9002(87)90386-X.

[187] E. Chardonnet. "The DUNE dual-phase liquid argon TPC". *JINST* 15.05 (2020), p. C05064. DOI: 10.1088/1748-0221/15/05/C05064.

[188] G. Charpak et al. "High-accuracy localization of minimum ionizing particles using the cathode-induced charge centre-of-gravity read-out". *NIM* 167.3 (1979), pp. 455–464. DOI: 10.1016/0029-554X(79)90227-1.

[189] M. Chefdeville et al. "An electron-multiplying 'Micromegas' grid made in silicon wafer post-processing technology". *NIMA* 556.2 (2006), pp. 490–494. DOI: 10.1016/j.nima.2005.11.065.

[190] V. Chepel and H. Araújo. "Liquid noble gas detectors for low energy particle physics". *JINST* 8.04 (2013), R04001. DOI: 10.1088/1748-0221/8/04/R04001.

[191] A. Chilingarov. "Temperature dependence of the current generated in Si bulk". *JINST* 8.10 (2013), P10003–P10003. DOI: 10.1088/1748-0221/8/10/p10003.

[192] V. Chmill et al. "Study of various high voltage protection structures for reduction of the insensitive region of silicon sensors designed for extreme radiation tolerance". *JINST* 6.01 (2011), pp. C01062–C01062. DOI: 10.1088/1748-0221/6/01/c01062.

[193] L. Christofek et al. SVX4 User's Manual. URL: https://lss.fnal.gov/archive/test-tm/2000/fermilab-tm-2318-e.pdf.

[194] A. Clark et al. "Design and test of a prototype silicon detector module for ATLAS Semiconductor Tracker endcaps". *NIMA* 538.1 (2005), pp. 265–280. DOI: 10.1016/j.nima.2004.09.001.

[195] CMS collaboration. "Transverse-momentum and pseudorapidity distributions of charged Hadrons in pp collisions at $\sqrt{s} = 7$ TeV". *PRL* 105 (2 2010), p. 022002. DOI: 10.1103/PhysRevLett.105.022002.

[196] CMS collaboration. "The CMS experiment at the CERN LHC". *JINST* 3.08 (2008), S08004. DOI: 10.1088/1748-0221/3/08/S08004.

[197] CMS collaboration. "Commissioning and performance of the CMS silicon strip tracker with cosmic ray muons". *JINST* 5.03 (2010), T03008–T03008. DOI: 10.1088/1748-0221/5/03/t03008.

[198] CMS collaboration. "Strategies and performance of the CMS silicon tracker alignment during LHC Run 2". *NIMA* 1037 (2022), p. 166795. DOI: 10.1016/j.nima.2022.166795.

[199] CMS collaboration. *The CMS Electromagnetic Calorimeter Project: Technical Design Report*. Geneva: CERN, 1997. URL: https://cds.cern.ch/record/349375.

[200] CMS collaboration. *The Phase-2 Upgrade of the CMS Endcap Calorimeter*. Geneva: CERN, 2017. DOI: 10.17181/CERN.IV8M.1JY2.

[201] CMS collaboration. Jet energy scale and resolution measurement with Run 2 Legacy Data Collected by CMS at 13 TeV. 2021. URL: http://cds.cern.ch/record/2792322.

[202] CMS collaboration. "Performance of the CMS muon detector and muon reconstruction with proton-proton collisions at $\sqrt{s} = 13$ TeV". *JINST* 13.06 (2018), P06015. DOI: 10.1088/1748-0221/13/06/p06015.

[203] CMS collaboration. "Performance of the CMS drift tube chambers with cosmic rays". *JINST* 5.03 (2010), T03015. DOI: 10.1088/1748-0221/5/03/T03015.

[204] CMS collaboration. "Performance of reconstruction and identification of τ leptons decaying to hadrons and ν_τ in pp collisions at $\sqrt{s} = 13$ TeV". *JINST* 13 (2018), P10005. DOI: 10.1088/1748-0221/13/10/P10005.

[205] CMS collaboration. "Reconstruction and identification of τ lepton decays to hadrons and ν_τ at CMS". *JINST* 11.01 (2016), P01019. DOI: 10.1088/1748-0221/11/01/p01019.

[206] CMS collaboration. "Performance of the CMS muon trigger system in proton-proton collisions at $\sqrt{s} = 13$ TeV". *JINST* 16.07 (2021), P07001. DOI: 10.1088/1748-0221/16/07/P07001.

[207] CMS collaboration. "The CMS trigger system". *JINST* 12.01 (2017), P01020. DOI: 10.1088/1748-0221/12/01/P01020.

[208] P. Coe, D. Howell, and R. Nickerson. "Frequency scanning interferometry in ATLAS: remote, multiple, simultaneous and precise distance measurements in a hostile environment". *Meas. Sci. Tech.* 15.11 (2004), p. 2175. DOI: 10.1088/0957-0233/15/11/001.

[209] V. Commichau et al. "A transition radiation detector for pion identification in the 100 GeV/c momentum region". *NIM* 176.1 (1980), pp. 325–331. DOI: 10.1016/0029-554X(80)90724-7.

[210] G. Contin et al. "The STAR MAPS-based PiXeL detector". *NIMA* 907 (2018), pp. 60–80. DOI: 10.1016/j.nima.2018.03.003.

[211] E. Conwell and V. Weisskopf. "Theory of impurity scattering in semiconductors". *Phys. Rev.* 77.3 (1950), pp. 388–390. DOI: 10.1103/PhysRev.77.388.

[212] A. Cookson and T. Lewis. "Variations in the Townsend first ionization coefficient for gases". *Br. J. Appl. Phys* 17.11 (1966), pp. 1473–1481. DOI: 10.1088/0508-3443/17/11/312.

[213] L. Costrell et al. "Standard NIM instrumentation system" (1990). DOI: 10.2172/7120327.

[214] C. Curceanu et al. "X-ray detectors for Kaonic atoms research at DAΦNE". *Condensed Matter* 4.2 (2019). DOI: 10.3390/condmat4020042.

[215] C. D'Ambrosio. A short overview on scintillators. CERN Academic Training Programme 2004/2005. 2005. URL: https://ph-dep-dt2.web.cern.ch/CAT2005_3a.pdf.

[216] M. Danilov et al. "Aging studies for the muon detector of HERA-B". *NIMA* 515.1 (2003), pp. 202–219. DOI: 10.1016/j.nima.2003.08.150.

[217] G. Darbo. "Experience on 3D silicon sensors for ATLAS IBL". *JINST* 10.05 (2015), pp. C05001–C05001. DOI: 10.1088/1748-0221/10/05/c05001.

[218] N. Das Gupta and S. Ghosh. "A report on the Wilson Cloud Chamber and its applications in physics". *Rev. Mod. Phys.* 18.2 (1946), pp. 225–290. DOI: 10.1103/RevModPhys.18.225.

[219] I. Dawson. *Radiation Effects in the LHC Experiments: Impact on Detector Performance and Operation*. CERN Yellow Reports. Geneva: CERN, 2021. DOI: 10.23731/CYRM-2021-001.

[220] O. de Aguiar Francisco et al. "Microchannel cooling for the LHCb VELO Upgrade I". *NIMA* 1039 (2022), p. 166874. DOI: 10.1016/j.nima.2022.166874.

[221] J. Del Peso and E. Ros. "On the energy resolution of electromagnetic sampling calorimeters". *NIMA* 276.3 (1989), pp. 456–467. DOI: 10.1016/0168-9002(89)90571-8.

[222] R. Devanathan et al. "Signal variance in gamma-ray detectors—A review". *NIMA* 565.2 (2006), pp. 637–649. DOI: 10.1016/j.nima.2006.05.085.

[223] P. Di Gangi. "The xenon road to direct detection of Dark Matter at LNGS: The XENON Project". *Universe* 7.8 (2021). DOI: 10.3390/universe7080313.

[224] A. Di Mauro et al. "Photoelectron backscattering effects in photoemission from CsI into gas media". *NIMA* 371.1 (1996), pp. 137–142. DOI: 10.1016/0168-9002(95)01146-3.

[225] A. Di Mauro et al. "Status of the HMPID CsI-RICH project for ALICE at the CERN/LHC". *IEEE TNS* 52.4 (2005), pp. 972–979. DOI: 10.1109/TNS.2005.852743.

[226] Y. Ding et al. "A new gadolinium-loaded liquid scintillator for reactor neutrino detection". *NIMA* 584.1 (2008), pp. 238–243. DOI: 10.1016/j.nima.2007.09.044.

[227] B. Dolgoshein. "Transition radiation detectors". *NIMA* 326.3 (1993), pp. 434–469. DOI: 10.1016/0168-9002(93)90846-A.

[228] Z. Drasal and W. Riegler. "An extension of the Gluckstern formulae for multiple scattering: Analytic expressions for track parameter resolution using optimum weights". *NIMA* 910 (2018), pp. 127–132. DOI: 10.1016/j.nima.2018.08.078.

[229] J. Eberth and J. Simpson. "From Ge(Li) detectors to gamma-ray tracking arrays–50 years of gamma spectroscopy with germanium detectors". *Prog. Part. Nucl. Phys.* 60.2 (2008), pp. 283–337. DOI: 10.1016/j.ppnp.2007.09.001.

[230] *Electron Tube Products - Condensed Catalog*. Hamamatsu Photonics K.K., 2020. URL: https://www.hamamatsu.com/content/dam/hamamatsu-photonics/sites/documents/99_SALES_LIBRARY/etd/ETDproducts_TOTH0016E.pdf.

[231] J. Engler. "Liquid ionization chambers at room temperatures". *J. Phys. G* 22.1 (1996), pp. 1–23. DOI: 10.1088/0954-3899/22/1/002.

[232] A. Ereditato. "The study of neutrino oscillations with emulsion detectors". *Adv. High Energy Phys.* 2013 (2013), p. 382172. DOI: 10.1155/2013/382172.

[233] V. Eremin, E. Verbitskaya, and Z. Li. "The origin of double peak electric field distribution in heavily irradiated silicon detectors". *NIMA* 476.3 (2002), pp. 556–564. DOI: 10.1016/S0168-9002(01)01642-4.

[234] G. Erskine. "Electrostatic problems in multiwire proportional chambers". *NIM* 105.3 (1972), pp. 565–572. DOI: 10.1016/0029-554X(72)90356-4.

[235] C. Escobar. "Track based alignment of the ATLAS Inner Detector ". *PoS(VERTEX 2008)*. Vol. 068. 2009, p. 026. DOI: 10.22323/1.068.0026.

[236] J. Evans. "The MINOS experiment: Results and prospects". *Adv. High Energy Phys.* 2013 (2013), p. 182537. DOI: 10.1155/2013/182537.

[237] C. Fabjan and D. Fournier. "Calorimetry". *Particle Physics Reference Library: Volume 2: Detectors for Particles and Radiation*. Ed. by C. Fabjan and H. Schopper. Springer International Publishing, 2020, pp. 201–280. ISBN: 978-3-030-35318-6. DOI: 10.1007/978-3-030-35318-6_6.

[238] C. Fabjan et al. "Iron liquid-argon and uranium liquid-argon calorimeters for hadron energy measurement". *NIM* 141.1 (1977), pp. 61–80. DOI: 10.1016/0029-554X(77)90747-9.

[239] F. Faccio and G. Cervelli. "Radiation-induced edge effects in deep submicron CMOS transistors". *IEEE TNS* 52.6 (2005), pp. 2413–2420. DOI: 10.1109/TNS.2005.860698.

[240] D. Falchieri. "Status and performance of the ALICE Silicon Drift Detector in pp and Pb–Pb collisions". *NIMA* 730 (2013), pp. 24–27. DOI: 10.1016/j.nima.2013.05.042.

[241] F. Fallavollita. "Aging phenomena and discharge probability studies of the triple-GEM detectors for future upgrades of the CMS muon high rate region at the HL-LHC". *NIMA* 936 (2019), pp. 427–429. DOI: 10.1016/j.nima.2018.10.180.

[242] U. Fano. "Ionization yield of radiations. II. The fluctuations of the number of ions". *Phys. Rev.* 72.1 (1947), pp. 26–29. DOI: 10.1103/PhysRev.72.26.

[243] U. Fano. "Penetration of protons, alpha particles, and mesons". *Ann. Rev. Nucl. Sci.* 13.1 (1963), pp. 1–66. DOI: 10.1146/annurev.ns.13.120163.000245.

[244] V. Fanti et al. "The beam and detector for the NA48 neutral kaon CP violation experiment at CERN". *NIMA* 574.3 (2007), pp. 433–471. DOI: 10.1016/j.nima.2007.01.178.

[245] J. Fast. "The Belle II imaging Time-of-Propagation (iTOP) detector". *NIMA* 876 (2017), pp. 145–148. DOI: 10.1016/j.nima.2017.02.045.

[246] L. Fayard. *Transition radiation*. LAL 88-55. Paris-Saclay: LAL, 1988. URL: https://inis.iaea.org/collection/NCLCollectionStore/_Public/21/034/21034408.pdf.

[247] T. Ferguson et al. "Aging studies of CMS muon chamber prototypes". *NIMA* 488.1 (2002), pp. 240–257. DOI: 10.1016/S0168-9002(02)00400-X.

[248] J. Fernández-Varea et al. "On the theory and simulation of multiple elastic scattering of electrons". *NIMB* 73.4 (1993), pp. 447–473. DOI: 10.1016/0168-583X(93)95827-R.

[249] H. Fesefeldt. *The Simulation of Hadronic Showers: Physics and Applications*. Tech. rep. Aachen: Aachen TU 3. Inst. Phys., 1985. URL: http://cds.cern.ch/record/162911.

[250] R. Forty and O. Ullaland. "Particle identification: Time-of-flight, Cherenkov and transition radiation detectors". *Particle Physics Reference Library: Volume 2: Detectors for Particles and Radiation*. Ed. by C. Fabjan and H. Schopper. Springer International Publishing, 2020, pp. 281–335. ISBN: 978-3-030-35318-6. DOI: 10.1007/978-3-030-35318-6_7.

[251] G. Fraser and E. Mathieson. "Monte Carlo calculation of electron transport coefficients in counting gas mixtures: I. Argon-methane mixtures". *NIMA* 247.3 (1986), pp. 544–565. DOI: 10.1016/0168-9002(86)90417-1.

[252] R. Frühwirth. "Application of Kalman filtering to track and vertex fitting". *NIMA* 262.2 (1987), pp. 444–450. DOI: 10.1016/0168-9002(87)90887-4.

[253] R. Frühwirth, E. Brondolin, and A. Strandlie. "Pattern recognition and reconstruction". *Particle Physics Reference Library: Volume 2: Detectors for Particles and Radiation*. Ed. by C. Fabjan and H. Schopper. Springer International Publishing, 2020, pp. 555–612. ISBN: 978-3-030-35318-6. DOI: 10.1007/978-3-030-35318-6_13.

[254] S. Fukuda et al. "The Super-Kamiokande detector". *NIMA* 501.2 (2003), pp. 418–462. DOI: 10.1016/S0168-9002(03)00425-X.

[255] T. Gabriel et al. "Energy dependence of hadronic activity". *NIMA* 338.2 (1994), pp. 336–347. DOI: 10.1016/0168-9002(94)91317-X.

[256] M. Garcia-Sciveres and N. Wermes. "A review of advances in pixel detectors for experiments with high rate and radiation". *Rep. Prog. Phys.* 81.6 (2018), p. 066101. DOI: 10.1088/1361-6633/aab064.

[257] E. Gatti and P. Rehak. "Semiconductor drift chamber — An application of a novel charge transport scheme". *NIM* 225.3 (1984), pp. 608–614. DOI: 10.1016/0167-5087(84)90113-3.

[258] E. Gatti et al. "Considerations for the design of a time projection liquid Argon ionization chamber". *IEEE TNS* 26.2 (1979), pp. 2910–2932. DOI: 10.1109/TNS.1979.4330558.

[259] E. Gatti et al. "Optimum geometry for strip cathodes or grids in MWPC for avalanche localization along the anode wires". *NIM* 163.1 (1979), pp. 83–92. DOI: 10.1016/0029-554X(79)90035-1.

[260] Geant4 - a simulation toolkit. URL: https://geant4.org/.

[261] A. Gelmi et al. "Longevity studies on the CMS-RPC system". *JINST* 14.05 (2019), pp. C05012–C05012. DOI: 10.1088/1748-0221/14/05/c05012.

[262] Y. Giomataris et al. "MICROMEGAS: a high-granularity position-sensitive gaseous detector for high particle-flux environments". *NIMA* 376.1 (1996), pp. 29–35. DOI: 10.1016/0168-9002(96)00175-1.

[263] Y. Giomataris et al. "Micromegas in a bulk". *NIMA* 560.2 (2006), pp. 405–408. DOI: 10.1016/j.nima.2005.12.222.

[264] R. Gluckstern. "Uncertainties in track momentum and direction, due to multiple scattering and measurement errors". *NIM* 24 (1963), pp. 381–389. DOI: 10.1016/0029-554X(63)90347-1.

[265] L. Gonella et al. "A serial powering scheme for the ATLAS pixel detector at sLHC". *JINST* 5.12 (2010), p. C12002. DOI: 10.1088/1748-0221/5/12/C12002.

[266] H. van der Graaf et al. "The ultimate performance of the Rasnik 3-point alignment system". 2021. DOI: 10.48550/ARXIV.2104.03601.

[267] V. Gratchev et al. "Double track resolution of cathode strip chambers". *NIMA* 365.2 (1995), pp. 576–581. DOI: 10.1016/0168-9002(95)00670-2.

[268] D. Green. *The physics of particle detectors*. CUP, 2000. ISBN: 9780521662260.

[269] M. Green and M. Keevers. "Optical properties of intrinsic silicon at 300 K". *Prog. Photovolt.* (1995), p. 189. DOI: 10.1002/pip.4670030303.

[270] D. Griffiths. *Introduction to elementary particles*. 2nd ed. New York, NY: Wiley-VCH, 2010. ISBN: 978-0-471-60386-3.

[271] D. Groom. Atomic and nuclear properties of materials. 2018. URL: https://pdg.lbl.gov/2019/AtomicNuclearProperties/.

[272] CERN Gas Detectors Development Group. *MSGC: Micro-Strip Gas Chambers - Discharges and aging of MSGCs*. URL: https://gdd.web.cern.ch/others-msgc.

[273] CMS Electromagnetic Calorimeter Group. "Radiation hardness qualification of PbWO4 scintillation crystals for the CMS Electromagnetic Calorimeter". *JINST* 5.03 (2010), P03010. DOI: 10.1088/1748-0221/5/03/P03010.

[274] C. Grupen and B. Shwartz. *Particle detectors*. CUP, 2008. ISBN: 9780511534966. DOI: 10.1017/CBO9780511534966.

[275] A. Güntherschulze. "Die Elektronengeschwindigkeit in Isolatoren bei hohen Feldstärken und ihre Beziehung zur Theorie des elektrischen Durchschlages". *Z. Phys.* 86.11 (1933), pp. 778–786. DOI: 10.1007/BF01337879.

[276] G. Hall. "Ionisation energy losses of highly relativistic charged particles in thin silicon layers". *NIM* 220.2 (1984), pp. 356–362. DOI: 10.1016/0167-5087(84)90296-5.

[277] R. Hall. "Electron-hole recombination in Germanium". *Phys. Rev.* 87.2 (1952), pp. 387–387.
       DOI: 10.1103/PhysRev.87.387.

[278] A. Hanson et al. "Measurement of multiple scattering of 15.7-Mev electrons". *Phys. Rev.* 84.4
       (1951), pp. 634–637. DOI: 10.1103/PhysRev.84.634.

[279] G. Harigel. "Die Groe Europäische Blasenkammer im CERN (Teil I)". *Physikalische Blätter*
       31.1 (1975), pp. 13–28. DOI: 10.1002/phbl.19750310105.

[280] N. Harnew. "Alternative particle identification techniques to Cherenkov detectors". *NIMA* 766
       (2014), pp. 274–282. DOI: 10.1016/j.nima.2014.05.103.

[281] R. Harper. "Radiation damage studies of silicon detectors and searching for an intermediate mass
       Higgs boson at ATLAS". PhD thesis. Sheffield University, 2001. URL: http://cds.cern.ch/
       record/2759942.

[282] F. Hasert et al. "Observation of neutrino-like interactions without muon or electron in the
       Gargamelle neutrino experiment". *Phys. Lett B* 46.1 (1973), pp. 138–140. DOI: 10.1016/0370-
       2693(73)90499-1.

[283] A. Heggelund et al. "Radiation hard 3D silicon pixel sensors for use in the ATLAS detector at the
       HL-LHC". *JINST* 17.08 (2022), P08003. DOI: 10.1088/1748-0221/17/08/P08003.

[284] D. Hellenschmidt. "Experimental studies on small diameter carbon dioxide evaporators for opti-
       mal Silicon Pixel Detector cooling". Dissertation. Universität Bonn, 2020. URL: https://cds.
       cern.ch/record/2748428.

[285] A. Herve et al. "Status of the construction of the CMS magnet". *IEEE Trans. Appl. Supercond.*
       14.2 (2004), pp. 542–547. DOI: 10.1109/TASC.2004.829715.

[286] V. Highland. "Some practical remarks on multiple scattering". *NIM* 129.2 (1975), pp. 497–499.
       DOI: 10.1016/0029-554X(75)90743-0. Erratum in: "Erratum". *NIM* 161.1 (1979), p. 171.
       DOI: 10.1016/0029-554X(79)90379-3.

[287] P. Hobson. "Precision coordinate measurements using holographic recording". *J. Phys. E* 21.2
       (1988), pp. 139–145. DOI: 10.1088/0022-3735/21/2/002.

[288] R. Holroyd and D. Anderson. "The physics and chemistry of room-temperature liquid-filled ion-
       ization chambers". *NIMA* 236.2 (1985), pp. 294–299. DOI: 10.1016/0168-9002(85)90164-0.

[289] P. Horowitz and W. Hill. *The art of electronics,* 3rd ed. CUP, 2015. ISBN: 978-0521809269.

[290] J. Jackson. *Classical Electrodynamics.* 3rd ed. New York, NY: Wiley, 1999. ISBN:
       9780471309321.

[291] A. Jeavons et al. "A proportional chamber positron camera for medical imaging". *NIM* 176.1
       (1980), pp. 89–97. DOI: 10.1016/0029-554X(80)90686-2.

[292] J. Jelley. *Čerenkov Radiation and Its Applications.* London: Pergamon Press, 1958. ISBN:
       9780343149239.

[293] W. Jesse and J. Sadauskis. "Alpha-particle ionization in mixtures of the noble gases". *Phys. Rev.*
       88.2 (1952), pp. 417–418. DOI: 10.1103/PhysRev.88.417.

[294] M. Jiang et al. "Atmospheric neutrino oscillation analysis with improved event reconstruction in
       Super-Kamiokande IV". *Prog. Theor. Exp. Phys.* 2019.5 (2019). DOI: 10.1093/ptep/ptz015.

[295] M. Johnson and M. Sheaff. "Use of de-randomizing buffers in a data acquisition system". *AIP
       Conf. Proc.* 422.1 (1998), pp. 313–323. DOI: 10.1063/1.55072.

[296] J. Kadyk. "Wire chamber aging". *NIMA* 300.3 (1991), pp. 436–479. DOI: 10.1016/0168-
       9002(91)90381-Y.

[297] T. Kajita. "Atmospheric neutrinos". *New J. Phys.* 6.1 (2004), p. 194. DOI: 10.1088/1367-2630/6/1/194.

[298] Y. Kalkan et al. "Cluster ions in gas-based detectors". *JINST* 10.07 (2015), P07004–P07004. DOI: 10.1088/1748-0221/10/07/p07004.

[299] I. Kawrakow. EGSnrc toolkit for Monte Carlo simulation of ionizing radiation transport. 2000. URL: https://doi.org/10.4224/40001303.

[300] M. Keil. "Upgrade of the ALICE inner tracking system". *JINST* 10.03 (2015), p. C03012. DOI: 10.1088/1748-0221/10/03/C03012.

[301] G. Keiser. *Optical fiber communications*. McGraw-Hill, 2000. ISBN: 978-0071088084.

[302] J. Kemmer. "Fabrication of low noise silicon radiation detectors by the planar process". *NIM* 169.3 (1980), pp. 499–502. DOI: 10.1016/0029-554X(80)90948-9.

[303] J. Kemmer and G. Lutz. "New detector concepts". *NIMA* 253.3 (1987), pp. 365–377. DOI: 10.1016/0168-9002(87)90518-3.

[304] J. Kemmer et al. "Experimental confirmation of a new semiconductor detector principle". *NIMA* 288.1 (1990), pp. 92–98. DOI: 10.1016/0168-9002(90)90470-Q.

[305] A. Khintchine. "Korrelationstheorie der stationären stochastischen Prozesse". *Math. Ann.* 109.1 (1934), pp. 604–615. DOI: 10.1007/BF01449156.

[306] Ch. Kim et al. "A review of inorganic scintillation crystals for extreme environments". *Crystals* 11.6 (2021). DOI: 10.3390/cryst11060669.

[307] J. Kirkby et al. "Role of sulphuric acid, ammonia and galactic cosmic rays in atmospheric aerosol nucleation". *Nature* 476.7361 (2011), pp. 429–433. DOI: 10.1038/nature10343.

[308] A. Kiryunin et al. "GEANT4 physics evaluation with testbeam data of the ATLAS hadronic end-cap calorimeter". *NIMA* 560.2 (2006), pp. 278–290. DOI: 10.1016/j.nima.2005.12.237.

[309] C. Klein. "Bandgap dependence and related features of radiation ionization energies in semiconductors". *J. Appl. Phys.* 39.4 (2003), pp. 2029–2038. DOI: 10.1063/1.1656484.

[310] O. Klein and Y. Nishina. "Über die Streuung von Strahlung durch freie Elektronen nach der neuen relativistischen Quantendynamik von Dirac". *Z. Phys.* 52.11 (1929), pp. 853–868. DOI: 10.1007/BF01366453.

[311] Sp. Klein. "Suppression of bremsstrahlung and pair production due to environmental factors". *Rev. Mod. Phys.* 71 (5 1999), pp. 1501–1538. DOI: 10.1103/RevModPhys.71.1501.

[312] G. Knoll. *Radiation detection and measurement*. 4th ed. John Wiley & Sons, Inc., 2010. ISBN: 9780470131480.

[313] E. Kobetich and R. Katz. "Electron energy dissipation". *NIM* 71.2 (1969), pp. 226–230. DOI: 10.1016/0029-554X(69)90019-6.

[314] K. Kodama et al. "Observation of tau neutrino interactions". *Phys. Lett. B* 504.3 (2001), pp. 218–224. DOI: 10.1016/S0370-2693(01)00307-0.

[315] A. Kok et al. "High aspect ratio deep RIE for novel 3D radiation sensors in high energy physics applications". *2009 IEEE Nuclear Science Symposium Conference Record (NSS/MIC)*. 2009, pp. 1623–1627. DOI: 10.1109/NSSMIC.2009.5402256.

[316] H. Kolanoski and N. Wermes. *Particle Detectors. Fundamentals and Applications*. OUP, 2020. ISBN: 9780198858362.

[317] S. Korpar and P. Krizan. "Photon detectors". *Handbook of Particle Detection and Imaging*. Ed. by I. Fleck et al. Springer International Publishing, 2021, pp. 353–370. ISBN: 978-3-319-93785-4. DOI: 10.1007/978-3-319-93785-4_13.

[318] G. Kramberger. "Advanced TCT setups". *PoS* Vertex2014 (2015), p. 032. DOI: 10.22323/1.227.0032.

[319] M. Krammer and F. Hartmann. "Introduction to silicon detectors". Talk given at EDIT 2011, Silicon strips and pixels technologies. 2011. URL: https://indico.cern.ch/event/124392/.

[320] O. Krasel. "Charge collection in irradiated silicon-detectors". Dissertation. Universität Dortmund, 2004. DOI: 10.17877/DE290R-14839.

[321] M. Krivda et al. "The ALICE trigger system performance for p-p and Pb-Pb collisions". *JINST* 7.01 (2012), p. C01057. DOI: 10.1088/1748-0221/7/01/C01057.

[322] M. Kubantsev et al. "Performance of the PrimEx electromagnetic calorimeter". *AIP Conference Proceedings* 867 (2006). DOI: 10.1063/1.2396938.

[323] K. Kumar, H. Skullerud, and R. Robson. "Kinetic theory of charged particle swarms in neutral gases". *Aust. J. Phys.* 33.2 (1980), pp. 343–448. DOI: 10.1071/PH800343B.

[324] L. Landau. "On the energy loss of fast particles by ionisation". *J. Phys. U.S.S.R.* 8 (1944), p. 201. English translation in "On the energy loss of fast particles by ionisation". *Collected Papers of L.D. Landau*. Ed. by D. Ter Haar. Pergamon, 1965, pp. 417–424. ISBN: 978-0-08-010586-4. DOI: 10.1016/B978-0-08-010586-4.50061-4.

[325] P. Lechner et al. "Silicon drift detectors for high resolution room temperature X-ray spectroscopy". *NIMA* 377.2 (1996), pp. 346–351. DOI: 10.1016/0168-9002(96)00210-0.

[326] P. Lecoq. "Scintillation detectors for charged particles and photons". *Particle Physics Reference Library: Volume 2: Detectors for Particles and Radiation*. Ed. by C. Fabjan and H. Schopper. Springer International Publishing, 2020, pp. 45–89. ISBN: 978-3-030-35318-6. DOI: 10.1007/978-3-030-35318-6_3.

[327] S. Lee, M. Livan, and R. Wigmans. "Dual-readout calorimetry". *Rev. Mod. Phys.* 90.2 (2018), p. 025002. DOI: 10.1103/RevModPhys.90.025002.

[328] W. Legler. "Die Statistik der Elektronenlawinen in elektronegativen Gasen, bei hohen Feldstärken und bei groer Gasverstärkung". *Z. Naturforsch. A* 16.3 (1961), pp. 253–261. DOI: 10.1515/zna-1961-0308.

[329] W. Leo. "The NIM standard". *Techniques for Nuclear and Particle Physics Experiments: A How-to Approach*. Berlin, Heidelberg: Springer, 1994, pp. 257–261. ISBN: 978-3-642-57920-2. DOI: 10.1007/978-3-642-57920-2_12.

[330] C. Leroy and P. Rancoita. *Principles of radiation interaction in matter and detection*. World Scientific, 2016. ISBN: 978-9814603188.

[331] H. Leutz. "Scintillating fibres". *NIMA* 364.3 (1995), pp. 422–448. DOI: 10.1016/0168-9002(95)00383-5.

[332] LHCb Collaboration. "The LHCb detector at the LHC". *JINST* 3.08 (2008), S08005. DOI: 10.1088/1748-0221/3/08/S08005.

[333] LHCb Collaboration. "The LHCb upgrade I" (2023). DOI: 10.48550/arXiv.2305.10515.

[334] LHCb Collaboration. LHCb Trigger and Online Upgrade Technical Design Report. 2014. URL: https://cds.cern.ch/record/1701361.

[335] Y.-C. Lin, M. Bettinelli, and M. Karlsson. "Unraveling the mechanisms of thermal quenching of luminescence in Ce3+-doped garnet phosphors". *Chem. Mater.* 31.11 (2019), pp. 3851–3862. DOI: 10.1021/acs.chemmater.8b05300.

[336] V. Lindstrom. *Displacement Damage in Silicon, On-line Compilation.* 2000. URL: https://rd50.web.cern.ch/niel/default.html.

[337] G. Lindström. "Radiation damage in silicon detectors". *NIMA* 512.1 (2003), pp. 30–43. DOI: 10.1016/S0168-9002(03)01874-6.

[338] C. Lippmann. "Detector physics of resistive plate chambers". PhD thesis. Frankfurt University, 2003. URL: http://cds.cern.ch/record/1303626/.

[339] C. Lippmann. "Particle identification". *NIMA* 666 (2012), pp. 148–172. DOI: 10.1016/j.nima.2011.03.009.

[340] G. Lipták et al. *Radiation Tests on Selected Electrical Insulating Materials for High-power and High-voltage Application.* CERN Yellow Reports. Geneva: CERN, 1985. DOI: 10.5170/CERN-1985-002.

[341] M. Livan, V. Vercesi, and R. Wigmans. *Scintillating-fibre Calorimetry.* CERN Yellow Reports. Geneva: CERN, 1995. DOI: 10.5170/CERN-1995-002.

[342] M. Livan and R. Wigmans. *Calorimetry for collider physics, an introduction.* Springer, 2020. ISBN: 978-3-030-23653-3. DOI: 10.1007/978-3-030-23653-3.

[343] X. Llopart et al. "Timepix, a 65k programmable pixel readout chip for arrival time, energy and/or photon counting measurements". *NIMA* 581.1 (2007), pp. 485–494. DOI: 10.1016/j.nima.2007.08.079.

[344] F. Di Lodovico. "The Hyper-Kamiokande experiment". *J. Phys. Conf. Ser.* 888.1 (2017), p. 012020. DOI: 10.1088/1742-6596/888/1/012020.

[345] M. Lucchini. "Crystal calorimetry". Talk given at the ECFA Detector R&D Roadmap Symposium of Task Force 6 Calorimetry. 2021. URL: https://indico.cern.ch/event/999820/contributions/4200695/attachments/2241036/3799740/2021_05_07_ECFA_TF6_Lucchini_CrystalCalorimetry.pdf/.

[346] P. Luke et al. "Low capacitance large volume shaped-field germanium detector". *IEEE TNS* 36.1 (1989), pp. 926–930. DOI: 10.1109/23.34577.

[347] E. Lund et al. "Track parameter propagation through the application of a new adaptive Runge-Kutta-Nyström method in the ATLAS experiment". *JINST* 4.04 (2009), P04001. DOI: 10.1088/1748-0221/4/04/P04001.

[348] M. Lupberger. "The Pixel-TPC: first results from an 8-InGrid module". *JINST* 9.01 (2014), pp. C01033–C01048. DOI: 10.1088/1748-0221/9/01/C01033.

[349] G. Lutz. *Semiconductor radiation detectors, device physics.* Springer, 2007. ISBN: 978-3-540-71678-5. URL: https://doi.org/10.1007/978-3-540-71679-2.

[350] G. Lutz and R. Klanner. "Solid state detectors". *Particle Physics Reference Library: Volume 2: Detectors for Particles and Radiation.* Ed. by C. Fabjan and H. Schopper. Springer International Publishing, 2020, pp. 137–200. ISBN: 978-3-030-35318-6. DOI: 10.1007/978-3-030-35318-6_5.

[351] G. Lutz et al. "The DEPFET sensor-amplifier structure: A method to beat 1/f noise and reach sub-electron noise in pixel detectors". *Sensors* 16.5 (2016), p. 608. DOI: 10.3390/s16050608.

[352] LXcat. URL: https://nl.lxcat.net/.

[353] G. Lynch and O. Dahl. "Approximations to multiple Coulomb scattering". *NIMB* 58.1 (1991), pp. 6–10. DOI: 10.1016/0168-583X(91)95671-Y.

[354] A. Machado and E. Segreto. "ARAPUCA a new device for liquid argon scintillation light detection". *JINST* 11.02 (2016), p. C02004. DOI: 10.1088/1748-0221/11/02/C02004.

[355] K. Majumdar and K. Mavrokoridis. "Review of liquid Argon detector technologies in the neutrino sector". *Appl. Sci.* 11.6 (2021). DOI: 10.3390/app11062455.

[356] L. Malter. "Thin film field emission". *Phys. Rev.* 50.1 (1936), pp. 48–58. DOI: 10.1103/PhysRev.50.48.

[357] R. Mankel. "A concurrent track evolution algorithm for pattern recognition in the HERA-B main tracking system". *NIMA* 395.2 (1997), pp. 169–184. DOI: 10.1016/S0168-9002(97)00705-5.

[358] A. Marchionni. "Status and new ideas regarding liquid argon detectors". *Ann. Rev. Nucl. Part. Sci.* 63.1 (2013), pp. 269–290. DOI: 10.1146/annurev.nucl.012809.104445.

[359] G. Marr and J. West. "Absolute photoionization cross-section tables for helium, neon, argon, and krypton in the VUV spectral regions". *Atomic Data and Nuclear Data Tables* 18.5 (1976), pp. 497–508. DOI: 10.1016/0092-640X(76)90015-2.

[360] S. Mathimalar et al. "Characterization of Neutron Transmutation Doped (NTD) Ge for low temperature sensor development". *NIMB* 345 (2015), pp. 33–36. DOI: 10.1016/j.nimb.2014.12.020.

[361] D. McCammon. "Thermal equilibrium calorimeters – An introduction". *Cryogenic Particle Detection*. Ed. by Christian Enss. Berlin, Heidelberg: Springer Berlin Heidelberg, 2005, pp. 1–34. ISBN: 978-3-540-31478-3. DOI: 10.1007/10933596_1.

[362] R. McIntyre. "Multiplication noise in uniform Avalanche diodes". *IEEE Trans. Electron Dev.* 13 (1966), p. 164. DOI: 10.1109/T-ED.1966.15651.

[363] A. McKemey. "The SLD CCD pixel vertex detector and its upgrade". *Il Nuovo Cimento A* 109.6 (1996), pp. 1027–1034. DOI: 10.1007/BF02823643.

[364] P. McNulty, V. Pease, and V. Bond. "Visual sensations induced by relativistic pions". *Radiation Research* 66.3 (1976), pp. 519–530. DOI: 10.2307/3574456.

[365] B. Mecking et al. "The CEBAF large acceptance spectrometer (CLAS)". *NIMA* 503.3 (2003), pp. 513–553. DOI: 10.1016/S0168-9002(03)01001-5.

[366] Ch. Meegan et al. "The Fermi gamma-ray burst monitor". *Astrophys. J.* 702.1 (2009), p. 791. DOI: 10.1088/0004-637X/702/1/791.

[367] C. Melcher. "Scintillators for well logging applications". *NIMB* 40-41 (1989), pp. 1214–1218. DOI: 10.1016/0168-583X(89)90622-8.

[368] M. Mentink et al. Superconducting detector magnets for high energy physics. 2022. DOI: 10.48550/ARXIV.2203.07799.

[369] F. Merritt et al. "Hadron shower punchthrough for incident hadrons of momentum 15, 25, 50, 100, 200 and 300 GeV/c". *NIMA* 245.1 (1986), pp. 27–34. DOI: 10.1016/0168-9002(86)90254-8.

[370] A. Meyer. Heavy Quark Production at HERA. Habilitationsschrift, Universität Hamburg. 2005. URL: https://www-h1.desy.de/psfiles/theses/h1th-409.pdf.gz.

[371] D. Michael et al. "The magnetized steel and scintillator calorimeters of the MINOS experiment". *NIMA* 596.2 (2008), pp. 190–228. DOI: 10.1016/j.nima.2008.08.003.

[372] MicroBooNE at Work - Event Displays. URL: https : / / microboone - exp . fnal . gov / public/approved_plots/Event_Displays.html.

[373] P. Miné. "Photoemissive materials and their application to gaseous detectors". *NIMA* 343.1 (1994), pp. 99–108. DOI: 10.1016/0168-9002(94)90538-X.

[374] MINOS Collaboration. The MINOS Detectors Technical Design Report. FERMILAB-DESIGN-1998-02. 1998. URL: https://lss.fnal.gov/archive/design/fermilab-design-1998-02.pdf.

[375] G. Molière. "Theorie der Streuung schneller geladener Teilchen I. Einzelstreuung am abgeschirmten Coulomb-Feld". *Z. Naturforsch. A* 2.3 (1947), pp. 133–145. DOI: 10.1515/zna-1947-0302.

[376] G. Molière. "Theorie der Streuung schneller geladener Teilchen II Mehrfach-und Vielfachstreuung". *Z. Naturforsch. A* 3.2 (1948), pp. 78–97. DOI: 10.1515/zna-1948-0203.

[377] M. Moll. "Radiation damage in silicon particle detectors: Microscopic defects and macroscopic properties". Dissertation. Universität Hamburg, 1999. DOI: 10.3204/PUBDB-2016-02525.

[378] M. Moll. "Displacement damage in silicon detectors for high energy physics". *IEEE TNS* 65.8 (2018), pp. 1561–1582. DOI: 10.1109/TNS.2018.2819506.

[379] L. Montanet et al. "Review of particle physics". *Phys. Rev.* D50 (1994), p. 1173. URL: https://pdg.lbl.gov/1995/.

[380] M. Moritz et al. "Performance study of new pixel hybrid photon detector prototypes for the LHCb RICH counters". *IEEE TNS* 5 (2004). DOI: 10.1109/TNS.2004.829450.

[381] N. Mott. "The wave mechanics of α-ray tracks". *Proc. R. Soc. A* 126.800 (1929), pp. 79–84. DOI: 10.1098/rspa.1929.0205.

[382] S. Mukhopadhyay and N. Majumdar. A nearly exact boundary element method. URL: https://nebem.web.cern.ch/nebem/.

[383] S. Mukhopadhyay et al. "Exact solutions to model surface and volume charge distributions". *J. Phys. Conf. Ser.* 759 (2016), p. 012073. DOI: 10.1088/1742-6596/759/1/012073.

[384] K. Murakami et al. "Systematic comparison of electromagnetic physics between Geant4 and EGS4 with respect to protocol data". *IEEE Symposium Conference Record Nuclear Science 2004*. Vol. 4. 2004, pp. 2120–2123. DOI: 10.1109/NSSMIC.2004.1462681.

[385] T. Nakamura et al. "The OPERA film: New nuclear emulsion for large-scale, high-precision experiments". *NIMA* 556.1 (2006), pp. 80–86. DOI: 10.1016/j.nima.2005.08.109.

[386] E. Nappi and J. Seguinot. "Ring imaging Cherenkov detectors: The state of the art and perspectives". *La Rivista del Nuovo Cimento* 28.8 (2005), pp. 1–130. DOI: 10.1393/ncr/i2006-10004-6.

[387] W. Newhauser and R. Zhang. "The physics of proton therapy". *Phys. Med. Biol.* 60.8 (2015), R155–R209. DOI: 10.1088/0031-9155/60/8/r155.

[388] C. Nielsen, T. Needels, and O. Weddle. "Diffusion cloud chambers". *Rev. Sci. Instr.* 22.9 (1951), pp. 673–677. DOI: 10.1063/1.1746033.

[389] I. Obodovskiy. "Chapter 6 - Interaction of gamma quanta with matter". *Radiation*. Ed. by I. Obodovskiy. Elsevier, 2019, pp. 137–150. ISBN: 978-0-444-63979-0. DOI: 10.1016/B978-0-444-63979-0.00006-9.

[390] A. Oed. "Position-sensitive detector with microstrip anode for electron multiplication with gases". *NIMA* 263.2 (1988), pp. 351–359. DOI: 10.1016/0168-9002(88)90970-9.

[391] A. Olszewski. "High energy physics experiments in GRID computing networks". *Comp. Sci.* 9.3 (2013), p. 97. URL: https://journals.agh.edu.pl/csci/article/view/191.

[392] OPAL collaboration. "The OPAL detector at LEP". *NIMA* 305.2 (1991), pp. 275–319. DOI: 10. 1016/0168-9002(91)90547-4.

[393] W. Orthmann. "Ein Differentialkalorimeter zur Absolutbestimmung kleinster Wärmemengen". *Z. Phys.* 60.3 (1930), pp. 137–142. DOI: 10.1007/BF01339818.

[394] A. Owens. "Spectral degradation effects in an 86 cm^3 Ge(HP) detector". *NIMA* 238.2 (1985), pp. 473–478. DOI: 10.1016/0168-9002(85)90487-5.

[395] A. Owens and A. Peacock. "Compound semiconductor radiation detectors". *NIMA* 531.1 (2004), pp. 18–37. DOI: 10.1016/j.nima.2004.05.071.

[396] V. Palladino and B. Sadoulet. "Application of classical theory of electrons in gases to drift proportional chambers". *NIM* 128.2 (1975), pp. 323–335. DOI: 10.1016/0029-554X(75)90682-5.

[397] A. Papanestis and C. D'Ambrosio. "Performance of the LHCb RICH detectors during the LHC Run II". *NIMA* 876 (2017), pp. 221–224. DOI: 10.1016/j.nima.2017.03.009.

[398] J. Parker and J. Lowke. "Theory of electron diffusion parallel to electric fields. I. Theory". *Phys. Rev.* 181.1 (1969), pp. 290–301. DOI: 10.1103/PhysRev.181.290. And J. Lowke and J. Parker. "Theory of electron diffusion parallel to electric fields. II. Application to real gases". *Phys. Rev.* 181.1 (1969), pp. 302–311. DOI: 10.1103/PhysRev.181.302.

[399] C. Parkes and R. Lidner. Framework TDR for the LHCb Upgrade II. CERN-LHCC-2021. 2021. URL: https://cds.cern.ch/record/2776420/files/LHCB-TDR-023.pdf.

[400] G. Parks. "Magnetosphere". *Encyclopedia of atmospheric sciences*. Ed. by G. North, J. Pyle, and F. Zhang. Second Edition. Oxford: Academic Press, 2015, pp. 309–315. ISBN: 978-0-12-382225-3. DOI: 10.1016/B978-0-12-382225-3.00211-5.

[401] H. Patton and B. Nachman. "The optimal use of silicon pixel charge information for particle identification". *NIMA* 913 (2019), pp. 91–96. DOI: 10.1016/j.nima.2018.10.120.

[402] M. Pelgrom. *Analog-to-digital conversion*. Springer, 2018. ISBN: 978-3-319-83175-6.

[403] L. Peralta. "Temperature dependence of plastic scintillators". *NIMA* 883 (2018), pp. 20–23. DOI: 10.1016/j.nima.2017.11.041.

[404] P. Phillips. "Design, development, characterisation and operation of ATLAS ITk strip staves". PhD thesis. University of Oxford, 2020. URL: https://ora.ox.ac.uk/objects/uuid: 16ea0c42-8bae-482d-bfbc-e1d898124461.

[405] *Photomultiplier Tubes - Basics and Applications*. Hamamatsu Photonics K.K., 2007.

[406] R. Pierret. *Field Effect Devices*. Addison-Wesley Publishing Company, 1990. ISBN: 9780201053234.

[407] L. Pitchford, S. ONeil, and J. Rumble. "Extended Boltzmann analysis of electron swarm experiments". *Phys. Rev. A* 23.1 (1981), pp. 294–304. DOI: 10.1103/PhysRevA.23.294.

[408] F. Piuz, R. Roosen, and J. Timmermans. "Evaluation of systematic errors in the avalanche localization along the wire with cathode strips read-out MWPC". *NIM* 196.2 (1982), pp. 451–462. DOI: 10.1016/0029-554X(82)90113-6.

[409] S. Plewnia et al. "A sampling calorimeter with warm-liquid ionization chambers". *NIMA* 566.2 (2006), pp. 422–432. DOI: 10.1016/j.nima.2006.07.051.

[410] K. Pretzl. "Cryogenic detectors". *Particle Physics Reference Library: Volume 2: Detectors for Particles and Radiation*. Ed. by C. Fabjan and H. Schopper. Springer International Publishing, 2020, pp. 871–912. ISBN: 978-3-030-35318-6. DOI: 10.1007/978-3-030-35318-6_19.

[411] PyBoltz GIT repository. URL: https://github.com/UTA-REST/PyBoltz.

[412] F. Quarati et al. "X-ray and gamma-ray response of a $2'' \times 2''$ LaBr$_3$:Ce scintillation detector". *NIMA* 574.1 (2007), pp. 115–120. DOI: 10.1016/j.nima.2007.01.161.

[413] V. Radeka. "Low-noise techniques in detectors". *Ann. Rev. Nucl. Part. Sci.* 38.1 (1988), pp. 217–277. DOI: 10.1146/annurev.ns.38.120188.001245.

[414] V. Radeka. "Signal processing for particle detectors". *Particle Physics Reference Library: Volume 2: Detectors for Particles and Radiation*. Ed. by C. Fabjan and H. Schopper. Springer International Publishing, 2020, pp. 439–484. ISBN: 978-3-030-35318-6. DOI: 10.1007/978-3-030-35318-6_10.

[415] V. Radeka and P. Rehak. "Second coordinate readout in drift chambers by charge division". *IEEE TNS* 25.1 (1978), pp. 46–52. DOI: 10.1109/TNS.1978.4329274.

[416] H. Raether. "Die Entwicklung der Elektronenlawine in den Funkenkanal". *Z. Phys.* 112.7 (1939), pp. 464–489. DOI: 10.1007/BF01340229.

[417] F. Ragusa and L. Rolandi. "Tracking at LHC". *New J. Phys.* 9.9 (2007), p. 336. DOI: 10.1088/1367-2630/9/9/336.

[418] S. Ramo. "Currents induced by electron motion". *Proceedings of the IRE* 27.9 (1939), pp. 584–585. DOI: 10.1109/JRPROC.1939.228757.

[419] V.-T. Rangel-Kuoppa et al. "Towards GaAs thin-film tracking detectors". *JINST* 16.09 (2021), P09012. DOI: 10.1088/1748-0221/16/09/P09012.

[420] M. Regler and R. Frühwirth. "Generalization of the Gluckstern formulas I: Higher orders, alternatives and exact results". *NIMA* 589.1 (2008), pp. 109–117. DOI: 10.1016/j.nima.2008.02.016.

[421] F. Reines and C. Cowan. "Detection of the free neutrino". *Phys. Rev.* 92.3 (1953), pp. 830–831. DOI: 10.1103/PhysRev.92.830.

[422] W. Riegler. "Limits to drift chamber resolution". PhD thesis. TU Wien, 1998. URL: https://cds.cern.ch/record/1274450.

[423] W. Riegler. "Extended theorems for signal induction in particle detectors". *NIMA* 535.1 (2004), pp. 287–293. DOI: 10.1016/j.nima.2004.07.129.

[424] W. Riegler. "Electric fields, weighting fields, signals and charge diffusion in detectors including resistive materials". *JINST* 11.11 (2016), P11002–P11002. DOI: 10.1088/1748-0221/11/11/p11002.

[425] W. Riegler, Ch. Lippmann, and R. Veenhof. "Detector physics and simulation of resistive plate chambers". *NIMA* 500.1 (2003). NIMA Vol 500, pp. 144–162. DOI: 10.1016/S0168-9002(03)00337-1.

[426] W. Riegler and P. Windischhofer. "Signals induced on electrodes by moving charges, a general theorem for Maxwell's equations based on Lorentz-reciprocity". *NIMA* 980 (2020), p. 164471. DOI: 10.1016/j.nima.2020.164471.

[427] W. Riegler et al. "Resolution limits of drift tubes". *NIMA* 443.1 (2000), pp. 156–163. DOI: 10.1016/S0168-9002(99)01014-1.

[428] A. Roberts. "Development of the spark chamber: A review". *Rev. Sci. Instrum.* 32.5 (1961), pp. 482–485. DOI: 10.1063/1.1717420; and other articles in this volume.

[429] G. Rochester and C. Butler. "Evidence for the existence of new unstable elementary particles". *Nature* 160.4077 (1947), pp. 855–857. DOI: 10.1038/160855a0.

[430] S. Roesler, R. Engel, and J. Ranft. "The Monte Carlo event generator DPMJET-III". *Advanced Monte Carlo for Radiation Physics, Particle Transport Simulation and Applications*. Ed. by Andreas Kling et al. Berlin, Heidelberg: Springer, 2001, pp. 1033–1038. ISBN: 978-3-642-18211-2.

[431] W. Röntgen. "Eine neue Art von Strahlen". *Sitzungsberichte der Würzburger physik.-medic. Gesellschaft* (1895). URL: https://wellcomecollection.org/works/avjgayxz. English translation in "On a New Kind of Rays". *Science* 3.59 (1896), pp. 227–231. DOI: 10.1126/science.3.59.227.

[432] B. Rossi. *High-energy Particles*. Prentice-Hall Physics Series. Prentice Hall, 1952. ISBN: 978-0133873245.

[433] B. Rossi and K. Greisen. "Cosmic-ray theory". *Rev. Mod. Phys.* 13.4 (1941), pp. 240–309. DOI: 10.1103/RevModPhys.13.240.

[434] B. Rossi and N. Nereson. "Experimental arrangement for the measurement of small time intervals between the discharges of Geiger-Müller counters". *Rev. Sci. Instr.* 17.2 (1946), pp. 65–71. DOI: 10.1063/1.1770435.

[435] L. Rossi et al. *Pixel detectors, from fundamentals to applications*. Berlin, Heidelberg: Springer, 2006. ISBN: 978-3-540-28332-4. DOI: 10.1007/3-540-28333-1.

[436] S. Roy, R. Apfel, and Y.-C. Lo. "Superheated drop detector: A potential tool in neutron research". *NIMA* 255.1 (1987), pp. 199–206. DOI: 10.1016/0168-9002(87)91101-6.

[437] Ö. Sahin and T. Kowalski. "Measurements and calculations of electron avalanche growth in ternary mixture of Ne-CO_2-N_2". *JINST* 11.11 (2016), P11012–P11012. DOI: 10.1088/1748-0221/11/11/p11012.

[438] Ö. Sahin, T. Kowalski, and R. Veenhof. "High-precision gas gain and energy transfer measurements in Ar–CO2 mixtures". *NIMA* 768 (2014), pp. 104–111. DOI: 10.1016/j.nima.2014.09.061.

[439] F. Sauli. "Principles of operation of multiwire proportional and drift chambers". CERN, Geneva, 1975 - 1976. CERN. Geneva: CERN, 1977, 92 p. DOI: 10.5170/CERN-1977-009.

[440] F. Sauli. "GEM: A new concept for electron amplification in gas detectors". *NIMA* 386.2 (1997), pp. 531–534. DOI: 10.1016/S0168-9002(96)01172-2.

[441] F. Sauli. *Gaseous radiation detectors: Fundamentals and applications*. Cambridge Monographs on Particle Physics, Nuclear Physics and Cosmology. CUP, 2014. ISBN: 9781107337701. DOI: 10.1017/CBO9781107337701.

[442] F. Sauli. "The gas electron multiplier (GEM): Operating principles and applications". *NIMA* 805 (2016), pp. 2–24. DOI: 10.1016/j.nima.2015.07.060.

[443] G. Saviano et al. "Properties of potential eco-friendly gas replacements for particle detectors in high-energy physics". *JINST* 13.03 (2018), P03012–P03012. DOI: 10.1088/1748-0221/13/03/p03012.

[444] P. Schade and J. Kaminski. "A large TPC prototype for a linear collider detector". *NIMA* 628.1 (2011), pp. 128–132. DOI: 10.1016/j.nima.2010.06.300.

[445] H. Schindler. "Microscopic simulation of particle detectors". PhD thesis. TU Wien, 2012. URL: https://cds.cern.ch/record/1500583.

[446] H. Schindler. Garfield++. URL: https://garfieldpp.web.cern.ch/garfieldpp/.

[447] H. Schindler, S. Biagi, and R. Veenhof. "Calculation of gas gain fluctuations in uniform fields". *NIMA* 624.1 (2010), pp. 78–84. DOI: 10.1016/j.nima.2010.09.072.

[448] S. Schmeling et al. "The detector safety system for LHC experiments". *IEEE TNS* 51.3 (2004), pp. 521–525. DOI: 10.1109/TNS.2004.828631.

[449] S. Schuh et al. "Long-term geometry stability of ATLAS MDT chambers studied with a high-precision X-ray tomograph". *IEEE Nuclear Science Symposium Conference Record, 2005*. Vol. 2. 2005, pp. 1024–1028. DOI: 10.1109/NSSMIC.2005.1596427.

[450] B. Sciascia. "LHCb Run 2 trigger performance". *PoS* BEAUTY2016 (2016), p. 029. DOI: 10.22323/1.273.0029.

[451] F. Sefkow et al. "Experimental tests of particle flow calorimetry". *Rev. Mod. Phys.* 88.1 (2016), p. 015003. DOI: 10.1103/RevModPhys.88.015003.

[452] F. Seitz. "On the theory of the bubble chamber". *Phys. Fluids* 1.1 (1958), pp. 2–13. DOI: 10.1063/1.1724333.

[453] P. Sellin and J. Vaitkus. "New materials for radiation hard semiconductor dectectors". *NIMA* 557.2 (2006), pp. 479–489. DOI: 10.1016/j.nima.2005.10.128.

[454] L. Shekhtman. "Micro-pattern gaseous detectors". *NIMA* 494.1 (2002), pp. 128–141. DOI: 10.1016/S0168-9002(02)01456-0.

[455] W. Shockley. "Currents to conductors induced by a moving point charge". *J. Appl. Phys* 9.10 (1938), pp. 635–636. DOI: 10.1063/1.1710367.

[456] W. Shockley and W. Read. "Statistics of the recombinations of holes and electrons". *Phys. Rev.* 87.5 (1952), pp. 835–842. DOI: 10.1103/PhysRev.87.835.

[457] A. Sirunyan et al. "Mechanical stability of the CMS strip tracker measured with a laser alignment system". *JINST* 12.04 (2017), P04023. DOI: 10.1088/1748-0221/12/04/P04023.

[458] T. Sjöstrand, St. Mrenna, and P. Skands. "PYTHIA 6.4 physics and manual". *J. High Energy Phys.* 2006.05 (2006), p. 026. DOI: 10.1088/1126-6708/2006/05/026.

[459] T. Sjöstrand, St. Mrenna, and P. Skands. "A brief introduction to PYTHIA 8.1". *Comp. Phys. Comm.* 178.11 (2008), pp. 852–867. DOI: 10.1016/j.cpc.2008.01.036.

[460] I. Smirnov. "Modeling of ionization produced by fast charged particles in gases". *NIMA* 554.1 (2005), pp. 474–493. DOI: 10.1016/j.nima.2005.08.064.

[461] G. Smith, J. Fischer, and V. Radeka. "Capacitive charge division in centroid finding cathode read-outs in MWPCs". *IEEE TNS* 35.1 (1988), pp. 409–413. DOI: 10.1109/23.12754.

[462] F. Spada. "AMS-02 on the International Space Station". *EPJ Web of Conferences* 70 (2014), p. 00026. DOI: 10.1051/epjconf/20147000026.

[463] H. Spieler. Analog and digital electronics for detectors. 1998. URL: https://www.desy.de/~garutti/LECTURES/ParticleDetectorSS12/spieler.pdf.

[464] H. Spieler. *Semiconductor Detector Systems*. OUP, 2005. ISBN: 0198527845.

[465] H. Spieler. Semiconductor detectors. URL: https://www-physics.lbl.gov/~spieler/SLAC_Lectures/PDF/Sem-Det-I.pdf.

[466] R. Sternheimer. "The density effect for the ionization loss in various materials". *Phys. Rev.* 88.4 (1952), pp. 851–859. DOI: 10.1103/PhysRev.88.851. Erratum in "The Density Effect for the Ionization Loss in Various Materials". *Phys. Rev.* 89.6 (1953), pp. 1309–1309. DOI: 10.1103/PhysRev.89.1309.2.

[467]   R. Sternheimer, M. Berger, and S. Seltzer. "Density effect for the ionization loss of charged particles in various substances". *Atomic Data and Nuclear Data Tables* 30.2 (1984), pp. 261–271. DOI: 10.1016/0092-640X(84)90002-0.

[468]   S. Sze and K. Kwok. "Physics and properties of semiconductors—A review". *Physics of Semiconductor Devices*. John Wiley & Sons, Ltd, 2006. ISBN: 9780470068328. DOI: 10.1002/9780470068328.ch1.

[469]   I. Tamm and I. Frank. "Coherent radiation of fast electrons in a medium". *Doklady Akad. Nauk SSSR* 14 (1937). English version reprinted in I. Frank and I. Tamm. "Coherent visible radiation of fast electrons passing through matter". *Selected Papers*. Ed. by Boris M. Bolotovskii, Victor Ya. Frenkel, and Rudolf Peierls. Berlin, Heidelberg: Springer, 1991, pp. 29–35. ISBN: 978-3-642-74626-0. DOI: 10.1007/978-3-642-74626-0_2.

[470]   M. Tavlet, A. Fontaine, and H. Schönbacher. *Compilation of Radiation Damage Test Data. Index des résultats d'essais de radiorésistance;* 2nd ed. CERN Yellow Reports. Geneva: CERN, 1998. DOI: 10.5170/CERN-1998-001.

[471]   M. Tavlet and H. Schönbacher. *Compilation of Radiation Damage Test Data*. CERN Yellow Reports. Geneva: CERN, 1989. DOI: 10.5170/CERN-1989-012.

[472]   P. Teng et al. "Radiation hardness and lifetime studies of the VCSELs for the ATLAS SemiConductor Tracker". *NIMA* 497.2 (2003), pp. 294–304. DOI: 10.1016/S0168-9002(02)01922-8.

[473]   M. Thomson. "Particle flow calorimetry and the PandoraPFA algorithm". *NIMA* 611.1 (2009), pp. 25–40. DOI: 10.1016/j.nima.2009.09.009.

[474]   M. Titov. "Radiation damage and long-term aging in gas detectors". *Innovative Detectors for Supercolliders*. World Scientific, 2004. DOI: 10.1142/9789812702951_0014.

[475]   T. Trippe. *Minimum Tension Requirement for Charpak Chamber Wires*. CERN-NP-INT-REP-69-18, CERN-NP-Internal-Report-69-18. Geneva: CERN, 1969. URL: https://cds.cern.ch/record/314008.

[476]   J. Troska, F. Vasey, and A. Weidberg. "Radiation tolerant optoelectronics for high energy physics". *NIMA* 1052 (2023), p. 168208. DOI: 10.1016/j.nima.2023.168208.

[477]   Y. Tsai. "Pair production and bremsstrahlung of charged leptons". *Rev. Mod. Phys.* 46.4 (1974), pp. 815–851. DOI: 10.1103/RevModPhys.46.815. Erratum in: "Erratum: Pair production and bremsstrahlung of charged leptons". *Rev. Mod. Phys.* 49.2 (1977), pp. 421–423. DOI: 10.1103/RevModPhys.49.421.

[478]   L. Urbán. *A Model for Multiple Scattering in GEANT4*. Geneva: CERN, 2006. URL: http://cds.cern.ch/record/1004190.

[479]   J. Va'vra. "Particle identification using the de/dx and the Cherenkov light detection methods in high energy physics". *IEEE TNS* 47.6 (2000), pp. 1764–1774. DOI: 10.1109/TNS.2000.914443.

[480]   J. Va'vra. "Particle identification methods in high-energy physics". *NIMA* 453.1 (2000), pp. 262–278. DOI: 10.1016/S0168-9002(00)00644-6.

[481]   J. Va'vra. "Summary of session 6: aging effects in RPC detectors". *NIMA* 515.1 (2003), pp. 354–357. DOI: 10.1016/j.nima.2003.09.023.

[482]   J. Va'vra and A. Sharma. "Single electron detection in quadruple-GEM detector with pad readout". *NIMA* 478.1 (2002), pp. 235–244. DOI: 10.1016/S0168-9002(01)01763-6.

[483]   M. Valentan, M. Regler, and R. Frühwirth. "Generalization of the Gluckstern formulas II: Multiple scattering and non-zero dip angles". *NIMA* 606.3 (2009), pp. 728–742. DOI: 10.1016/j.nima.2009.05.024.

[484] A. Van Lysebetten, B. Verlaat, and M. van Beuzekom. "CO_2 cooling experience (LHCb)". *PoS* Vertex 2007 (2008), p. 009. DOI: 10.22323/1.057.0009.

[485] P. Vavilov. "Ionization losses of high-energy heavy particles". *JETP* 4 (1957), pp. 749–751. URL: http://jetp.ras.ru/cgi-bin/e/index/e/5/4/p749?a=list.

[486] R. Veenhof. "Numerical methods in the simulation of gas-based detectors". *JINST* 4.12 (2009), P12017–P12017. DOI: 10.1088/1748-0221/4/12/p12017.

[487] R. Veenhof. Garfield - simulation of gaseous detectors. URL: https://garfield.web.cern.ch/garfield/.

[488] B. Verlaat and J. Noite. "Design considerations of long length evaporative CO2 cooling lines". *10th IIR-Gustav Lorentzen Conference on Natural Working Fluids (GL2012)*. Vol. 10. 2012, GL–209. ISBN: 9782913149908.

[489] G. Viehhauser. "Thermal management and mechanical structures for silicon detector systems". *JINST* 10.09 (2015), P09001–P09001. DOI: 10.1088/1748-0221/10/09/p09001.

[490] C. Vignoli. "The ICARUS T600 liquid argon purification system". *Phys. Procedia* 67 (2015), pp. 796–801. DOI: 10.1016/j.phpro.2015.06.135.

[491] A. Walenta. "The time expansion chamber and single ionization cluster measurement". *IEEE TNS* 26.1 (1979), pp. 73–80. DOI: 10.1109/TNS.1979.4329616.

[492] B. Wang et al. "Operational experience of the Belle II pixel detector". *NIMA* 1032 (2022), p. 166631. DOI: 10.1016/j.nima.2022.166631.

[493] M. White and J. Bernstein. Microelectronics Reliability: Physics-of-Failure Based Modeling and Lifetime Evaluation. 2008. URL: https://nepp.nasa.gov/files/16365/08_102_4_%20JPL_White.pdf.

[494] N. Wiener. "Generalized harmonic analysis". *Acta Math.* 55 (1930), pp. 117–258. DOI: 10.1007/BF02546511.

[495] R. Wigmans. "On the energy resolution of uranium and other hadron calorimeters". *NIMA* 259.3 (1987), pp. 389–429. DOI: 10.1016/0168-9002(87)90823-0.

[496] R. Wigmans. "Calorimetry". *Scientifica Acta* 2.1 (2008), pp. 18–55. URL: http://siba.unipv.it/fisica/ScientificaActa/volume_2_1/Wigmans.pdf.

[497] R. Wigmans. *Calorimetry: Energy measurement in particle physics*. OUP, 2017. ISBN: 9780198786351. DOI: 10.1093/oso/9780198786351.001.0001.

[498] E. Williams. "Correlation of certain collision problems with radiation theory". *Det Kgl. Danske Videnskabernes Selskab. Mathematisk-fysiske Meddelelse* XIII, 4 (1935). URL: http://gymarkiv.sdu.dk/MFM/kdvs/mfm%2010-19/mfm-13-4.pdf.

[499] C. Wilson. "On a method of making visible the paths of ionising particles through a gas". *Proc. R. Soc. Lond.* A85 (1911), pp. 285–288. URL: http://doi.org/10.1098/rspa.1911.0041.

[500] P. Windischhofer and W. Riegler. "The statistics of electron–hole avalanches". *NIMA* 1003 (2021), p. 165327. DOI: https://doi.org/10.1016/j.nima.2021.165327.

[501] S. Withington. "Quantum electronics for fundamental physics". *Contemp. Phys.* 63.2 (2022), pp. 116–137. DOI: 10.1080/00107514.2023.2180179.

[502] J. Woithe. S'Cool Lab, do it yourself manual, cloud chamber. 2020. URL: https://scoollab.web.cern.ch/sites/default/files/documents/20200521_JW_DIYManual_CloudChamber_v7.pdf.

[503] M. Wolf et al. "Flip chip bumping technology—Status and update". *NIMA* 565.1 (2006), pp. 290–295. DOI: 10.1016/j.nima.2006.05.046.

[504] F. Wooten. *Optical Properties of Solids*. New York and London: Academic Press, 1972. ISBN: 978-0-12-763450-0. DOI: 10.1016/C2013-0-07656-6.

[505] R. Workman et al. "Review of Particle Physics". *PTEP* 2022 (2022), p. 083C01. DOI: 10.1093/ptep/ptac097.

[506] A. Wright. "The optical interface to PMTs". *The Photomultiplier Handbook*. OUP, 2017. ISBN: 9780199565092. DOI: 10.1093/oso/9780199565092.003.0003.

[507] S. Yamada et al. "New online system without hardware trigger for the Super-Kamiokande experiment". *2007 IEEE Nuclear Science Symposium Conference Record*. Vol. 1. 2007, pp. 111–114. DOI: 10.1109/NSSMIC.2007.4436298.

[508] A. Yamamoto et al. "The ATLAS central solenoid". *NIMA* 584.1 (2008), pp. 53–74. DOI: 10.1016/j.nima.2007.09.047.

[509] H. Yasuda. "New insights into aging phenomena from plasma chemistry". *NIMA* 515.1 (2003), pp. 15–30. DOI: 10.1016/j.nima.2003.08.125.

[510] T. Ypsilantis and J. Seguinot. "Theory of ring imaging Cherenkov counters". *NIMA* 343.1 (1994), pp. 30–51. DOI: 10.1016/0168-9002(94)90532-0.

[511] ZEUS collaboration. *The ZEUS Detector, status report*. Tech. rep. 1993. URL: https://www-zeus.desy.de/bluebook/bluebook.html.

[512] R.-Y. Zhu. "Radiation damage effects". *Handbook of Particle Detection and Imaging*. Ed. by Claus Grupen and Irène Buvat. Berlin, Heidelberg: Springer, 2012, pp. 535–555. ISBN: 978-3-642-13271-1. DOI: 10.1007/978-3-642-13271-1_22.

[513] M. Zolotorev and K. McDonald. "Classical radiation processes in the Weizsacker-Williams approximation" (2000). DOI: 10.48550/arXiv.physics/0003096.

[514] G. Zunica. "The upgrade of LHCb VELO". *NIMA* 1047 (2023), p. 167804. DOI: 10.1016/j.nima.2022.167804.

[515] P. Zyla et al. Data files and plots of cross-sections and related quantities in the 2021 Review of Particle Physics. 2021. URL: https://pdg.lbl.gov/2021/hadronic-xsections/hadron.html.

Index